Mathematics of the 19th Century

Edited by
A.N. Kolmogorov
A.P. Yushkevich

Translated from the Russian
by A. Shenitzer, H. Grant
and O. B. Sheinin

Mathematical Logic

Algebra

Number Theory

Probability Theory

Second Revised Edition

Birkhäuser Verlag
Basel · Boston · Berlin

Editors' addresses

A.N. Kolmogorov †
Moscow State University
Faculty of Mathematics and Mechanics
117 234 Moscow
Russia

A.P. Yushkevich †
Institute of History of Science and Technology
Staropanski Pereulok 1/5
103 012 Moscow
Russia

Originally published as:
Matematika XIX veka: Matematicheskaya logika, algebra, teoriya chisel, teoriya veroyatnostei
© Izdatel'stvo «Nauka», Moskva 1978.
All illustrations are taken from the original Russian edition.

A set of the three volumes of the Mathematics of the 19th Century is also available and includes the following:

Volume 1, 2nd rev. ed. (ISBN 3-7643-6441-6)
Mathematical Logic – Algebra – Number Theory – Probability Theory

Volume 2 (ISBN 3-7643-5048-2)
Geometry – Analytic Function Theory

Volume 3 (ISBN 3-7643-5845-9)
Function Theory According to Chebyshev – Ordinary Differential Equations – Calculus of Variations –
Theory of Finite Differences.

ISBN 3-7643-5868-8 (HC Set)

A CIP catalogue record for this book is available from the Library of Congress, Washington D.C., USA

Deutsche Bibliothek Cataloging-in-Publication Data
Mathematics of the 19th century / ed. by A. N. Kolmogorov ; A. P.Yushkevich. -
Basel ; Boston ; Berlin : Birkhäuser
    Einheitssacht.: Mathematika XIX veka ‹engl.›

Mathematical logic, algebra, number theory, probability theory. - 2., rev. ed. - 2001
    ISBN 3-7643-6442-4
    ISBN 3-7643-6441-6

ISBN 3-7643-6441-6 (Hardcover) Birkhäuser Verlag, Basel – Boston – Berlin
ISBN 3-7643-6442-4 (Softcover) Birkhäuser Verlag, Basel – Boston – Berlin

© 2001 Birkhäuser Verlag, P.O.Box 133, CH-4010 Basel, Switzerland
Printed on acid-free paper produced from chlorine-free pulb. TCF ∞
Cover design: Markus Etterich, Basel
Cover Illustration: Gauss, Dedekind, De Morgan, Markov, Boole, Minkowski
Printed in Germany

9 8 7 6 5 4 3 2 1

Contents

Preface IX

Introduction to the English Translation XIII

Chapter One
Mathematical Logic (by Z.A. Kuzicheva) 1

    The Prehistory of Mathematical Logic 1
    Leibniz's Symbolic Logic 2
    The Quantification of a Predicate 9
    The "Formal Logic" of A. De Morgan 10
    Boole's Algebra of Logic 14
    Jevons' Algebra of Logic 20
    Venn's Symbolic Logic 24
    Schröder's and Poretskiĭ's Logical Algebra 27

    Conclusion 33

Chapter Two
Algebra and Algebraic Number Theory
(by I.G. Bashmakova and A.N. Rudakov with the
assistance of A.N. Parshin and E.I. Slavutin) 35

1 Survey of the Evolution of Algebra and of the Theory
    of Algebraic Numbers During the Period of 1800–1870 35

2 The Evolution of Algebra 41
    Algebraic Proofs of the Fundamental Theorem
    of Algebra in the 18th Century 41
    C.F. Gauss' First Proof 43
    C.F. Gauss' Second Proof 44
    The Kronecker Construction 47

The Theory of Equations 50
Carl Friedrich Gauss 50
Solution of the Cyclotomic Equation 52
Niels Henrik Abel 55
Evariste Galois 57
The Algebraic Work of Evariste Galois 58
The First Steps in the Evolution of Group Theory 63
The Evolution of Linear Algebra 68
Hypercomplex Numbers 72
William Rowan Hamilton 74
Matrix Algebra 77
The Algebras of Grassmann and Clifford 78
Associative Algebras 79
The Theory of Invariants 80

3 The Theory of Algebraic Numbers and the
   Beginnings of Commutative Algebra 86
   Disquisitiones Arithmeticae of C.F. Gauss 86
   Investigation of the Number of Classes of Quadratic Forms 92
   Gaussian Integers and Their Arithmetic 94
   Fermat's Last Theorem. The Discovery of E. Kummer 99
   Kummer's Theory 102
   Difficulties. The Notion of an Integer 106
   The Zolotarev Theory. Integral and p-Integral Numbers 108
   Dedekind's Ideal Theory 116
   On Dedekind's Method. Ideals and Cuts 123
   Construction of Ideal Theory in Algebraic Function Fields 125
   L. Kronecker's Divisor Theory 131

   Conclusion 133

Chapter Three
Problems of Number Theory (by E.P. Ozhigova with
the assistance of A.P. Yushkevich) 137

1 The Arithmetic Theory of Quadratic Forms 137
   The General Theory of Forms; Ch. Hermite 137
   Korkin's and Zolotarev's Works on
   the Theory of Quadratic Forms 144
   The Investigations of A.A. Markov 151

2 Geometry of Numbers 154
   Origin of the Theory 154
   The Work of H.J.S. Smith 159
   Geometry of Numbers: Hermann Minkowski 161
   The Works of G.F. Voronoĭ 166

3 Analytic Methods in Number Theory 171
    Lejeune-Dirichlet and the Theorem on Arithmetic Progressions 171
    Asymptotic Laws of Number Theory 177
    Chebyshev and the Theory of Distribution of Primes 182
    The Ideas of Bernhard Riemann 189
    Proof of the Asymptotic Law of Distribution
    of Prime Numbers 192
    Some Applications of Analytic Number Theory 194
    Arithmetic Functions and Identities.
    The Works of N.V. Bugaev............................... 196

4 Transcendental Numbers 201
    The Works of Joseph Liouville............................ 201
    Charles Hermite and the Proof of the Transcendence of
    the Number e; The Theorem of Ferdinand Lindemann 205

    Conclusion 209

Chapter Four
The Theory of Probability (by B.V. Gnedenko and O.B. Sheĭnin) ... 211

    Introduction 211
    Laplace's Theory of Probability.......................... 212
    Laplace's Theory of Errors 222
    Gauss' Contribution to the Theory of Probability 226
    The contributions of Poisson and Cauchy 230
    Social and Anthropometric Statistics 242
    The Russian School of the Theory
    of Probability. P.L. Chebyshev........................... 247
    New Fields of Application of the Theory of Probability.
    The Rise of Mathematical Statistics 268
    Works of the Second Half of the 19th Century
    in Western Europe 276

    Conclusion 280

Addendum (by O.B. Sheĭnin)............................... 283
    1. French and German Quotations 283
    2. Notes 285
    3. Additional Bibliography 286

Bibliography (by F.A. Medvedev) 289

Abbreviations 302

Index of Names 304

Preface

This multi-authored effort, *Mathematics of the nineteenth century* (to be followed by *Mathematics of the twentieth century*), is a sequel to the *History of mathematics from antiquity to the early nineteenth century*, published in three volumes from 1970 to 1972.[1] For reasons explained below, our discussion of twentieth-century mathematics ends with the 1930s. Our general objectives are identical with those stated in the preface to the three-volume edition, i.e., we consider the development of mathematics not simply as the process of perfecting concepts and techniques for studying real-world spatial forms and quantitative relationships but as a social process as well. Mathematical structures, once established, are capable of a certain degree of autonomous development. In the final analysis, however, such immanent mathematical evolution is conditioned by practical activity and is either self-directed or, as is most often the case, is determined by the needs of society. Proceeding from this premise, we intend, first, to unravel the forces that shape mathematical progress. We examine the interaction of mathematics with the social structure, technology, the natural sciences, and philosophy. Through an analysis of mathematical history proper, we hope to delineate the relationships among the various mathematical disciplines and to evaluate mathematical achievements in the light of the current state and future prospects of the science.

The difficulties confronting us considerably exceeded those encountered in preparing the three-volume edition. The history of nineteenth-century and especially of twentieth-century mathematics has been much less studied than that of preceding periods. In the past 150–175 years, mathematics has splintered into a number of distinct and highly specialized fields. The relevant

---

1 From here on, this work will be referred to by HM, followed by volume and page reference.

primary sources are virtually limitless. Although more authors worked on this volume than on *The history of mathematics*, we have not been able to do justice to all the ramifications of the mathematical achievements of the period. Our presentation is more a collection of essays than a connected history of nineteenth and twentieth-century mathematics — we have not succeeded in avoiding gaps, and have omitted several topics, for example, some chapters of the theory of differential equations and certain classes of special functions. In some cases we were not able to find specialists willing to undertake the difficult task of studying recent history. The treatment of some areas is incomplete in places because of scant historical material. Thus, in the essays on the history of nineteenth-century mathematics, the history of computational methods, which at the time were not accorded the status of a separate branch of mathematics and were relegated to algebra or to one of the analytic disciplines, is presented in relatively less detail. We have, nevertheless, sought to discuss the most important mathematical developments in their entirety.

We should add that we could not avoid occasional repetitions, first, because some ideas belong to several mathematical disciplines and second, because we have tried in each case to provide a self-contained presentation of the relevant issue.

The plan of the book is similar to that of the three-volume edition. As before, our primary objective has been to treat the evolution of mathematics as a whole. We have concentrated on the essential concepts, methods, and algorithms. As before, we present short biographies of the most distinguished mathematicians while providing only the sketchiest of biographical material — in many cases the vital statistics only — of others.

The historical period covered extends from the early nineteenth century to the end of the 1930s. The discussion falls naturally into two parts: the nineteenth century and the first four decades of this century. Of course, neither 1801 nor 1900 are, in themselves, turning points in the history of mathematics, although each date is notable for a remarkable event: the first for the publication of Gauss' *Disquisitiones arithmeticae*, the second for Hilbert's *Mathematical problems*. We do not abide strictly by these dates, and while discussing the history of a particular area we may deviate from 1801 or 1900 in either direction. We seldom bring the discussion up to the very end of the nineteenth century, since in a number of cases the natural theoretical watershed lies somewhere in the 1870s and 1880s. The nineteenth century, as a whole, differs radically from the eighteenth, which is characterized by the direct development of the basic ideas of the mathematics of Descartes, Fermat, Newton and Leibniz — ideas, many of which existed in rudimentary form in ancient Greece. Beginning in the second quarter of the nineteenth century mathematics underwent a revolution as crucial and profound in its consequences for the general world outlook as the mathematical revolution at the beginning of the modern era. The latter consisted in the creation of the infinitesimal calculus which, as Euler noted, "revolves" around the concept of a function. The basic concepts and principles of specific areas of mathematics were constantly changing throughout the nineteenth century. This development in no way implied the rejection of ideas inherited from the eighteenth century. The main changes included a new statement of the problem of the existence of mathematical objects, particularly in the calculus, and soon thereafter the formation of non-standard (i.e. non-Euclidean) structures in geometry, arithmetic, and algebra. The foundations for these developments were laid by men educated for the most part in the first quarter of the nineteenth century, while the most creative work was accomplished during the second and third quarters of the century. These were men with a new kind of mathematical mentality: Cauchy, and somewhat earlier Bolzano, in analysis; Lobachevskiĭ and Bolyai in geometry; Galois, Hamilton, and Grassmann in algebra; and their predecessor Gauss. This became apparent as the new methods spread to all of mathematics. At first, the nineteenth century appeared to its contemporaries as a time of brilliant flowering of mathematical analysis and its physical applications, but, as we said earlier, its forms were fundamentally different from earlier ones. These radical changes in mathematics occurred in the sharply changing economic and political environment of the nineteenth century, which engendered changes in the social role of mathematics, in the training of specialists and the organization of research, including the creation of a new type of mathematician. The role of mathematics in technology, and even in the social sciences, increased enormously in this period of developing capitalism, which established itself in the most advanced countries of the world. This led to major changes in secondary education and even more so in higher education. Creative mathematical activity in the seventeenth and eighteenth centuries was to a significant extent the province of amateur mathematicians, many of whom were self-taught, and the State Academies of the Sciences were the chief centers of mathematical activity in eighteenth-century France, Russia, and Germany. As mathematics developed into a primarily university science in the nineteenth century, the number of university-based schools of mathematics grew apace.

The upper bound of nineteenth-century mathematics may be fixed in a variety of ways. It is indisputable, though, that early twentieth-century mathematics is clearly distinguishable from its mid-nineteenth-century predecessor. As far as mathematical analysis is concerned, its complete subordination to the generalizing ideas of functional analysis was becoming ever more pronounced, even for problems pertaining mainly to mathematical physics. The possibility of non-standard geometries, which had been clearly anticipated in earlier periods, led to geometry becoming, in effect, a theory of classes of spaces far more general than 3-dimensional Euclidean space. Set theory, and later the logical investigation of the construction of completely formalized deductive theories, became the twin basis of all mathematics.

All of these trends had been noted by the end of the nineteenth century and received legitimate expression by the 1930s. Our last series of essays, devoted to the study of the mathematics of the first four decades of the twentieth century, deals with the restructuring of the entire science of mathematics on a greatly expanded basis, as was briefly suggested above. The only late-nineteenth-century works included in these essays are works that can be fully covered in this broader perspective.

The frenetic growth of postwar mathematics lies beyond the scope of our book. The scholars who fashioned prewar mathematics have either died, retired from active work, or are now engaged in an evaluation of their own efforts. Their achievements can be judged with some degree of objectivity. This, however, would be difficult to do for postwar mathematics.

Given our chronological bounds, it is only natural to discuss broadly the increased interest in computational mathematics, to which we devote a chapter. But the influence of computers on mathematics exceeds the chronological limits we have adopted here.

Our essays are thus divided into two sets: *Mathematics of the nineteenth century*, in four books, and *Mathematics of the twentieth century*, in two.

Because of the vast scope of each subject, there will be several books with explanatory subtitles. Book I includes essays on the history of nineteenth-century mathematical logic, algebra, number theory, and probability theory. Book II contains the history of nineteenth-century geometry, differential and integral calculus, and computational mathematics. Books III and IV deal with other areas of analysis, from methods of mathematical physics to historiography, the history of mathematical education, and the organization of mathematical research. Books V and VI deal with the twentieth century. Each book contains a bibliography and name index. Titles of periodicals are abbreviated, giving the issue number and year of publication; if the actual year of publication is different from the stated year, the latter is given in parentheses; a list of abbreviations is given after the bibliography.

Chapter One was read by Professor V.A. Uspenskiĭ, Chapters Two and Three by D.K. Fadeev, Corresponding Member of the USSR Academy of Sciences. The entire book was read by S.Kh. Siragdinov, member of the Academy of Sciences of the Uzbek SSR, to whom the authors and editors express their deep gratitude for many valuable comments and recommendations. D.K. Faddeev is also the author of several passages in the book. In deference to his wish we have not listed him as a coauthor.

Moscow, June 1, 1977.   A.N. Kolmogorov, A.P. Yushkevich

Introduction to
the English Translation

I must take it upon myself to write the introduction to this volume of the work "Mathematics of the 19th Century" since, regrettably, my colleague and collaborator, Academician Andrei Nikolaevich Kolmogorov (April 25, 1908 – October 20, 1987), passed away several years ago.

In the preface of this book we mentioned that, due to special features of the development of certain branches of the mathematical sciences, some chapters of our exposition of the history of mathematics in the 19th century stop short of 1900. This is because the 1870s and 1880s turned out to be a natural chronological boundary. Some readers have raised objections. While we see no reason to make changes in this first book of this work, we will try, in the future, to consider reader remarks. Specifically, it has been more natural to take the geometry chapter in the second volume as far as Hilbert's work on the foundations of geometry and Poincaré's work on topology.

For reasons over which the editors have no control, the plan of the work had to be modified: it turned out that the order in which various chapters were completed by their authors differed from the originally anticipated order. In view of the autonomous character of the chapters, we changed the structure of the second volume to have it comprise just two chapters. One chapter deals with all geometric disciplines, and the other with the theory of analytic functions, including some of its special topics. The author of the latter chapter, A.I. Markushevich, passed away in 1979.

The third volume of this work was published in 1987. It consists of the following chapters:

1. Chebyshev's Approach to the Theory of Functions (N.I. Akhiezer, † 1980);

2. Ordinary Differential Equations (S.S. Demidov, with the collaboration of S.S. Petrova and N.I. Simonov, † 1979);

3. The Calculus of Variations (A.V. Dorofeeva);

4. The Calculus of Finite Differences (S.S. Petrova and A.D. Solov'ev).

The fourth volume is essentially ready and deals with the theory of partial differential equations. Part of the fifth volume, which deals with different approaches in computational mathematics, is also practically ready. All these volumes do not really cover all of the areas of mathematics that were successfully developed in the 20th century. It is my hope that in the future additional people will become involved in this enterprise and that the Editorial Board will create a plan for further work. Personally, I would like the final volume to contain a synthetical overview of the achievements of 19th-century mathematics and of its "social history", including such topics as mathematical education, the issue of mathematical journals, 19th-century Congresses, etc. (Kolmogorov shared my view in this matter). Finally, a new Editorial Board might also discuss the possibility of realizing one of my and Kolmogorov's dreams, the publication of several volumes treating the History of 20th-Century Mathematics.

We wish to thank Birkhäuser Verlag for undertaking the publication of an English translation of the present book as well as all individuals who have contributed to various phases of translation and correction.

Moscow, July 1991 A.P. Yushkevich

Additional Note

The second edition of this book gives me the opportunity to correct misprints and errors. In this connection I want to acknowledge the help of H. Grant, S. Mykytiuk, and S. Shenitzer. I am especially grateful to S. Mykytiuk, who checked the essay on logic (Chapter One) and corrected errors some of which were carried from the Russian original.

A. Shenitzer

Editor of the
second edition

Chapter One
Mathematical Logic

The Prehistory of Mathematical Logic

Mathematical logic was not treated in the three-volume *History of Mathematics*. We shall, therefore, briefly survey its prehistory before analyzing its development in the nineteenth century.

The first surviving systematic construction and exposition of logic is contained in the treatise of Aristotle (384–322 B.C.), organized by his ancient commentators under the title *Organon*. The *Organon* includes the *Categoriae* (on categories), the *De interpretatione* (on propositions), the *Analytica priora* (on inference), the *Analytica posteriora* (on proofs), the *Topica* (on a proof based on likely premises) and the related *De sophisticis elenchis* (on fallacies). In the *Analytica posteriora* Aristotle presents his theory of proof and formulates the basic requirements for a "demonstrative science" and for mathematics in particular. Noting the rigor of Aristotle's logical propositions, Leibniz observed that "Aristotle was the first to write in a mathematical way on non-mathematical subjects."[1]

A different style of logic, a distinctive logic of propositions, was developed by the philosophers of the Megarian School, founded by Socrates' pupil, Euclid of Megara (c. 450–380 B.C.). Euclid, in turn, taught Eubulides of Miletus (fourth century B.C.), to whom the logicians are indebted for certain paradoxes, including the celebrated "liar paradox" and "heap paradox". Philo of Megara (c. 300 B.C.) was the last representative of the Megarian school. About the same time, however, Zeno of Citium (c. 336–264 B.C.), a student of Philo, founded the Stoic school, which inherited the Megarians' basic ideas and style. The most outstanding representative of Stoicism was Chrisippus

---

1 Leibniz, G.W., *Fragmente zur Logik*. Berlin 1960, p. 7.

of Soli (c. 281–208 B.C.), of whom it was said that if the gods needed logic, it would be his. Fragments of the logical teachings of the Megarians and the Stoics show that these teachings anticipated the modern calculus of propositions to a remarkable extent.

The period of late Greco-Roman antiquity has usually been regarded as a time in which the study of logic made little headway and has scarcely been studied in terms of the history of logic. Nevertheless, the well-known commentaries of Alexander of Aphrodisias (A.D. 2nd–3rd centuries) on Aristotle's treatises, *The Introduction to "Categoriae"* by Porphirius (c. 232–c. 304), and works by Boethius (c. 480–c. 524), the translator of and commentator on both Aristotle and Porphirius, date from that time.

In the early Middle Ages, logic developed as an autonomous discipline only in the Arab world. In this connection, we note the series of logical treatises by Abū Naṣr al-Fārābī (c. 870–950), the logical component of *Ash-Shifa* (*The Recovery*, i.e., of the soul from ignorance) called *Sufficientia* in Latin versions by Abū 'Ali Al-Husain Ibn 'Abdallāh Ibn Sīnā (Avicenna, 980–1037), commentaries on Aristotle by Ibn Rushd (Averroës, 1126–1198), and the treatise on logic by al-Tūsī (Abū Nāṣīr al-dīn, a native of Tûs, in Korâsân, 1201–1274), who continued the traditions of Fārābi and Ibn Sīnā. These works present and comment on Aristotle's *Organon*, and especially on his doctrine of syllogisms.

Late medieval Europe saw the development of scholastic logic, which adapted Aristotelian logic to the needs of the Church. During that time the "new logic", or *logica modernorum*, took form in the works of Pierre Abelard (1079–1142), Petrus Hispanus (c. 1215–1277), John Duns Scotus (c. 1270–1308), and William of Occam (c. 1300–c. 1350). These works contain many laws of the propositional calculus (for example, De Morgan's laws) as well as anticipations of the idea of a "universe" and of other ideas.

The notion of mechanizing the process of logical inference, expressed by the Spanish philosopher Raymundus Lullius (Ramon Lull or Lully, c. 1235–1315) in his *Ars magna et ultima*, also belongs to this period.

We can, without much loss of completeness, pass to the 17th century, when issues of logic and of its rebuilding on a new algebraic foundation attracted the attention of scientists, of whom G.W. Leibniz was the most prominent (HM, v. 2, pp. 251–252).

Leibniz's Symbolic Logic

Leibniz conceived of logic in the broadest sense of the word. To him it was not only the analytics of Aristotle — the art of reasoning, and of proving known truths — but also the art of inventing and discovering new truths.

The study of the works of Aristotle made a very strong impression on the young Leibniz and decisively influenced his view of logic. Leibniz thought highly of Aristotle's theory of syllogisms. He wrote,

... the invention of the syllogistic form is one of the finest and indeed one of the most important, to have been made by the human mind. It is a kind of universal mathematics whose importance is too little known.[2]

But there are forms of inference more complex than Aristotelian syllogistic. Leibniz regarded Euclid's rules of addition, multiplication, and permutation of terms of a proportion as instances of such complex forms. What is obtained by operating according to these rules is true, and the process of obtaining the results constitutes a proof (*argumenta in forma*).[3] The following is Leibniz's scheme for the improvement and formulation of logic.

First, one must analyze all concepts, reducing these to combinations of the simplest possible notions. The list of these simple, undefinable notions would constitute an "alphabet of human thought". All remaining notions would be obtained by combining these initial ones.

Leibniz believed that such an analysis of ideas would enable all known truths to be proved, and thus yield a kind of "encyclopedia of proofs".

Finally, it would be necessary to introduce appropriate symbols to represent simple and complex notions and propositions, that is, to create a universal symbolism or *characteristica universalis*.

Leibniz attached great importance to symbolism. He emphasized that apt symbolism could become a decisive factor — he noted that the successes of contemporary algebra were to a large extent due to the apt symbolism of François Viète and René Descartes.

Leibniz suggested that such a universal symbolism ought to serve as an international language with which to express all existing and potential knowledge as well as a tool for discovering and proving new truths, or, as he put it, as the art of invention. For this purpose, a logical calculus completing the construction of the new logic was required. Leibniz believed that this new logic made possible the easy resolution of human conflicts. Opponents would be able to pick up their pens and say: "Let us calculate".

Leibniz thus began his refinement of traditional logic with the establishment of this "alphabet of human thought". He thought that there exist finitely many simple notions into which all other notions can be factored and from which new notions can be obtained through combination. The task will be solved, he said,

> when we will be able to find a small number of thoughts from which there arise in succession infinitely many other thoughts. Thus from the few numbers from 1 to 10 there arise in succession all other numbers.[4]

---

2 Leibniz, G.W., *New Essays on Human Understanding*, transl. and ed. by P. Remnant and J. Bennett. Cambridge, 1981, p. 479.

3 Leibniz, G.W., *Fragmente zur Logik*. Berlin 1960, pp. 7–9.

4 Leibniz, G.W., *Opuscules et fragments inédits de Leibniz*, L. Couturat (Ed.). Paris 1903, pp. 429–432.

Leibniz calls the simple notions "first-order terms"* and assigns them to the first class. He assigns second-order terms, i.e., combinations of two first-order terms, to the second class.** The order of the two terms is not significant. The third class consists of third-order terms: combinations of three first-order terms, or of one first-order and one second-order term, and so on.

Depending on the manner of combination, each compound term may have several expressions. In order to determine whether the expressions are "the same", it is sufficient to break the expressions down into simple, that is, first-order terms; if the first-order terms coincide, then these expressions represent the same compound term.

Thus the analysis of a notion consists of reducing it to its constituent first-order terms.

Leibniz expresses each proposition in the subject-predicate form a est b (a is b). By a "truth" he means a proposition whose predicate is contained in the notion of the subject. In order to analyze the truth of such a proposition one should reduce both of its terms to simple notions and then compare them.

However, there is a distinction between the analysis of a truth and that of a concept: while it is intended that the analysis of a concept must (and can) reduce it to simple notions, the analysis of a truth need not necessarily be complete. "For the analysis of truth is complete — writes Leibniz — when we have found its proof, whereas it is not always necessary to complete the analysis of a subject or predicate to find a proof of a proposition. In most cases, the beginnings of an analysis of things suffices for the analysis or complete knowledge of the truth known about a thing."[5]

Leibniz treated the analysis of relations as a translation from a language that includes polyvalent relations to one in which all propositions are expressed as subject and predicate. Leibniz based his arithmetical interpretation of logic on the analogy between the classification of notions into first-order terms and the factoring of composite numbers into primes. His writings contain several variants of such an interpretation. One of the first is found in his *Elementa calculi* (1679).[6] In this representation, a characteristic number is assigned to each term. The characteristic numbers are natural numbers; in particular primes are assigned to the simple notions (that is, notions from the "alphabet"). Numbers are assigned to complex terms according to the

---

* Today we would call these terms "atomic formulas". (Transl.)

** These second-order terms have no connection with what are now known as second-order expressions, i.e., expressions with quantifiers over sets and not over elements. (Transl.)

5 Leibniz, G.W., *Die philosophischen Schriften*, Band 7, hsg. von C. Gerhardt. Berlin 1890, pp. 82–85.

6 Leibniz, G.W., *Opuscules et fragments inédits de Leibniz*, L. Couturat (Ed.). Paris 1903, pp. 49–57.

following rule: the characteristic number of a complex term is the product of the characteristic numbers of its components terms; the definition thus obtained is unique (up to order, which is irrelevant for numbers as well as for component terms).

This calculus did not include the notion of negation, without which it was impossible to represent negative propositions. Leibniz uses this notion in a later scheme in which he denotes negation by the sign "−" and assigns a positive or negative number to each term. He does not use the signs "+" and "−" as signs of arithmetic operations: all divisors of a characteristic number having the sign "−" are assigned the sign "−"; similarly, all divisors of a characteristic number having the sign "+" are assigned the sign "+".

Leibniz defines the conditions under which a term P is contained in a term S, and interprets the classical forms of categorical propositions. He employs the interpretation of these inference forms to verify the modes of syllogisms.

Leibniz also used geometric diagrams to analyze the logical relations between notions. In his philosophical works published by Gerhardt,[7] we find diagrams in which notions are represented by segments of parallel lines; dotted lines indicating whether the notions include each other completely, partially, or not at all, are drawn through the end-points of the segments. As Couturat discovered, Leibniz, in his paper "On the proof of Logical Form by the Drawing of Lines" (*De formae logicae comprobatione per linearum ductus*),[8] uses circle diagrams similar to Euler's.

Figure 1 shows Leibniz's representation of the four classical propositions, *a*, *e*, *i* and *o*, by means of two types of line diagrams and by circle diagrams. Leibniz notes that the circle diagrams can lead to ambiguous interpretations. Thus the diagram for the proposition SaP can also be interpreted as "Some P are not S", and the diagrams for SiP and SoP are practically identical. *Figure 2* shows some examples of line and circle diagrams used by Leibniz to represent the modes of the syllogisms of types *aaa* and *eee*, commonly known as Barbara and Cesare respectively. This work remained unknown until 1903, when it was published by Couturat. As a result, Leonhard Euler is usually assumed to be the founder of geometric methods of logic. Euler's interpretation of Aristotelian syllogistic with the aid of circles is given in the second volume of his "Letters to a German Princess" (*Lettres à une princesse d'Allemagne sur divers sujets de physique et de philosophie*. T. 2, St.-Pétersbourg 1768).[9]

As noted above, Leibniz's program for logic envisaged the creation of logical calculus. He sketched a number of approaches to this project, but un-

7 Leibniz, G.W., *Die philosophischen Schriften*, Band 7, hsg. von C. Gerhardt. Berlin 1890.

8 Leibniz, G.W., *Opuscules et fragments inédits de Leibniz*. Paris 1903, pp. 293–321.

9 These *Letters* have been published: *Leonhardi Euleri Opera Omnia*, Ser. 3, vol. 11–12.

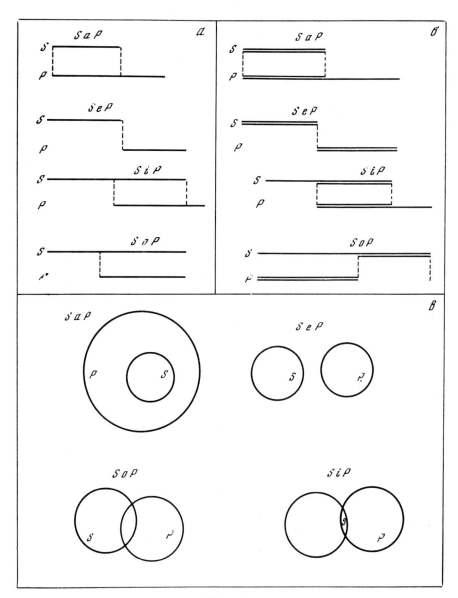

Figure 1.

fortunately not one of them was developed in detail or carried out to its conclusion.

Leibniz denotes terms by Latin lowercase letters, a, b, c, ... He uses one binary operation, that of adding a symbol on the right: ab. Of the rules of inference, the rule of substitution, permitting the replacement of a variable by a term or another variable in a true proposition, is explicitly formulated.

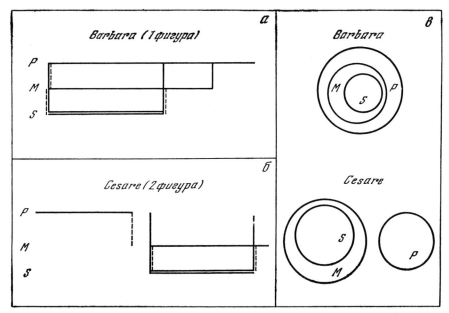

Figure 2.

As a copula, he uses *est* for inclusion, *sunt idem* or *eadem sunt* for equality, and *diversa sunt* for inequality. He denotes by a non-*a* the negation of *a*.

Although Leibniz distinguished between axioms of logical calculus (*propositiones per se verae*) and propositions deduced from axioms (*verae propositiones*), his calculus shows no clear distinction between axioms and theorems; he makes a number of assertions but proves only a few of them.

It seems that the following assertions may be regarded as examples of axioms:

$$(1)\ ab = ba, \quad (2)\ aa = a, \quad (3)\ a \text{ est } a, \quad (4)\ ab \text{ est } a.$$

Examples of theorems are:

(1)   if *a* est *b*, then *ac* est *bc*,

(2)   if *a* est *b* and *b* est *a*, then $a = b$,

(3)   if *a* est *c* and *b* est *c*, then *ab* est *c*.

A variety of circumstances delayed the publication of most of Leibniz's writings on logic until the beginning of the 20th century. Proper evaluation of Leibniz's contribution in this field is due to the efforts of the French philosopher of mathematics Louis Couturat (1868–1914). After studying Leibniz's papers on logic preserved in Hannover, Couturat wrote *La logique de Leibniz* (Paris, 1901), and later published the previously cited *Opuscules et fragments inédits de Leibniz* (1903), which contained more than two hundred previously unknown essays.

It would, however, be wrong to assume that Leibniz's ideas played no role at all in the creation of mathematical logic. They exercised a notable influence on various 18th-century scientists including, for instance, J.A. von Segner (1704-1777), a student of mathematics and the natural sciences (HM, v. 3, p. 97) as well as logic, in which he used symbolism widely; and Gottfried Ploucquet (1716–1790), professor of logic and philosophy at Tübingen, who attempted to develop an entire system of "logical calculus" (... *des logischen Kalküls*). The distinguished mathematician, J.H. Lambert, who is repeatedly mentioned in Volume III of HM, introduced a number of truly original ideas to logic. Without going into a detailed discussion of Lambert's work, which is particularly relevant to the calculus of propositions, or his occasional dispute with Ploucquet, we note that he broached the idea of the quantification of a predicate, a concept later expanded by 19th-century English scholars.

Mathematical logic arose as a consequence of the application of mathematical, and, particularly, algebraic methods for the solution of logical problems. Elements of mathematical logic had appeared in the works of mathematicians and logicians prior to the 19th century. However, mathematical logic established itself as an independent discipline only in the nineteenth century, when the term "mathematical logic" acquired common currency. The deduction of consequences from premises has always remained one of the basic problems of logic. As will be shown, the quantitative definition of the predicate provides the theoretical possibility for expressing the initial propositions (premises) in terms of equalities. Analogies with algebra suggest techniques for transforming these equalities by introducing operations similar to algebraic operations. As noted, these analogies had been noticed quite some time ago, though it was only Leibniz who really advanced the algebraization of logic. These ideas were most fully realized by representatives of the new approach to logic in the 19th century. Efforts to adapt algebra to the needs of logic resulted in the development of the "Boolean" algebraic structure, which is nowadays defined as follows:

A Boolean algebra is a set A with two distinguished elements denoted by 0 and 1, respectively, with two binary operations $+$, $\cdot$ and one unary operation $^{-}$ such that

$$(1) \qquad x \cdot y = y \cdot x, \qquad\qquad\qquad x + y = y + x,$$
$$(2) \qquad x \cdot (y \cdot z) = (x \cdot y) \cdot z, \qquad x + (y + z) = (x + y) + z,$$
$$(3) \qquad x \cdot (x + y) = x, \qquad\qquad x + x \cdot y = x,$$
$$(4) \qquad x \cdot (y + z) = x \cdot y + x \cdot z,$$
$$(5) \qquad x \cdot \bar{x} = 0,$$
$$(6) \qquad x + \bar{x} = 1.$$

An example of a Boolean algebra is provided by the algebra of classes, whose elements are classes — subsets of a fixed set U referred to as the universe. 0 denotes the empty class, + the union, · the intersection, and ¯ the complement. Since a class is often thought of as the content of a term, the algebra of classes is sometimes called the logic of classes.

The first systems of this type were formulated by Boole and De Morgan, and were later developed by 19th-century logicians. The quantification of a predicate (quantitative definition of a predicate) had been developed some time earlier.

The Quantification of a Predicate

In traditional categorical propositions the word "is" ("are") is regarded as a non-strict inclusion. For instance, the proposition "all x is y" can mean "x equals y" as well as "x is a proper part of y". The quantification of a predicate consists in determining whether the full content of a predicate or only part of it coincides (or does not coincide) with the content of the subject. This can be done by adding the words "all" or "some" to the predicate noun. Consequently, instead of the four traditional forms of a categorical proposition

  a: All x is y,

  e: No x is y,

  i: Some x is y,

  o: Some x is not y,

there are eight forms:

  All x is all y,

  All x is some y,

  Some x is all y,

  Some x is some y,

  Any x is not any y,

  Any x is not some y,

  Some x is not some y,

  Some x is not any y.

The traditional meaning of "all" and "some" changes. Whereas tradition-ally "some" was treated as "some and possibly all", it is now treated as "at least some but not all". "All" is used in its collective sense, in the sense of the whole class. Thus in a proposition of the type "all x is all y" the subject is identical with the predicate, that is, x equals y, and can be written as $x = y$. If vy denotes a proper subclass of the class y, the proposition "all x is some y", that is, the inclusion of x in y, can be written as the identity $x = vy$. All other propositions can be written down similarly.

The interpretation of simple propositions of the traditional syllogistic in terms of the equivalence (equality) of subject and predicate is due to George

Bentham, an English botanist (1800–1884), who published this interpretation in his *Outline of a New System of Logic* (London 1827). This view, however, went unnoticed for quite some time. The eight forms listed above were obtained, independently of Bentham, by William Hamilton, a Scottish philosopher (1788–1856) and professor of logic and metaphysics at the University of Edinburgh (not to be confused with his younger contemporary, the Irish mathematician William Rowan Hamilton (1805–1865)). A detailed presentation of Hamilton's results is found in his *Lectures on Metaphysics and Logic* (Edinburgh-London 1860).

Syllogistic ideas analogous to Hamilton's were conceived somewhat later by A. De Morgan, who developed logic much further. When Hamilton learned of a paper which De Morgan had sent to the Cambridge Philosophical Society, he forwarded a résumé of his logical concepts to De Morgan. A prolonged and fruitless debate over priority soon sprang up between them. We will not, however, dwell on this, and proceed to a discussion of De Morgan's logic.

The "Formal Logic" of A. De Morgan

Augustus De Morgan (1806–1871) was born in India to the family of a British colonel; he received his education at Trinity College, Cambridge. From 1828 to 1831 and from 1836 to 1866 he was professor of mathematics at London University College; Isaac Todhunter and James Joseph Sylvester were among his pupils. In 1865, as first President of the London Mathematical Society, he delivered the opening address of its inaugural session. De Morgan made several discoveries in algebra and analysis, which we will discuss later. In an 1838 article he introduced the term "mathematical induction", which became widely known through Todhunter's textbook on algebra. In the study as well as in the teaching of mathematics De Morgan was most interested in basic mathematical principles and their strict logical development. He referred to mathematics and logic as "the eyes" of precise knowledge and regretted that mathematicians cared less for logic than logicians cared for mathematics. De Morgan tried to narrow the gap between the two disciplines. His main scientific contribution was his construction of logic along mathematical lines.

In his *Formal Logic, or the Calculus of Inference, Necessary and Probable* (London, 1847), De Morgan begins with the premise that logic must serve the task of accurately expressing thoughts, thereby eliminating common, everyday linguistic ambiguity.

Every language contains "affirmative" and "negative" terms. This classification is, to a large extent, random — a certain term may have a paired negated term in one language but not in another. Not all terms in every language are provided with paired negated terms. Nevertheless, every term divides a "collection of entities" into two groups: those having characteristics produced by the particular term and those not having them. Therefore, every name expression contains its own negation, for example, "perfect" – "imperfect", "human" – "non-human". Logic must not only consider this fact but

A. De Morgan

express it as well. However, negation has always been a problem for logic. As William Minto said, in logic, by negating a given quality, we simply assert that the quality does not belong to a given object; the negation only removes, eliminates, but does not permit of an implication:[10] "'Not-b' is something entirely indefinite: it may cover anything except b." To remove this ambiguity of negation, De Morgan introduces the notion of the "whole" or "universe", which is to be selected depending on the subject of the investigation. Thus, if X stands for a class of things (in fact, De Morgan more often refers to name expressions which denote things), everything in the "whole" (the "universe") which is not X is to be regarded as non-X. If x stands for a non-X, then it becomes obvious that there is no essential distinction between affirmative and negative propositions: "No X is Y" means the same as "All X is non-Y", that is, "All X is Y".

Having established that X and x have the same status from the indicated standpoint, De Morgan considers instead of a pair of terms X and Y of the traditional syllogistic four pairs, $X, Y; x, y; X, y;$ and x, Y, generating sixteen

10 Minto, W., *Logic, Inductive and Deductive.* London 1893, p. 37.

logically possible combinations, of which eight are distinct. For the latter De Morgan introduces two kinds of symbolism: A_1, E_1, I_1, O_1, A', E', I', O', indicating the connection with the symbolism of the simple propositions of the traditional syllogistic, and $X)Y$; $X \cdot Y$; $X : Y$; XY; $x)y$, etc.:

A_1	Every X is a Y	$X)Y$;	A'	Every x is a y	$x)y$;
E_1	No X is a Y	$X \cdot Y$;	E'	No x is a y	$x \cdot y$;
I_1	Some X's are Y's	XY;	I'	Some x's are y's	xy;
O_1	Some X's are not Y's	$X : Y$;	O'	Some x's are not y's	$x : y$.

For these propositions, which he calls simple, De Morgan establishes the following equivalences:

$$X)Y = X \cdot y = y)x; \qquad x)y = x \cdot Y = Y)X;$$

$$X \cdot Y = X)y = Y)x; \qquad x \cdot y = x)Y = y)X;$$

$$XY = X : y = Y : x; \qquad xy = x : Y = y : X;$$

$$X : Y = Xy = y : x; \qquad x : y = xY = Y : X;$$

(Here De Morgan used the sign "=" in the sense of equivalence; he used this sign in a different sense in the expression $X)Y + Y)Z = X)Z$, where "=" means "consequently".)

Using the sign "+" for the conjunction of simple propositions, De Morgan constructs complex propositions or relations:

$$D = A_1 + A' = X)Y + x)y; \qquad C = E_1 + E' = X \cdot Y + x \cdot y;$$

$$D_1 = A_1 + O' = X)Y + x : y; \quad C_1 = E_1 + I' = X \cdot Y + xy;$$

$$D' = A' + O_1 = xy + X : Y; \qquad C' = E' + I_1 = x \cdot y + XY;$$

$$P = I_1 + I' + O_1 + O'.$$

In commonly used symbolic vocabulary, D is "$X = Y$", D_1 is "$X \subset Y$", D' is "$X \supset Y$", C is "$X = y$", C_1 is "$X \subset y$", C' is "$X \supset y$", and P means that each of the classes $X \cap Y$, $X \cap y$, $x \cap Y$, $x \cap y$ is nonempty.

De Morgan employs the relations thus obtained in order to construct his theory of syllogisms. He refers to a syllogism whose premise is a complex proposition as a complex syllogism. In his view, both complex propositions and complex syllogisms are simpler than the traditional "simple" ones.

Without going into the details of De Morgan's theory of syllogisms, we note the following. For his syllogistic vocabulary De Morgan first uses "simple" name expressions (simple terms), X, Y, Z, etc. Then he considers complex name expressions (complex terms): PQ is the name of that "which is both P

and Q" (that is, $P \cap Q$ in the symbolism of set theory), $P * Q^{11}$ is the name expression for "P or Q, or both P and Q" (that is, $P \cup Q$). U is the name for the universe and u denotes the negation of the universe (that is, the empty set). Introducing what we now call operations of intersection and union De Morgan establishes properties such as

$$XU = X; \quad Xu = u; \quad X * U = U; \quad X * u = X,$$

formulates what are now known as De Morgan's laws:

the negation of PQ is $p * q$; the negation of $P * Q$ is pq,

proves distributivity:

$$(P * Q)(R * S) = PR * PS * QR * QS,$$

and constructs a theory of syllogisms using complex name expressions as subjects and predicates of the premises.[12]

In this way De Morgan employed a system in his theory of syllogisms later to be known as "Boolean algebra". This system was first explicitly formulated by William Stanley Jevons.

De Morgan later introduced the general concept of a relation and of operations on relations, thereby laying the foundation for the modern theory of relations. The latter was to be developed in various directions and from different perspectives by Charles Sanders Peirce, Ernst Schröder, Giuseppe Peano, Georg Cantor, Gottlob Frege, and Bertrand Russell.

De Morgan was frequently accused of dissipating his brilliant and subtle ideas by overloading his presentation with details and new notations. By emphasizing various nuances of the meaning of terms he made his ideas still more difficult to grasp, diluted their impact and obscured the picture of the system as a whole. However, the profundity and accuracy of his observations, combined with the diversity of his ideas, suggest that De Morgan's contribution has not yet been sufficiently explored.

Boole's Algebra of Logic

George Boole's (1815–1864) work on mathematical logic appeared almost simultaneously with that of De Morgan. Boole's father, a shoemaker, took a great deal of interest in mathematics and the crafting of optical instruments. His business talents, though, were rather mediocre and he was not able to

---

11 De Morgan denotes such a complex name by "P,Q".

12 De Morgan, A., *Formal logic* ... , Ch. 6. London 1847.

provide his son with a higher education. Young Boole completed elementary school and was otherwise self-taught. In this way he learned several languages.

Early in life he decided to devote himself entirely to mathematics. By himself, he exhaustively studied Newton's *Principia Mathematica*, Lagrange's *Mécanique analytique*, and other works. He earned his living as a teacher. Boole began his mathematical research by developing operator methods in analysis and in the theory of differential equations. Later Boole, like his good friend De Morgan, became involved in mathematical logic.

Boole's youngest daughter Ethel-Lilian, Voinich by marriage, achieved fame as author of *The Gadfly*, a novel about the struggle of the Carbonari for Italian independence.

Two of Boole's basic works bear characteristic titles. They are *The Mathematical Analysis of Logic, Being an Essay Towards a Calculus of Deductive Reasoning* (Cambridge, 1847), and *An Investigation Into the Laws of Thought, on Which are Founded the Mathematical Theories of Logic and Probabilities* (London, 1854), in which he laid the foundations of modern mathematical logic.[13]

It was Boole who provided a distinct, "quantitative" interpretation of logical objects and consistently applied this new approach to the solution of logical problems.

This approach inevitably demanded a change and extension of the scope of traditional logic as well as of algebraic symbolism: the choice of the appropriate symbolism, operations, and laws defining these operations and reflecting the specific nature of the objects of investigation, that is, in effect, the creation of a new calculus.

Boole pointed out that the essence of the new calculus lay in the specific nature of the operations and the laws which these operations obey:

> They who are acquainted with the present state of the theory of symbolical Algebra, are aware, that the validity of the process of analysis does not depend upon the interpretation of the symbols which are employed, but solely upon the laws of their combination. Every system of interpretation which does not affect the truth of the relations supposed, is equally admissible, and it is thus that the same process may, under one scheme of interpretation, represent the solution of a question on the properties of numbers, under another, that of a geometrical problem, and under a third, that of a problem of dynamics or optics. This principle is indeed of fundamental importance...[14]

---

13 Concerning De Morgan's and Boole's interest in the principles of probability, see pp. 279–280.

14 Boole, G., *The mathematical analysis of logic, being an essay towards a calculus of deductive reasoning.* Cambridge 1847, p. 3.

14

George Boole

Boole's contribution to the calculus includes a clear understanding of its abstract nature and of the fact that the calculus obeys the same laws as operations.

As initial symbols Boole takes[15]

1. x, y, z as symbols of classes;

2. $+$, $-$, $\times$ as operational symbols; instead of the symbol $\times$ he often uses the symbol $\cdot$ or omits the symbol altogether.

3. $=$ as the symbol of identity.

Boole denotes by $x \cdot y$ the class of all those and only those members which are members of both classes x and y.

If the classes x and y have no member in common, then $x + y$ represents the class consisting of the members of x and the members of y.

---

15 Boole did not construct his system as a calculus in the modern sense of the term; he did not separate postulated properties of operations from deducible ones.

If each member of the class y is a member of the class x then $x - y$ represents the class consisting of all those and only those members of x which are not members of y.

Thus Boole introduces three operations on the set of classes: $\cdot$, $+$, $-$; the last two operations are not defined on every pair of classes. Boole notes the following elementary properties of these operations:

1. $x \cdot y = y \cdot x$;
2. $x + y = y + x$; here equality is "two-directional": if $x + y$ is admissible, so is $y + x$, and both represent the same class, and conversely;
3. $x \cdot x = x$;
4. more generally, if x is a subclass of y, $x \cdot y = x$, and this condition is sufficient for x to be a subclass of y;
5. $z \cdot (x + y) = z \cdot x + z \cdot y$; here equality is "one-directional": If $z \cdot (x + y)$ is admissible, that is, if the classes x and y do not intersect, then $z \cdot x + z \cdot y$ is also admissible and represents the same class, but not conversely — the fact that $z \cdot x$ does not intersect $z \cdot y$ does not, in general, imply that x does not intersect y. Although Boole does not stipulate this, he always uses, so to say, distributive classes;
6. $z \cdot (x - y) = z \cdot x - z \cdot y$. The comment in 5. applies to this equality.

Boole notes that the condition "$0 \cdot y = 0$ for any class y" will be *a priori* satisfied if the symbol "0" represents "Nothing", the empty class. Furthermore, the condition "$1 \cdot y = y$ for any class y" will be satisfied *a priori* if "1" represents a class such that any class is its subclass. Boole calls this class the universal class (the "universe").

If the system of classes includes the universe 1, then we have the law of contradiction: for any class x

$$x(1 - x) = 0. \tag{1}$$

In fact, $x = x \cdot x$, therefore $x - x \cdot x = 0$. Since x is a subclass of 1, the class $x(1 - x)$ is defined. But by property 5,

$$x(1 - x) = x - x \cdot x = 0.$$

The operations $+$ and $-$ introduced by Boole, are not, in general, defined for all pairs of classes. In view of the particular nature of Boole's definitions, there are two ways of extending these operations to all pairs of classes in line with commonly used set-theoretic operations:

(a) either $x + y$ is the union of x and y no matter whether they intersect or not; or

(b) $x + y$ is the symmetric difference of x and y; in the latter case $x - y$ must also denote the symmetric difference since it is its own inverse.

16

Because of the presence of the universe 1 in the system, case (a) leads, in essence, to the construction of a Boolean algebra, and case (b) to a Boolean ring. The Boolean ring representation is due to I.I. Zhegalkin.[16]

We note that although the operations introduced by Boole are not everywhere defined, they constitute a "complete system of connectives", that is, all set-theoretic operations can be expressed in terms of these connectives, namely, the intersection of x and y can be expressed as $x \cdot y$, the complement of x as $1 - x$, and the union of x and y as $xy + x(1-y) + y(1-x)$. Boole gives this expression for the union of classes in terms of his operation of addition.[17]

Boole uses the terms "logical equation" and "logical function" to denote an equation or a function containing class symbols $x, y, \ldots$. For example,

$$f(x) = x, \quad f(x) = \frac{1+x}{1-x}, \quad f(x,y) = \frac{x+y}{x-2y}.$$

If the equality $x(1-x) = 0$ is regarded as an algebraic equation, that is, if it is assumed that x takes on numerical values, then this equation has the roots 0 and 1. Boole writes:

> Hence, instead of determining the measure of formal agreement of the symbols of Logic with those of Number generally, it is more immediately suggested to us to compare them with symbols of quantity *admitting only of the values* 0 *and* 1.[18]

Proceeding from this notion, Boole adopts the following general method for treating logical functions and logical equations:

> But as the formal processes of reasoning depend only upon the laws of the symbols, and not upon the nature of their interpretation, we are permitted to treat the above symbols, x, y, z, as if they were quantitative symbols of the kind described above. *We may in fact lay aside* the logical interpretation of the symbols in the given equation; convert them into quantitative symbols, susceptible only of the values 0 and 1; perform upon them as such all the requisite processes of solution; and finally restore to them their logical interpretation.[19]

In accordance with this convention, one can substitute 0 and 1 in the "logical functions" for the symbols $x, y, z, \ldots$: if $f(x) = \frac{a+x}{a-2x}$, then $f(0) = \frac{a}{a}$

---

16 Zhegalkin, I.I., "On the technique of the propositional calculus in symbolic logic". Matem. sb. **34**, 1927, vyp. 1 (in Russian).

17 Boole, G., *An investigation of the laws of thought...*. London 1854, p. 62.

18 Ibid., p. 41.

19 Ibid., p. 76.

and $f(1) = \frac{a+1}{a-2}$. Boole repeatedly points out that the intermediate results do not have to have a logical interpretation.

Each function $f(x)$ can be written as

$$f(x) = ax + b(1 - x).$$

In fact, if $x = 1$, then $f(1) = a$, and if $x = 0$, then $f(0) = b$, that is,

$$f(x) = f(1)x + f(0)(1 - x).$$

Similarly,

$$f(x, y) = f(1, y)x + f(0, y)(1 - x).$$

But, since

$$f(1, y) = f(1, 1)y + f(1, 0)(1 - y), \quad f(0, y) = f(0, 1)y + f(0, 0)(1 - y),$$

it follows that

$$f(x, y) = f(1, 1)xy + f(1, 0)x(1 - y) + f(0, 1)(1 - x)y$$
$$+ f(0, 0)(1 - x)(1 - y).$$

Boole refers to x and $(1-x)$ in the development of $f(x)$ and to xy, $x(1-y)$, $(1-x)y$, $(1-x)(1-y)$ in the development of $f(x, y)$ as constituents, and formulates a general rule for developing a function into constituents:

1. form a complete system of constituents of the required development,
2. find the coefficients of the development,
3. multiply each constituent by the corresponding coefficient and add the results.

To obtain the constituents in the case of several variables — for example, x, y, z — Boole suggests the following procedure: take as the first constituent the product xyz of the given symbols; change any symbol in this product, say z to $(1 - z)$, thus obtaining a new constituent $xy(1 - z)$. In both of these change any initial symbol, for example y, into $(1 - y)$, then in the four constituents thus obtained change the remaining initial symbol x to $(1 - x)$. This yields a complete system of constituents for x, y, z:

$$xyz, \quad xy(1 - z), \quad x(1 - y)z, \quad x(1 - y)(1 - z), \quad (1 - x)yz,$$
$$(1 - x)y(1 - z), \quad (1 - x)(1 - y)z, \quad (1 - x)(1 - y)(1 - z).$$

A similar procedure yields a system of constituents for any finite number of initial symbols.

18

To determine the coefficient of a constituent in the development of a given function we must replace by 1 those symbols in the representation of this function which enter as factors in this constituent and replace by 0 those symbols whose complements enter as factors in the given constituent. Thus for the function $f(x,y) = \frac{x+y}{x-2y}$, the constituent $(1-x)y$ has the coefficient $f(0,1) = -\frac{1}{2}$, and the constituent $x(1-y)$ has the coefficient $f(1,0) = 1$.

The sum of all the constituents of an expansion is 1. The product of any two constituents is 0. Therefore, the constituents of the development of a logical function can be interpreted as a partition of the universe into pairwise disjoint classes. Thus the class x splits the universe into x and $1-x$; the classes x and y yield xy, $x(1-y)$, $(1-x)y$, $(1-x)(1-y)$, etc.

Boole introduces operations on classes in order to use them for solving logical equations.

Let an equation $w = v$ be given, where w designates a class and v a logical function. To solve this equation with respect to any class in v means to express this particular class in terms of the remaining classes. To do this, Boole recommends expressing this class formally using algebraic transformations. If the result is not interpretable in logical terms, for example, if the result is a quotient, then we must develop it into the constituents of all the symbols contained in it, and interpret it according to the following rules:

1. all constituents with coefficient 1 enter the final expression;

2. all constituents with the coefficient 0 are discarded;

3. the coefficient $0/0$ is replaced by the symbol for the indeterminate class v;

4. all constituents whose coefficients are different in form than those given in 1.–3. are equated to 0.

Suppose we are required to solve the equation

$$0 = ax + b(1-x) \tag{2}$$

(where a and b contain no x) with respect to x, that is, to express x in terms of a and b. We write (2) as $(a-b)x + b; = 0$, whence $x = \frac{b}{b-a}$. Since a quotient is not interpretable, we develop the right side into constituents:

$$\frac{b}{b-a} = \frac{1}{0}ab + \frac{0}{-1}a(1-b) + 1(1-a)b + \frac{0}{0}(1-a)(1-b).$$

The result is

$$x = (1-a)b + v(1-a)(1-b), \quad ab = 0.$$

Boole calls the expression $ab = 0$ a necessary and sufficient condition for the relation (2) to hold. When it is satisfied, x is expressed in terms of a and b as

$$x = (1-a)b + v(1-a)(1-b).$$

Boole's system was perfected by later logicians through the simplification of operations, interpretation of non-interpretable expressions, and refinement of the concept of "solving a logical equation".

Jevons' Algebra of Logic

The main logical works of William Stanley Jevons (1835–1882), professor of logic, philosophy, and political economy at Manchester (1866–1876) and London (1876–1880) are: *Pure Logic* (London, 1863), *The Substitution of Similars* (1869), and *The Principles of Science* (London, 1874).

Jevons defines the operations $+$, $\cdot$, $\bar{}$ on a set of classes. By $x \cdot y$ he means the intersection of x and y. He extends the operation $+$ to any two classes. He denotes by $x + y$ the union of x and y (to distinguish this operation from that of addition in Boole's sense, Jevons calls the latter alternation and denotes it by the symbol $\cdot|\cdot$). He does not introduce the operation of subtraction. He denotes by $\bar{x}$ the complement of x, by 0 the empty class, by 1 the universe, and by $x = y$ the identity of classes; he regards classes as identical if they consist of the same members. To express the inclusion of x in y, he writes $x = xy$; he calls such identities partial. Jevons notes the following elementary properties of these operations:

$$xy = yx, \quad x + y = y + x, \qquad x(yz) = (xy)z$$

$$x + (y + z) = (x + y) + z, \quad x(y + z) = xy + xz.$$

In constructing his theory, Jevons employs the law of identity $x = x$, the law of contradiction $x\bar{x} = 0$, the law of the excluded middle $x + \bar{x} = 1$, and the principle of substitution.

Jevons calls the inferences obtained by application of the principle of substitution, direct. For example, the following is a direct inference: from the identity $x = y$ one infers that $xz = yz$, since $xz = xz$, and it is possible to replace x on the left side by an equal y.

Jevons calls the inferences employing the law of contradiction and the law of the excluded middle, indirect. For example: Let $x = xy$, $y = yz$. By the law of the excluded middle, $x = xy + x\bar{y}$, $x = xz + x\bar{z}$. Further,

$$x = xyz + xyz\bar{z} + xy\bar{y}z + xy\bar{y}z\bar{z},$$

yielding $x = xyz$.

To solve the problem of characterizing a class using the specified conditions or logical premises, Jevons suggests the following method, based on the use of the complete disjunctive normal form of the algebra of classes.

Consider for definiteness, three classes x, y and z. First, Jevons writes down the constituents of the three classes; he calls the system of constituents a

W.S. Jevons

logical alphabet. Comparison of the dependence relations between classes and constituents given in the premises shows that some constituents contradict the premises; in this case, according to the law of contradiction, they are to be removed. "The remaining terms can be equated to the required term, and this will yield the desired characteristic."[20]

Example. The premises $x = xy$, $y = yz$ are given. Express the class $\bar{z}$ in terms of the remaining classes. To do this, one needs to associate with the premises the constituents containing $\bar{z}$: $xy\bar{z}$, $x\bar{y}\bar{z}$, $\bar{x}y\bar{z}$, $\bar{x}\bar{y}\bar{z}$. Of these, only $\bar{x}\bar{y}\bar{z}$ does not contradict the premises, therefore $\bar{z} = \bar{x}\bar{y}\bar{z}$, that is, $\bar{z}$ is a subclass of $\bar{x}\bar{y}$. Jevons solves inverse problems as well: given combinations of classes, find the conditions from which these combinations can be derived. The solution of these problems also involves the enumeration of many combinations. Solving problems by means of development into constituents and enumeration is rather tedious, although, as a rule, it can be reduced to the repetition of similar and, for the most part, simple operations: writing down constituents, comparing them with the premises, and removing those which

---

20 Jevons, W.S., *The principles of science*. London 1874.

contradict the initial propositions. The tediousness of these techniques led Jevons to the idea of performing them mechanically.

Since the system of distinct constituents depends only on the number of classes in the problem, it is possible to solve various problems involving the same number of classes by using a list of constituents prepared in advance. This idea is the basis of Jevons' logical board. Logical alphabets of problems (involving up to six variable statements) are written on an ordinary school blackboard. Premises of specific problems are written on the free portion of the board and are compared with the corresponding column of constituents. The contradictory combinations are simply erased.

Jevons noticed that the erasing procedure could also be mechanized. He devised a logical abacus: a slant board with four horizontal ledges and a stack of wooden disks with inscribed constituents something like a tangram alphabet. All constituents of the required number of variable statements are placed on one ledge, and those in the premise on another. The constituents contradicting the premises are removed from the upper row, leaving on it the "logical unit" of the problem. This done, it is possible to solve problems pertaining to individual classes.

Envisaging the possibilities of mechanizing the processes of selecting and comparing constituents, and even the possibility of composing logical equalities, Jevons, after ten years of efforts, devised a logical machine which became known as Jevons' logical machine. He exhibited it in 1870 at the London Royal Society, of which he was a member. He described this machine in his article "On the Mechanical Performance of Logical Inference" (Philos. Trans., 1870, 160, 497–518), and in *The Principles of Science* in 1874.

Jevons' machine resembles a small upright piano or organ (*Figure 3*). The symbols of "machine" operations, "finis" (clearance) and "full stop" (output) as well as the symbol for the logical operation ·|· are inscribed on the keyboard, from left to right, followed by the symbols of the four classes and their complements, repeated twice — to the left and right of the central key representing the copula "is" or "are". The operation labelled "finis" prepares the machine for the solution of the problem; the operation labelled "full stop" fixes a certain state of the solution of the problem, for example, the termination of composing the logical unit, that is, readiness to answer questions concerning classes. Moreover, the machine contains a vertical board with slits and a collection of moving plates with inscribed constituents of the four classes.

The premises of the problem are reduced to the form a is b and the copula "is" or "are" corresponds to inclusion. To feed the identity $a = b$ into the machine, $a = b$ is represented as two inclusions: a is ab, and b is ab. The premises are fed into the machine by pressing keys corresponding to classes. The keys to the left of the key "is" or "are" convey the left side of the relation, the keys to the right convey the right side of the relation. After

22

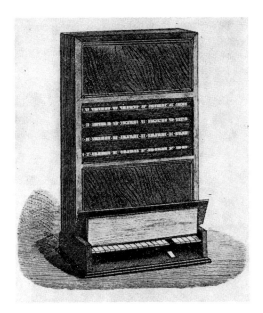

Figure 3. Jevons' machine.

the premises of the problem have been fed into the machine, only the logical unit of the problem is left on the vertical board. This done, it is possible to solve problems concerning individual classes. For instance, by pressing the key "d", we see constituents on the vertical board whose sum is the class d. By pressing the key "full stop", we restore the logical unit of the problem and can now "inquire" about another class. If the conditions relative to a particular class are contradictory, the machine "refuses" to answer questions about the class — this class does not appear on the vertical board at all. In order to bring the machine into the starting position, the "finis" key has to be pressed. As P.S. Poretskiĭ wrote,

> In this position the machine once again imitates the mind that is capable of thinking about classes of things but has no knowledge about them. As the mind receives this knowledge in the form of premises, it starts functioning, either rejecting familiar alternatives or — once again — bringing them back.[21]

Jevons' logical machine (now on display in the Oxford Museum of the History of Science) was, in some respects, a predecessor of the computer. "Playing" this machine as one plays a piano, it was possible to solve some kinds of logical problems more quickly than otherwise. "The logical piano", as Jevons

---

21 Poretskiĭ, P.S., "On ways of solving logical equations...". Kazan' 1884, p. 109 (in Russian).

himself called it, aroused great interest among his contemporaries. In Russia, P.S. Poretskiĭ described this machine in his book (see above) published in 1884. In 1893, I.V. Sleshinskiĭ (1854–1931), a professor at the University of Novorosiĭsk (Odessa), presented a report entitled "Jevons' Logical Machine" to the Novorosiĭsk Society of Natural Scientists. This report was published the same year in *Herald of Experimental Physics and Elementary Mathematics* (Vestnik opytnoĭ fiziki i elementarnoĭ matematiki) in Odessa (Vol. 175, No. 7, pp. 145–154).

Venn's Symbolic Logic

John Venn (1834–1923), son of a clergyman, graduated from a college at Cambridge University in 1858. Following the family tradition, he became a clergyman, but his inclination for scientific studies led him back to Cambridge, where he began teaching logic and moral sciences.

One of Venn's most important works, devoted to the substantiation and development of new methods in mathematical logic, is *Symbolic Logic* (London 1881, 2nd ed. London 1894).

Like Boole and Jevons, Venn regarded as the first task of symbolic logic the creation of a special language "to carry out the design of securing the widest possible extension of our logical processes by the aid of symbols".[22]

Venn uses Latin letters as symbols for classes. Like all logicians of his time, he denoted by 0 and 1 the empty and universal classes, respectively, and by $\bar{x}$ the complement of x. In addition to the operations $\cdot$ and $+$ (by $x + y$ he signifies the union of classes), Venn introduces the operations of subtraction and division. By $x - y$ he signifies the exclusion of the class y from x, provided y is a part of x. He treats this operation as the inverse of the union of classes; he shows that it can be dispensed with, since, by definition,

$$x - y = x \cdot \bar{y}.$$

Venn regards the operation of division as the inverse of the intersection of classes and interprets it as follows: x/y represents a class z such that $x = yz$; one can take as z an arbitrary class of the form $x + v \cdot \bar{y}$.[23]

According to Venn, $x > 0$ means that the class x is not empty; $x > 0$ is regarded as the negation of $x = 0$. Venn interprets the equality $x = y$ to mean that "there exists no member belonging to x and not belonging to y,

---

22 Venn, J., *Symbolic Logic*. London 1894, p. 2.

23 Division is not an operation in the strict sense of the word, since its "outcome" is not unique. It is clear that division reduces to addition and multiplication and can be dispensed with. Venn retains the operations of subtraction and division to emphasize the analogy between his system and arithmetic.

and no member belonging to y and not belonging to x", and writes it in the "null form"

$$x\bar{y} + \bar{x}y = 0.$$

Venn formulates some properties of the operations, for example, $x(y+z) = xy + xz$, $x = x + xy$, $x = x(x + y)$. He restricts himself to illustrating these properties by means of examples, and makes no attempt to separate postulated properties from those to be proved.

Like other 19th-century algebraic logicians, Venn regarded the solution of equations and the exclusion of unknowns as the most important problems of symbolic logic. In fact, he solved these problems in the same way as other logicians, and we shall not dwell upon his methods for solving equalities. Unlike Boole and Jevons, Venn solved both equations and logical inequalities. A relevant problem follows.

Let

$$ax + b\bar{x} + c > 0. \tag{3}$$

Find the necessary and sufficient condition for the existence of x such that the class $ax + b\bar{x} + c$ is not empty.

The required necessary and sufficient condition is

$$a + b + c > 0. \tag{4}$$

Indeed, it follows from $a + b + c = 0$ that $a = 0$, $b = 0$, $c = 0$, that is, $ax + b\bar{x} + c = 0$ for all x. Let $a + b + c > 0$. This means that at least one of the classes, a, b, c is not empty. If $a > 0$, put $x = a$. Then $aa + b\bar{a} + c > 0$. If $b > 0$, put $x = \bar{b}$. Then $a\bar{b} + bb + c > 0$. If $c > 0$, then $ax + b\bar{x} + c > 0$ for all x. Venn calls condition (4) the result of exclusion of x from (3).

Venn also considers cases where the premises include equations and inequalities. For example,

$$ax + b\bar{x} + c > 0, \quad dx + e\bar{x} + f = 0. \tag{5}$$

The necessary and sufficient conditions for the existence of x such that $ax + b\bar{x} + c$ is not empty and the class $dx + e\bar{x} + f$ is empty is the following:

$$de + f = 0, \quad a\bar{d} + b\bar{e} + c > 0. \tag{6}$$

In fact, it follows from $dx + e\bar{x} + f = 0$ that $dx = 0$, $e\bar{x} = 0$, $f = 0$, $dex = 0$, $de\bar{x} = 0$, $de(x + \bar{x}) = 0$, $de = 0$. If $dx = 0$, $e\bar{x} = 0$, then $x \subset \bar{d}$, $\bar{x} \subset \bar{e}$, whence $ax \subset a\bar{d}$, $b\bar{x} \subset b\bar{e}$. If $ax + b\bar{x} + c > 0$, then a fortiori, $a\bar{d} + b\bar{e} + c > 0$.

Therefore, (5) implies (6). Let (6) be satisfied. Then $de = 0$, $f = 0$, and at least one of the classes $a\bar{d}$, $b\bar{e}$, c is not empty. If $a\bar{d} > 0$, put $x = \bar{d}$. Then

$$ax + b\bar{x} + c = a\bar{d} + bd + c > 0,$$

$$dx + e\bar{x} + f = d\bar{d} + ed + f = 0.$$

If $b\bar{e} > 0$, put $x = e$. Then

$$ax + b\bar{x} + c = ae + b\bar{e} + c > 0,$$

$$dx + e\bar{x} + f = de + e\bar{e} + f = 0.$$

For $c > 0$ the class $ax + b\bar{x} + c$ is not empty for any x; if $x = e$, then $de + e\bar{e} + f = 0$. Therefore, (6) is sufficient for the existence of a class x satisfying (5).

In solving logical problems Venn used algebraic methods as well as what are now known as Venn diagrams.

Venn sketches his diagrams by first dividing the plane into 2^n regions by means of closed curves; here n is the number of classes specified in the statement of the problem. In his problems, Venn considers only the cases where $n \leq 5$. As the number of variables increases, the figures become very complicated. Hence, to graphically represent problems with many classes, Venn uses tables consisting of 2^n cells, i.e., Venn tables. The tables and figures are in one-to-one correspondence: there is a figure based on each table, and, conversely, each figure is the basis for a table. *Figure 4* shows the tables for the cases where $n = 1, 2, 3$. It is possible to go from a table with n variables to one with $n + 1$ variables by doubling the former with respect to a contour line.

Venn illustrated the method of drawing figures and tables by many examples. However, he gives no general definition of the concept of a table. Analyzing these examples, we can say that an n-variable Venn diagram is a figure or a table with n variables whose cells are shaded, empty, or marked with an asterisk. The shaded cells correspond to classes which contradict the statement of the problem, that is, they are subject to removal in accordance with Jevons' method. The unshaded cells constitute the logical unity of the problem. The diagrams without asterisks correspond to formulas in propositional calculus. The asterisks enable one to represent certain special propositions. It should be noted that Venn uses an asterisk in only one example. *Figure 5* represents a Venn diagram with four variables, in which the cells $ab\bar{c}\bar{d}$, $abc\bar{d}$, $\bar{a}bcd$, $ab\bar{c}d$, $\bar{a}bcd$, $\bar{a}\bar{b}cd$, are shaded. This diagram represents the proposition

$$ab\bar{c}\bar{d} + abc\bar{d} + a\bar{b}cd + ab\bar{c}d + \bar{a}bcd + \bar{a}\bar{b}cd = 0.$$

Occasionally, the method of Venn diagrams brings us to the goal much sooner than the analytical method of solving the problem. In his "Lectures on the Algebra of Logic (Exact Logic)" (*Vorlesungen über die Algebra der Logik (exakte Logik)*, Bd. 1–3. Leipzig 1890–1905), Schröder cites Jevons' problem of simplifying the premises:

$$a = b + c, \quad b = \bar{d} + \bar{c}, \ \bar{c}\bar{d} = 0, \quad ad = bcd.$$

Shading all the cells which are supposed to be empty in accordance with the statement of the problem, one obtains from *Figure 6* that $a = b = c = 1$, $d = 0$. The same result can be obtained analytically.

It is often said that Venn took over the notion of Euler's circles and merely perfected Euler's method somewhat. We cannot, however, agree with this opinion.

Although Euler's and Venn's methods both involve the representation of the content of concepts in the plane, the methodological foundation of Venn diagrams, i.e., the development of logical functions into "constituents", one of the key ideas of logical algebra, is not part of Euler's method. Diagrams drawn with the aid of development into constituents are not only more intuitive, but also enable us to obtain more information from the statement of the problem. In addition, and this is another crucial distinction between Euler's and Venn's methods, Venn diagrams are intended not merely for the illustration of solutions but as a tool for actually solving logical problems.

Schröder's and Poretskiĭ's Logical Algebra

The works of Ernst Schröder and P.S. Poretskiĭ were published contemporaneously with Venn's writings.

Ernst Schröder (1841–1902), a German algebraist and logician, served from 1874 as professor at the Polytechnic in Darmstadt, and from 1876 at the Technical University in Karlsruhe. Schröder continued the development of logical algebra, which he called "logical calculus" (*Logikkalkül*): it was Schröder who originated the term "propositional calculus" (*Aussagenkalkül*). Schröder, like Jevons, believed that the operations of subtraction and division were superfluous in logic. He constructed a system involving the operations $\cdot$, $+$, $^-$, the identity $=$, and the constants 0 and 1. Unlike Boole, Jevons, and Venn, Schröder clearly indicated which properties of the operations he accepted as axioms and which he obtained as theorems. Schröder's first work in mathematical logic, "The Class of Operations of the Logical Calculus" (*Der Operationskreis des Logikkalküls*), published in Leipzig in 1877,[24] contained for the first time the formulation of the duality principle. He summarized the results of his research in this area in his three-volume "Lectures on the Algebra of

---

24 See Bobynin, V.V., "Attempt to give a mathematical exposition of logic. The works of Ernst Schröder". *The physical-mathematical sciences in the past and present, 1886–1894*, 2, pp. 65–72, 178–192, 438–458 (in Russian).

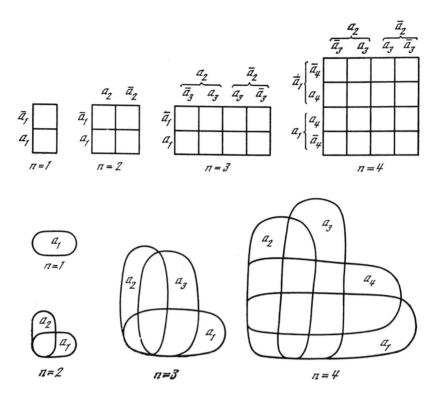

Fig. 4

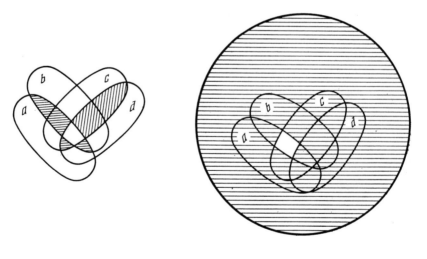

Fig. 5

Fig. 6

Logic (Exact Logic)" (*Vorlesungen über die Algebra der Logik*, Leipzig 1890–1905). The third volume is subtitled "Algebra and the Logic of Relations" (*Algebra und Logik der Relative*), and is a comprehensive discussion of the calculus of relations. Schröder also studied general properties of "calculi" and devised a version of what is now known as the theory of quasi-groups.

Schröder, like other mathematicians in this discipline, regarded the solution of logical equations to be one of the central problems of logical algebra. We will pay particular attention to his treatment of this matter.

Schröder reduces each equality to the form

$$ax + b\bar{x} = 0. \tag{7}$$

This can be done, since any equality $y = z$ is equivalent to the equality $y\bar{z} + \bar{y}z = 0$, and development of the left side of the latter into constituents with respect to x, y, z yields

$$(y\bar{z} + \bar{y}z)x + (y\bar{z} + \bar{y}z)\bar{x} = 0,$$

that is, an equality of the form $ax + b\bar{x} = 0$.

Among equalities of this form, Schröder distinguishes between "analytic" or identically true, for example, $x\bar{x} + \bar{x}x = 0$, and "synthetic", or true only for some x. Analytic equalities could not, according to Schröder, be equations because he regarded an equation as a condition which x must satisfy, that is, turn this equation into an identity. According to Schröder, solving an equation means discovering whether it has a solution, that is, discovering whether there exist expressions which when substituted for x in the equation turn it into an identity, and, if such expressions exist, finding some general form from which all solutions of equation (7) can be obtained.

The answer to the first question is provided by the result of exclusion of x from equation (7), or by the resolvent $ab = 0$. As Boole noted earlier (see above), the latter is a necessary and sufficient condition for Eq. (7) to have a solution, and the general form of the solution is

$$x = b\bar{u} + \bar{a}u \tag{8}$$

for any value u (see Venn's solution).

If the condition $ab = 0$ is satisfied, then the substitution $x = b\bar{u} + \bar{a}u$ turns Eq. (7) into
$$a(b\bar{u} + \bar{a}u) + b\overline{(b\bar{u} + \bar{a}u)} = 0,$$

that is
$$ab\bar{u} + a\bar{a}u + b\bar{b}a + b\bar{b}\bar{u} + bua + bu\bar{u} = 0.$$

E. Schröder

If $ab = 0$ is identically zero, then

$$0\bar{u} + 0u + 0a + 0\bar{u} + 0u + b0 = 0,$$

that is, a true identity. Here it is essential that u is an arbitrary class. Conversely, if some x satisfies (7), then, for that x, the relation $x = b\bar{u} + \bar{a}u$ holds with $u = x$ for the same x.

In other words, Schröder actually showed that for each x the equality $ax + b\bar{x} = 0$ is equivalent to the fact $ab = 0$ holds and there exists a class u such that $x = b\bar{u} + \bar{a}u$.

The assertion concerning the existence of u such that x determined by the equality (8) satisfies equation (7), is equivalent to the assertion that any u satisfying (8) yields a solution of (7).

Thus, according to Schröder, solving equation (7) is equivalent to replacing it by the following two conditions:

1. $ab = 0$ and

2. "there exists a class u such that $x = b\bar{u} + \bar{a}u$."

Naturally, the relation (8) is a logical consequence of (7) only if (8) is taken together with the assertion of the existence of u, although, conversely, for each u the relation (7) is a logical consequence of (8).

An important role in both the development of logical algebra and in its dissemination throughout Russia was played by the distinguished Russian scholar Platon Sergeevich Poretskiĭ (1846–1907). The son of a military physician, Poretskiĭ graduated from the Department of Physics and Mathematics of the University of Kharkov in 1870, with a major in astronomy. In 1876 he began his career as an astronomer at the University of Kazan, where he defended his doctoral dissertation in astronomy in 1886 and taught courses on astronomy and mathematics. Poretskiĭ taught the first course in mathematical logic in Russia at Kazan. From 1881 to 1904 a number of his works devoted to problems of logical algebra were published. We touch only on some aspects of his research in this area, presented in his first major work, "On Methods of Solving Logical Equalities and an Inverse Method of Mathematical Logic" (Kazan, 1884). This work served as the basis for his previously mentioned lecture course. We will not discuss his subsequent writings since they either systematized and developed the ideas of this book, or only referred in a limited way to the calculus of propositions.

Poretskiĭ regarded a logical equation not as a condition which needs to be satisfied but as a premise from which it is necessary to derive all or some logical conclusions of a certain type. In view of this, he proposed a different definition of what is meant by solving a logical equation.

Poretskiĭ reduces the premise given by an equality of the form $x = y$ to $x\bar{y} + \bar{x}y = 0$, which he calls the logical zero of the premise, or to $xy + \bar{x}\bar{y} = 1$, which he calls the logical unit of the premise. He calls the sum of the logical zeros of premises the logical zero of the problem, and denotes it by $N(x, y, z, \ldots)$. He calls the product of the logical units of premises the logical unit of the problem, and denotes it by $M(x, y, z, \ldots)$. He thus formulates a method of solving the problem: in order to define a class x in terms of the remaining classes, it suffices to multiply this class by a function obtained from the logical unit of the problem by substituting 1 for the class to be determined and 0 for its complement:

$$x = x \cdot M(1, y, z, \ldots),$$
$$\bar{x} = \bar{x} \cdot M(0, y, z, \ldots).$$

Using De Morgan's law, we obtain from the second equality $x = x + \bar{M}(0, y, \ldots)$, and therefore two expressions for x, namely,

$$x = x \cdot M(1, y, z, \ldots)$$

and

$$x = x + \bar{M}(0, y, \ldots).$$

(9)

P.S. Poretskiĭ

Poretskiĭ calls the last two expressions a complete solution, because "fragmentary information suggested in the premises can be condensed in two simple formulas intended for drawing a complete picture of the role played by each given class".[25]

Furthermore, Poretskiĭ poses the question as to whether it is possible to characterize the class x using not all the information contained in the premises and relevant as well to other classes, but only the information concerning the class x. This characteristic of the class x is the solution

$$x = x \cdot M(1, y, \ldots),$$
$$x = x + \bar{M}(0, y, \ldots) \cdot M(1, y, \ldots), \tag{10}$$

which Poretskiĭ calls an exact solution.

---

25 Poretskiĭ, P.S., "On methods of solution of logical equalities...". Kazan' 1884, p. 65 (in Russian).

As mentioned earlier, Poretskiĭ meant by the solution of an equation, the deduction of consequences from primary information. An exact solution is obtained by leaving out some terms equal to zero. It is sometimes said that it is obtained by the exclusion of information.

In his "Algebra of Logic" (*L'algèbre de la logique*, Paris 1905), which summarized the progress of 19th-century logical algebra and was strongly influenced by Poretskiĭ, Louis Couturat wrote:

> In logic, the difference between known and unknown terms is artificial and virtually useless: all terms are essentially known, and what one wants to do is deduce from given relations that hold for them, new relations (that is, unknown or only implicitly known relations). This is the aim of Poretskiĭ's method...

We note that I.V. Sleshinskiĭ, mentioned earlier in this chapter and noted for making Couturat's derivation of basic formulas more precise in his appendices to his Russian translation of Couturat's book (1909), became, after Poretskiĭ, one of the initiators of the study of mathematical logic in Russia and in particular in Odessa. In Odessa, as early as 1896–1899, E.L. Bunitskiĭ (1874–1921) published his articles on logical algebra and its applications in arithmetic. In 1901 S.O. Shatunovskiĭ (1859–1929) delivered a report to the Society of Natural Scientists at Novorossiĭsk on the applicability of the law of the excluded middle to infinite sets. Shatunovskiĭ's ideas on the subject, however, were published only in 1917. In essence, Shatunovskiĭ claimed that the application of the law of the excluded middle required complete accuracy in each individual case in the definition of both the set and the element under consideration, though he did not reject, as did Brouwer, the possibility of such application in the general case. This problem, however, lies far beyond the scope of the present chapter.

Conclusion

Mathematical logic developed in the 19th century primarily in the form of logical algebra. The analogy leading to the creation of logical algebra lay in the fact that each solution of a problem, *via* the setting up and solution of an equation, is in essence the derivation of consequences from the statement of the problem. The idea, developed by Boole, was to extend algebraic methods beyond quantitative problems. It was necessary, of course, to find a way of expressing information in terms of equalities or inequalities and, in addition, of establishing rules for processing this information. The search for such rules of operation led to the creation of an algebraic system now called Boolean algebra, which first appeared in Jevons' works and was later improved upon by Venn, Schröder, and Poretskiĭ. The relations obtained were, as a rule, interpreted in the language of classes — the content of concepts. At the same time, logical relations among propositions were made more precise, thus facilitating the process whereby the investigation of these relations was made into a "calculus of propositions" by Frege at the end of the 19th century. By that time, the introduction of quantifiers — abbreviations for "all" and "some" — created a veritable revolution in logic. The present chapter does not, however, deal with these issues.

Chapter Two
Algebra and Algebraic
Number Theory

1 Survey of the Evolution of Algebra and of the Theory of Algebraic Numbers During the Period of 1800–1870

The 19th century was an age of deep qualitative transformations and, at the same time, of great discoveries in all areas of mathematics, including algebra. The transformation of algebra was fundamental in nature. Between the beginning and the end of the last century, or rather between the beginning of the last century and the twenties of this century, the subject matter and methods of algebra and its place in mathematics changed beyond recognition.

It is difficult to describe precisely what algebra was at the end of the 18th century. Of course, it was no longer the art of computing with numbers, letters, and mysterious magnitudes, or an art involving a handful of rules and formulas and the skill of their correct interpretation. Complex numbers were virtually accepted by all, there existed something like a theory of linear equations, some principles were being outlined and so were the beginnings of a theory of equations of one variable of arbitrary degree. But next to the majestic edifice of analysis all this paled into insignificance. Algebra was on the periphery of mathematics. By the beginning of the 20th century all this had radically changed. Algebra had grown prodigiously in content, had been enriched by remarkable concepts and theories and, moreover, its new concepts and spirit began to penetrate virtually the whole of mathematics. There was a manifest tendency toward the algebraization of mathematics. Remarkable new disciplines, such as algebraic number theory, algebraic geometry, the theory of Lie groups, combining algebra with number theory, geometry and analysis respectively, came into being and flourished. Much like the fundamentals of analysis, so too the fundamentals of new algebraic theories, such as groups, fields, and vector spaces, became indispensable components of general mathematical education not only in the universities but also in engineering and technical schools.

The period up to the seventies of the last century was a basically latent period of the explosive growth of algebra. Here the historian of science is faced with the task of reconstructing the mysterious process of birth, growth and interaction of mathematical ideas on the basis of notes, papers, letters and recollections. What makes this task particularly difficult is that in the history of every great mathematical thought there is a period of implicit existence when, unbeknownst to contemporaries, it begins to turn up here and there in the masquerade garb of special cases and applications. And when it suddenly appears in all its beauty, it is not immediately possible — and at times altogether impossible — to determine who helped it make the decisive step of appearing on the stage.

During the period under consideration only some of the already-created new algebraic theories had explicitly entered mathematics. They included the elements of the theories of groups, fields and algebras, linear algebra (the by then significantly advanced theory of linear algebraic equations, linear transformations, and quadratic forms), and, around the seventies, Galois theory, linking new and old algebra, and algebraic number theory, a theory at once flourishing and rich in algebraic ideas.

Before embarking on a more detailed investigation of the evolution of algebra during the period 1800–1870 we will describe in general terms the main stages and paths of this evolution.

The first major event of this period was the appearance, in 1801, of C.F. Gauss' *Disquisitiones Arithmeticae*. Of the seven parts of the book only one is devoted to an algebraic issue, namely the cyclotomic equation $x^n - 1 = 0$. But the author's brilliant algebraic thinking is apparent in all the other parts as well. *Disquisitiones*, an epoch-making work in algebraic number theory, was for a long time a handbook and source of ideas in algebra. In the course of his study of the cyclotomic equation Gauss shows that it is solvable for every n in the sense that the solutions are expressible in terms of radicals, gives a method for explicitly finding these expressions, and singles out the values of n for which the solutions are expressible in quadratic radicals and thus the values of n for which it is possible to construct a regular n-gon by means of ruler and compass. As always, his investigations are strikingly profound and detailed. They were continued by N.H. Abel, who proved the insolvability by radicals of the general quintic and singled out a class of equations, now named after him, that are solvable by radicals. The new notions of field (domain of rationality) and group (group of an equation) turned up in Abel's papers with greater definiteness. The next step in this direction that completed the theory was the papers of the young E. Galois, published in fragmentary form between 1830 and 1832, and, after his death, in more complete form by Liouville in 1846.

36

The papers of Abel, and especially of Galois, already belong to the radically new trend of ideas now generally accepted in algebra. In his study of the ancient problem of solution of equations by radicals Galois shifted the center of gravity from the problem to the methods of its solution: he gave clear-cut definitions of the concepts of a field and of the group of an equation, established the correspondence between the subgroups of the group of an equation and the subfields of the splitting field of the polynomial on the left side of that equation, and, finally, singled out the normal subgroups of a group and studied its composition series. These were completely new and extremely fruitful methods of investigation and yet they were apprehended by mathematicians only in the 70s. The one exception was groups of substitutions. Such groups were considered by Galois and their investigation began already in the 40s.

Another source of group theory was Gauss' theory of composition of classes of forms. In this theory one applied an operation analogous to addition (or multiplication) of numbers to objects very different from numbers. Gauss' study of forms of the same discriminant was in effect a study of the fundamental properties of cyclic and general abelian groups.

The two parts of Gauss' remarkable paper "The theory of biquadratic residues" appeared in 1828 and 1832 respectively. In it Gauss not only gave a geometric interpretation of the complex numbers (this was done before him) but also — and this is very important — transferred to complex numbers the notion of a whole number, a concept that had seemed inseparable from the rational integers for more than 2000 years.

Gauss constructed an arithmetic of complex integers entirely analogous to the usual arithmetic and used the new numbers to formulate the law of biquadratic reciprocity. This opened for arithmetic boundless new horizons. Soon Eisenstein and Jacobi formulated and proved the law of cubic reciprocity and used for this purpose numbers of the form $K + m\rho$, $\rho^3 = 1$, $\rho \neq 1$, and in 1846 P. Lejeune-Dirichlet found all units (that is invertible elements) of the ring of integers of the field $\mathbb{Q}(\theta)$, where θ is a root of

$$x^n + a_1 x^{n-1} + \ldots + a_n = 0,$$

$a_i \in \mathbb{Z}$.[1] This paper, with its deep results in the theory of algebraic numbers, is also of interest from the point of view of group theory: in it Dirichlet constructed the first nontrivial example of an infinite abelian group and investigated its structure.

Further progress in algebraic number theory was linked to reciprocity laws and to Fermat's last theorem. Attempts to prove this theorem brought E. Kummer to the study of the arithmetic of fields $\mathbb{Q}(\zeta)$, $\zeta^p = 1$, $\zeta \neq 1$. In

---

1 $\mathbb{Z}$ is the ring of integers and $\mathbb{Q}$ is the field of rational numbers.

1844–1847 Kummer discovered that if one defines a "prime" number to be an indecomposable integer in a field $\mathbb{Q}(\zeta)$, then the law of unique factorization into prime factors fails for the integers in $\mathbb{Q}(\zeta)$. To "save the day" and restore the possibility of constructing an arithmetic analogous to the usual (arithmetic) he introduced ideal factors. In so doing, he laid the foundations for the subtlest and most abstract theories of algebraic number theory. Kummer's methods were local. They were further developed by E.I. Zolotarev, K. Hensel, and others, and now form the core of commutative algebra.

Linear algebra continued to develop in the first half of the 19th century. In this connection, the first thing to be noted is that whereas no part of Gauss' *Disquisitiones* deals directly with linear algebra, its advance was bound to be furthered by the detailed study of integral quadratic forms in two variables contained in that work. A. Cauchy's "On an equation for the determination of the secular inequalities of planetary motions" (1826) dealt implicitly with the eigenvalues of matrices of arbitrary order. Somewhat later, in 1834, there appeared C.G.J. Jacobi's "On the transformation of two arbitrary homogeneous functions of the second order by means of linear substitutions into two others containing only squares of the variables; together with many theorems on the transformation of multiple integrals" in which he explicitly studied quadratic forms and their reduction to canonical form. Jacobi also perfected the theory of determinants (1841). What was still lacking in this theory was geometric features and, above all, the all-important and fundamental notion of a linear space. The first, none-too-clear definition of a linear space was given by H. Grassmann in his *Die lineale Ausdehnungslehre* of 1844. This work, rich in new ideas but written in a muddled manner, first attracted attention when its author published a reworked and improved version in 1862. In particular, the work contains a construction of exterior products and the now famous Grassmann algebra. In 1843 there appeared A. Cayley's *Chapters in the analytical geometry of (n) dimensions*, a work less rich in ideas but better known to contemporary mathematicians. There is a close connection between the development of linear algebra and the theory of hypercomplex numbers (now known as the theory of algebras) which elicited considerable interest at the time. Years of fruitless attempts to generalize the complex numbers were crowned with success in 1843 by W.R. Hamilton's discovery of the quaternions. Hamilton studied the quaternions for over 20 years, for the rest of his life. His researches are summarized in two fundamental works: *Lectures on quaternions* (1853) and *Elements of the theory of quaternions* (1866). Their subsequent significance is due not so much to quaternions but to the new notions and methods of "vector calculus" introduced in this connection.

To resume our account of the further development of group theory we mention the series of A. Cauchy's papers, published between 1844 and 1846, in which he proves a great variety of theorems on groups of substitutions (subgroups of the symmetric group), including the famous theorem of Cauchy to the effect that a group whose order is divisible by a prime p contains an element of order p. A further major event in the history of group theory was the publication — in threee parts (1854, 1854, 1859) — of Cayley's paper *On the theory of groups, as depending on the symbolic equation* $\theta^n = 1$. Following the spirit of the English school, Cayley views a group as an abstract set of symbols with a given law of composition and defines a number of fundamental notions of abstract group theory, chief among them being the notions of a group and of isomorphism. This was a notable step in the evolution of the new abstract mathematical thinking.

Of crucial importance for the further development of group theory was the appearance, in 1870, of C. Jordan's fundamental *Traité des substitutions et des équations algébriques*. This work contained the first systematic and complete exposition of Galois theory as well as a detailed presentation of results in group theory up to that time, including Jordan's own significant results in these areas. In it Jordan also introduced what is now known as the Jordan canonical form of matrices of linear transformations. The publication of Jordan's work was a major event in all of mathematics.

Mention must be made of the flourishing, in the middle of the 19th century, of an area of algebra, intermediate between linear algebra and algebraic geometry, known as the theory of invariants. On the one hand, its content consists in the generalization and development of topics in linear algebra such as reduction to canonical form of quadratic forms and matrices of linear transformations. On the other hand, it is the study, in concrete situations, of the answer to the following question: "Given a geometric object determined in some coordinate system by certain algebraic conditions, find a way of obtaining from the algebraic conditions geometric characteristics of the object that are invariant with respect to coordinate transformations." Between 1840–1870 many of the works of various mathematicians dealt with the determination of systems of invariants in different concrete situations. The best known are the works of Cayley, Eisenstein, Sylvester, Salmon and Clebsch. In this connection one must single out two papers by Hesse, published in 1844 and 1851, respectively, in which he introduced the notion of a hessian and applied it to geometry, and P. Gordan's famous 1868 paper in which he proved a general algebraic theorem on the existence of a finite system of base invariants. An important paper close to these investigations is Cayley's *A sixth memoir upon quantics* (1859). In it Cayley showed how to consider the metric properties of geometric figures from the single viewpoint of the theory of invariants. This paper was one of the sources of F. Klein's Erlangen Program that resulted in revolutionary changes in geometry.

At that time, an important achievement in linear algebra was Sylvester's 1852 proof of the law of inertia of quadratic forms, presented in the paper *Proof of the theorem that every homogeneous quadratic polynomial can be reduced by means of a real orthogonal substitution to the form of a sum of positive and negative squares*. It was proved, but not published, somewhat earlier by Jacobi. In 1858 there appeared Cayley's *Memoir on the theory of matrices.* In it Cayley introduced the algebra of square matrices and established the isomorphism between the algebra of quaternions and a certain algebra of second-order matrices (a subalgebra of the algebra of all square second-order complex matrices). This work was of great importance for the clarification of the relation between the theory of algebras and linear algebra.

In the sixties, the activities of K. Weierstrass had an important influence on the development of mathematics. He published virtually nothing but included the results of his investigations in his lectures at Berlin University. In his 1861 lectures Weierstrass introduced the notion of a direct sum of algebras and showed that every (finite dimensional) commutative algebra (over the field of real numbers) without nilpotent elements is the direct sum of copies of the fields of real and complex numbers. This was one of the earliest classification results in algebra.

One of the main problems of algebraic number theory in the sixties and seventies was the extension of Kummer's divisibility theory from cyclotomic fields to general algebraic number fields. This was accomplished in three different constructions due, respectively, to E.I. Zolotarev, R. Dedekind, and L. Kronecker. Of the three, it was Dedekind's work — the Xth Supplement to Dirichlet's lectures on number theory published in 1871 and the XIth Supplement to subsequent editions — that was accepted by all mathematicians as the solution of the problem. Dedekind's clear, algebraically transparent, account became the model of mathematical style for many decades to come. By this and other works Dedekind laid the foundations of the contemporary axiomatic presentation of mathematical theories.

In our survey of the evolution of algebra we have not touched on the theory of elliptic and abelian functions — one of the central lines of development of 19th-century mathematics, an area in which Gauss, Abel, Jacobi, Clebsch, Gordan, Weierstrass and many others invested great efforts. In the 19th century this area belonged primarily to analysis, more specifically to the theory of functions of a complex variable, and it was only gradually, especially at the end of the 19th century, that the role of algebraic ideas in it became very significant.

The algebraization of the area began with Dedekind's transfer of his theory, in a joint work with H. Weber (1882), to the field of algebraic functions. This established the deep parallelism between the theories of algebraic numbers and algebraic functions and was the decisive step for an abstract definition of the concepts of field, module, ring, and ideal. From the end of the last century ideas began to flow in the opposite direction, from the theory of algebraic functions to number theory. This resulted in the introduction of p-adic numbers and topology by means of p-adic metrics. But this is already part of the mathematics of the present century.

The evolution of the ideas, methods, and theories just described resulted in the creation of abstract "modern algebra" and, later, of algebraic geometry whose flourishing we witness today.

2 The Evolution of Algebra

Algebraic Proofs of the Fundamental Theorem of Algebra in the 18th Century

The fundamental theorem of algebra was first stated by P. Rothe, A. Girard and R. Descartes in the first half of the 17th century. Their formulations were very different from the modern one. Thus Girard claimed that an nth degree equation has exactly n roots, real or imaginary, without assigning a clear meaning to the latter term and Descartes merely stated that the number of roots of an algebraic equation can be the same as its degree.[2]

In the 40s of the 18th century Maclaurin and Euler gave a formulation of the fundamental theorem equivalent to the modern one: every polynomial with real coefficients can be written as a product of factors of first and second degree with real coefficients; in other words, an equation of degree n has n real and complex roots.

The first proof of the fundamental theorem was given in 1746 by d'Alembert. While 18th-century scientists saw no flaws in the proof, they felt that it smacked of analysis. Mathematicians tried to justify the fundamental theorem in a purely algebraic manner, relying solely on the theory of equations. We now know that, while it is possible to reduce the use of continuity to a minimum, we cannot produce a proof that eschews the use of such properties altogether. The first "maximally algebraic" proof of the fundamental theorem is due to Leonhard Euler.

Euler's "Investigations of imaginary roots of equations" (*Recherches sur les racines imaginaires des équations*), which contained a proof of the fundamental theorem of algebra, was published in 1751 in the *Memoirs* of the Berlin Academy of Sciences for 1749. The Latin variant of the paper (*Theoremata de radicibus aequationum imaginariis*) was presented by Euler to the Berlin Academy of Sciences as early as November 10, 1746. This means that Euler and d'Alembert carried out their respective researches almost simultaneously. Nevertheless their starting principles were completely different.

We will not consider d'Alembert's proof. For one thing, it is quite well known. For another, it has no elements in common with the works of Euler. Unlike d'Alembert's proof, Euler's proof is almost forgotten today. And yet this proof is based on an idea that was subsequently used, with or without modifications, in all so-called algebraic proofs of the fundamental theorem. The proofs might be short or long, pedestrian or refined, flawless or flawed, but in all of them the basic idea remained the same.

One more thing. In the process of proving the theorem Euler made first use of methods of investigation of equations that were subsequently developed by Lagrange and became fundamental in his papers devoted to the question of

---

2 See HM, vol. 2, pp. 24–25, 42; vol. 3, pp. 74–76.

solution of equations by radicals. Also, these methods later entered Galois theory as indispensable constituent components.

The modern "algebraic proofs" of the fundamental theorem can be divided into three parts: (1) the topological proposition to the effect that every algebraic equation $f(x) = 0$ of odd degree with real coefficients has a real root; (2) the construction of the splitting field of the polynomial $f(x)$, that is, of the ["minimal"] field in which $f(x)$ can be written as a product of linear factors; and (3) the step that reduces the problem of finding a root of the equation $f(x) = 0$ of degree $m = 2^k r$, r odd, to that of finding a root of an equation $F(x) = 0$ of degree $2^{k-1} r_1$, r_1 odd.

All three parts can be found in Euler's proof: he states the topological assertion and regards it as obvious. He then assumes that every polynomial with real coefficients can be written as a product

$$f_m(x) = (x - \alpha_1)(x - \alpha_2)\ldots(x - \alpha_m), \tag{1}$$

where $\alpha_1, \ldots, \alpha_m$ are certain symbols or imaginary quantities of which we know nothing beforehand other than that we may apply to them the usual arithmetical operations in accordance with the same rules that apply to ordinary numbers (in other words, addition and multiplication of these symbols are commutative, multiplication is distributive over addition, and so on). Operating with the symbols $\alpha_1, \ldots, \alpha_m$ Euler carried out the reduction process for equations of degree 4, 8, and 16 and sketched it for equations of degree $m = 2^k$. The latter reduction was carried out with all necessary rigor by Lagrange in the paper "On the form of imaginary roots of equations" (*Sur la forme des racines imaginaires des équations.* Nouveaux Mémoires de l'Academie Royale des Sciences et Belles-Lettres de Berlin (172, 1774)). Lagrange made use of theorems on symmetric and similar[3] functions. He proved that the α_i are real or complex numbers.

In his mathematical lectures at the Ecole Normale in 1795 (*Leçons de Mathématiques données à l'Ecole Normale en 1795*) Laplace, relying on the existence of the decomposition (1), described a radical simplification of Euler's reduction procedure. He considers the polynomials

$$F_g(x) = \prod_{1 \le i < j \le m} [x - (\alpha_i + \alpha_j) - s\alpha_i\alpha_j], \tag{2}$$

where s is some real number. He notes that if the degree of the polynomial (1) is $m = 2^i r$, r odd, then the degree of the polynomial (2) is $g = 2^{i-1} r_1$.

---

3 Functions $\varphi(x_1,\ldots,x_n)$ and $\psi(x_1,\ldots,x_n)$ of the roots of an equation of degree n are called similar if they belong to the same subgroup H of the group S_n of permutations of the roots of this equation, that is, they are unchanged under the permutations in H and are changed by all other permutations in S_n.

Laplace analyzes the case $i = 1$. In that case g is odd and $F_g(x)$ has a real root. Taking different real s he obtains

$$\alpha_i + \alpha_j - s_1 \alpha_i \alpha_j = \ell_1,$$

$$\alpha_i + \alpha_j - s_2 \alpha_i \alpha_j = \ell_2,$$

where ℓ_1 and ℓ_2 are real numbers. Then $\alpha_i + \alpha_j$ and $\alpha_i \alpha_j$ will be real and the polynomial (1) will have the real quadratic factor

$$x^2 - (\alpha_i + \alpha_j)x - \alpha_i \alpha_j.$$

Now let i be arbitrary. Then Laplace shows that if equation (2) of degree $2^{i-1} r_1$ has a real quadratic factor, the given equation has a 4th-degree factor that splits into a product of two real quadratic factors. The latter fact can be verified by direct computation.

We see that neither Euler nor Lagrange nor Laplace made an attempt to rigorize the second part of the proof. They regarded it as obvious. Also, all the proofs of the theorems on symmetric functions, similar functions, and so on, were based on the implicit assumption of the existence of the splitting field of an arbitrary polynomial.

C.F. Gauss' First Proof

This way of looking at the problem, shared by all 18th-century mathematicians, was sharply criticized by the young C.F. Gauss. In his "New proof that every rational integral algebraic function of one variable can be decomposed into real factors of first and second degree" (*Demonstratio nova theorematis omnem functionem algebraicam rationalem integram unius variabilis in factores reales primi vel secundi gradus resolvi posse.* Helmstadii, 1799) he wrote:

> Since, apart from real and imaginary magnitudes $a + b\sqrt{-1}$, it is not possible to represent any other forms of magnitudes, it is not quite clear in what sense what is to be proved differs from what is assumed as a fundamental proposition; and even if one could think of other forms of magnitudes, some F, F', F'', ..., one could not accept without proof that every equation is satisfied by a real value x, or a value of the form $a + b\sqrt{-1}$, or of the form F, or F', and so on. That is why this fundamental proposition can have only the following meaning. Every equation can be satisfied by a real value of the unknown, or by an imaginary value of the form $a + b\sqrt{-1}$, or, possibly, by a value of another, yet unknown, form, or by a value that is not subsumed under any form. How these magnitudes of which we can form no idea whatever

— these shadows of shadows — are to be added or multiplied cannot be understood with the kind of clarity required in mathematics.[4]

Gauss' first proof was completely analytic and we won't consider it here. In 1815 Gauss returned to the fundamental theorem of algebra and gave an algebraic proof in the paper "A second new proof of the theorem to the effect that every integral rational function of one variable can be decomposed into real factors of first and second degree" (*Demonstratio nova altera theorematis omnem functionem algebraicam rationalem integram unius variabilis in factores reales primi vel secundi gradus resolvi posse*, 1815. Commentationes societatis regiae scientiarum Gottingensis recentiores, 1816). On this occasion he again criticized the reasoning of 18th-century scholars. This time he wrote:

> This proposition [about the possibility of decomposing a polynomial into linear factors — I.B.] is, at least at the point dealing with the general proof of such decomposability, nothing other than *petitio principii*.[5]

The charge that Euler's proof involved circular reasoning was unfair. This is most readily seen by analyzing Gauss' second proof. In this proof Gauss carries out essentially the same reduction as Euler, Lagrange, and Laplace without using the proposition of the existence of a splitting field anywhere. How could he have done this? Of course, only by operating with congruences modulo a polynomial, that is, essentially, by constructing the required splitting field. We will try so show how Gauss did this.

C.F. Gauss' Second Proof

Gauss begins his paper with a new proof of the theorem on symmetric functions. To avoid operating with the roots of a polynomial he proceeds as follows. He considers m independent magnitudes $\alpha_1, \alpha_2, \ldots, \alpha_m$, puts

$$\alpha_1 + \alpha_2 + \cdots + \alpha_m = \sigma_1,$$

$$\alpha_1\alpha_2 + \alpha_2\alpha_3 + \cdots + \alpha_{m-1}\alpha_m = \sigma_2$$

$$\cdots\cdots\cdots\cdots\cdots\cdots\cdots\cdots\cdots\cdots$$

$$\alpha_1\alpha_2\cdots\alpha_m = \sigma_m,$$

and proves the following theorem: If ρ is an integral rational symmetric function of $\alpha_1, \alpha_2, \ldots, \alpha_m$, then it is possible to find an integral rational function of the same number of indeterminates $s_1, s_2, \ldots, s_m$ that goes over into ρ

---

4 Gauss, C.F., *Werke*, Bd. 3. Göttingen 1866, pp. 1–30.

5 Ibid., pp. 31–56.

following the substitutions $s_i = \sigma_i$ $(i = 1, 2, \ldots, m)$, and this can be done *in one way only.*

The latter assertion is of great importance for the rest of the theory. It is easy to see that it means that the elementary symmetric functions $\sigma_1, \sigma_2, \ldots, \sigma_m$ are algebraically independent. It follows that any algebraic relation of the form

$$\Phi(\sigma_1, \sigma_2, \ldots, \sigma_m) = 0$$

is necessarily an identity. This means that we can replace the σ_i with indeterminates s_i. If, in turn, we put in place of the s_i the coefficients $a_1, \ldots, a_m$ of an arbitrary equation

$$f(x) = x^m - a_1 x^{m-1} + \cdots \pm a_m = 0 \tag{3}$$

about whose roots we make no assumption whatever, then we obtain the relation

$$\Phi(a_1, a_2, \ldots, a_m) = 0.$$

Thus the relation deduced with respect to the coefficients of a decomposable polynomial turns out to be valid for any polynomial.

Gauss was the first to discover this important method of proof. It is now known as Gauss' principle or the principle of continuation of identities.

Using this principle Gauss introduces the discriminant $\Delta(f)$ of a polynomial and proves its fundamental properties. In particular, he proves that $\Delta(f)$ vanishes if and only if $f(x)$ and $f'(x)$ have a common factor.

Gauss assumes that the fundamental theorem has been proved for all polynomials of degree $2^{n-1}r$, r odd, and proves it for equations of degree $m = 2^n r_1$, r_1 odd. He takes m arbitrary magnitudes $\alpha_1, \ldots, \alpha_m$ and constructs the auxiliary polynomial

$$F(u, x) = \prod_{1 \leq i < j \leq m}^{m} [u - (\alpha_i + \alpha_j)x + \alpha_i \alpha_j] = F(u, x, \sigma_1, \ldots \sigma_m), \tag{4}$$

of degree $m(m-1)/2 = 2^{n-1}r_2$, r_2 odd, so that the induction assumption holds for this polynomial.

We note that if we think of $\alpha_1, \alpha_2, \ldots, \alpha_m$ as the roots of the equation (1), then the polynomial (4) differs little from the Laplace polynomial (2). The principle of their construction is the same.

If Gauss had allowed himself to use the splitting field of the polynomial $f(x)$, then the rest of the proof would have been easy. But he had to give a construction that would make it unnecessary to assume the existence of such a field. To this end Gauss establishes the identity: if the polynomial $F(u, x) = F(u, x, \sigma_1, \ldots, \sigma_m)$ is a product of linear factors, then

$$F\left(u + w\frac{\partial F}{\partial x}, x - w\frac{\partial F}{\partial u}\right) = F(u, x)\varphi(u, x, w) \tag{5}$$

where $\varphi(u, x, w)$ is an entire rational function of all its arguments.

According to Gauss' principle, in this identity one can replace $\sigma_1, \ldots, \sigma_m$ by the coefficients $a_1, \ldots, a_m$ of $f(x)$. This gives

$$F\left(u + w\frac{\partial \bar{F}}{\partial x}, x - w\frac{\partial \bar{F}}{\partial u}\right) = \bar{F}(u, x)\bar{\varphi}(u, x, w), \tag{5'}$$

where the bar indicates that one has replaced in the function in question $\sigma_1, \ldots, \sigma_m$ by $a_1, \ldots, a_m$.

Now let $x = x_0$ be a real number such that the discriminant $\Delta(\bar{F}) \neq 0$. Since the discriminant vanishes for only finitely many values of x, x_0 can be chosen in infinitely many ways. Then $\bar{F}(u, x_0)$ has real coefficients and $\bar{F}(u, x_0)$ and $\frac{\partial \bar{F}}{\partial u}(u, x_0)$ have no common factors.

We choose $\bar{u}$ from the values of u for which $\frac{\partial \bar{F}}{\partial u}\big|_{\substack{u=\bar{u}\\x=x_0}} \neq 0$ (this excludes only finitely many values). Gauss puts

$$\frac{\partial \bar{F}}{\partial u}\bigg|_{\substack{u=\bar{u}\\x=x_0}} = U' \neq 0, \quad \frac{\partial \bar{F}}{\partial x}\bigg|_{\substack{u=\bar{u}\\x=x_0}} = X',$$

makes the substitution

$$w = (x_0 - x)/U',$$

and obtains

$$\bar{F}\left(\bar{u} + \frac{x_0 - x}{U'}X', x\right) = \bar{F}(\bar{u}, x_0) \cdot \bar{\varphi}\left(\bar{u}, x_0, \frac{x_0 - x}{U'}\right). \tag{6}$$

By the induction assumption the polynomial $\bar{F}(u, x_0)$ has the real or complex root $\bar{u} = u_0$ (that is, $\frac{\partial \bar{F}}{\partial u}\big|_{\substack{u=u_0\\x=x_0}} \neq 0$). The identity (6) shows that the polynomial $\bar{F}(u, x)$ vanishes for $u = u_0 + \frac{x_0 - x}{U'}X'$, that is, it is algebraically divisible by the difference $u - (u_0 + \frac{x_0 - x}{U'}X')$.

This established, Gauss puts $u = x^2$. Then

$$\bar{F}(x^2, x) = [f(x)]^{m-1}$$

is divisible by the trinomial

$$x^2 + \frac{X'}{U'}x - \left(u_0 + \frac{X'}{U'}x_0\right)$$

with real coefficients. This proves the fundamental theorem.

We see that in his proof Gauss does not work with the roots of the equation. Instead, he establishes certain identities and makes use of the laws of divisibility of polynomials. By means of this method he essentially constructs a field κ in which $\bar{F}(u, x_0)$ has a linear factor and the initial polynomial $f(x)$ has a quadratic factor.

Gauss presented his method in very veiled form and it took about 60 years before L. Kronecker was able to make it explicit. To obtain the splitting field by Gauss' method we proceed as follows.

The relation (6) shows that the polynomial $\bar{F}(u, x)$ is divisible by the trinomial $u - (u_0 + \frac{x_0 + x}{U'} X')$ if u_0 is a root of the equation $\bar{F}(u, x_0) = 0$, that is

$$F(u, x) = \left[u - \left(u_0 + \frac{x_0 - x}{U'}X'\right)\right]\psi(u, x, u_0, x_0), \tag{7}$$

where ψ is an entire rational function of all its arguments. The difference

$$\bar{F}(u, x) - \left[u - \left(v + \frac{x_0 - x}{U'}X'\right)\right]\psi(u, x, v, x_0)$$

vanishes if we replace v by a root of the polynomial $\bar{F}(u, x_0)$. Hence

$$[f(x)]^{m-1} \equiv \left[x^2 + \frac{X'}{U'}x - \left(v + \frac{x_0}{U'}X_0'\right)\right]\psi(x^2, x, v, x_0)[\mathrm{mod}\, F(v, x_0)]. \tag{8}$$

This means that it is possible to obtain the coefficients of the trinomial from the congruence (8), without consideration of the roots of equation (1). This fact saves the algebraic proof of the fundamental theorem from the vicious-circle syndrome.

The Kronecker Construction

L. Kronecker analyzed Gauss' "second proof" on a number of occasions. He states that he presented it in his course of lectures of 1870/71 as well as in his lectures on the theory of algebraic equations. He devoted to this proof a section of his paper "The foundations of an arithmetic theory of algebraic magnitudes" (*Grundzüge einer arithmetischen Theorie der algebraischen Grössen.* Berlin, 1882). Finally, in his "A fundamental theorem of general arithmetic" (*Ein Fundamentalsatz der allgemeinen Arithmetik.* — J. für Math., 1887) he generalized Gauss' idea by posing the problem: find an irreducible polynomial $f(x)$ such that a given polynomial splits into linear factors in the field of residues mod $f(x)$.

In other words, given a polynomial

$$F(x) = x^n + A_1 x^{n-1} + \cdots + A_n, \tag{9}$$

Kronecker poses the problem of constructing a field κ that is a finite extension of the fundamental domain of rationality $\mathbb{Q}(A_1, \ldots, A_n)$ such that the polynomial (9) splits in κ into a product of linear factors. Of course, κ must be constructed without relying on the determination of roots of the equation $F(x) = 0$.

Kronecker seeks a polynomial $G(x)$ such that

$$F(x) \equiv (x - \alpha_1)(x - \alpha_2) \ldots (x - \alpha_n) \,[\mathrm{mod}\, G(x)].$$

To solve the problem, Kronecker considers a factorable polynomial $P(x)$ of degree n

$$P(x) = (x - x_1)(x - x_2) \ldots (x - x_n) = x^n + S_1 x^{n-1} + \cdots + S_n \tag{10}$$

and constructs the polynomial $G(z)$ of degree $n!$

$$G(z, u_1, \ldots, u_n, S_1, \ldots, S_n) = \prod_i (x - u_1 x_{i_1} - u_2 x_{i_2} - \cdots - u_n x_{i_n}), \tag{11}$$

where $u_1, \ldots, u_n$ are indeterminates and the product is taken over all permutations $i_1, \ldots, i_n$, that is, it has $n!$ factors. Put $\theta = u_1 x_1 + \cdots + u_n x_n$. If we permute $x_1, \ldots, x_n$ in all possible ways, then θ takes on $n!$ different values and can be taken as a primitive element of the field $\kappa(x_1, \ldots, x_n) = K$, that is $K = \kappa(\theta)$, $x_k = \varphi_k(\theta, S_1, \ldots, S_n)$, $k = 1, \ldots, n$. Kronecker replaces the latter equalities by the congruences

$$x_k \equiv \varphi_k(z, S_1, \ldots, S_n) \,[\mathrm{mod}\,(z - \theta)], \ k = 1, 2, \ldots, n.$$

Juxtaposition of these and the equality (10) yields

$$P(x) \equiv \prod_{k=1}^n [x - \varphi_k(z, S_1, \ldots, S_n)] \,[\mathrm{mod}\,(z - \theta)].$$

Since this congruence contains only symmetric functions of $x_1, \ldots, x_n$, it holds not only for θ but also for $\theta_2, \ldots, \theta_{n!}$ and therefore also for the product of the corresponding moduli, that is for $G(z, S_1, \ldots, S_n)$:

$$P(x) \equiv \prod_k [x - \varphi_k(z, S_1, \ldots, S_n)] \,[\mathrm{mod}\, G(z, S_1, \ldots, S_n)].$$

48

Using Gauss' principle, Kronecker puts $S_1 = A_1, \ldots, S_n = A_n$. Then

$$F(x) \equiv \prod_k [x - \varphi_k(z, A_1, \ldots, A_n)] \, [\mathrm{mod} \, G(z, A_1, \ldots, A_n)]. \qquad (12)$$

This shows that an arbitrary polynomial $F(x)$ splits into linear factors relative to its corresponding Galois normal polynomial. But Kronecker wants to construct a *field* in which the polynomial $F(x)$ splits into linear factors, that is, a domain in which $AB = 0$ implies that either A or B is congruent to zero mod G. That is why he splits $G(z, A_1, \ldots, A_n)$ over κ into irreducible factors. If we denote one of these as $G_1(z)$, then

$$F(x) \equiv \prod_k [x - \varphi_k(z, A_1, \ldots, A_n)] \, [\mathrm{mod} \, G_1(z)]. \qquad (12')$$

This completes the construction.

We see that Kronecker constructed the splitting field of an arbitrary polynomial in the form of the field of residues with respect to an appropriate irreducible polynomial without assuming the existence of the field $\mathbb{R}$ of real numbers or the field $\mathbb{C}$ of complex numbers. This purely algebraic construction has found numerous applications in algebra as well as in algebraic number theory.

We note that a special case of a similar construction is due to Cauchy. It appeared in 1847 in Cauchy's "Memoir on a new theory of imaginaries and on symbolic roots of equations and equivalences" (*Mémoire sur une nouvelle théorie des imaginaires, et sur les racines symboliques des équations et des équivalences. – C. r. Acad. sci. Paris, 1847*), and thus before Kronecker's construction and, apparently, independently of the relevant work of Gauss. Cauchy constructed the field $\mathbb{C}$ of complex numbers in the form of the field of residues of the ring $\mathbb{R}[x]$ of real polynomials in x modulo $x^2 + 1$. We have no reason to suppose that Kronecker knew of Cauchy's paper. On the other hand, on various occasions Kronecker stressed the similarity between his ideas and those of Gauss in the latter's second proof.

Kronecker's proof also attests the influence of Galois theory. This is especially clear in his definition of a primitive element of the field κ (for details on Galois theory see below). The construction shows that Kronecker had a profound understanding of Galois' ideas and could use them in a masterly manner.

Viewed as an elementary proposition of the theory of functions of a complex variable, the fundamental theorem of algebra is of little interest. And yet mathematicians as great as Euler, Lagrange, Laplace, and Gauss worked on it, and Gauss gave four different proofs of it. What was interesting about this theorem? We can now answer this question in so far as it pertains to the algebraic proofs of the fundamental theorem. As it turned out, there was an intimate connection between such proofs and the general theory of equations.

The connection between algebraic proofs of the theorem and the theory of symmetric and similar functions of the roots of an equation became apparent already in the proofs of Euler and Lagrange. The study of the latter functions is an essential part of Galois theory. To give a proof not based on the existence of a splitting field Gauss used his "principle of continuation of identities" and may be said to have constructed a field in which a given polynomial had a quadratic factor. Gauss' method of construction of this field was subsequently developed by Kronecker and became one of the most powerful tools of algebra. In this way the fundamental theorem of algebra stimulated the creation of new algebraic methods.

We gave the analyzed proofs the tentative name of algebraic proofs but now we see that they are indeed essentially algebraic.

The Theory of Equations

At the end of the 18th century one of the most important problems of algebra was the problem of solution of equations by radicals. Algebra was known as the science of solution of equations. More specifically, the problem was that of finding methods for expressing the solutions of equations of the form

$$a_n x^n + \cdots + a_1 x + a_0 = 0$$

in terms of the coefficients using the four operations of arithmetic and extraction of roots of arbitrary degree. Here $a_0, \ldots, a_n$ are arbitrary, or are connected by some relations, or are concrete numbers. Of course, the greatest achievement would have been to find methods of solution of equations of arbitrary degree with arbitrary coefficients, but all attempts to find such methods failed. In fact, until 1770 no one knew how to solve by radicals the equation $x^n - 1 = 0$ for $n > 10$, and Vandermonde's paper of 1770 in which he analyzed the case $n = 11$ was regarded as an important advance (see HM, vol. 3, p. 93).[6]

The solution of the equation $x^n - 1 = 0$ for arbitrary n was a remarkable achievement of the young Gauss.

Carl Friedrich Gauss

Carl Friedrich Gauss was born in Braunschweig on April 30, 1777 into the poor family of a master waterpipe installer. The boy amazed his elementary

---

6 We note that in solving equations by radicals we use only the operation of extraction of arithmetic roots. This being so, writing $x = \sqrt[n]{1}$ gives just one root of the equation $x^n - 1 = 0$, namely the number 1, whereas it is required to find all roots of this equation, cf. pp. 52–53.

school teacher by his unusual mathematical ability. The teacher called the attention of the duke of Braunschweig to the boy and the duke made possible his continued education. Between 1795 and 1798 Gauss studied at the University of Göttingen. He attended lectures on mathematics and philology and could not decide which of the two disciplines to follow. What helped him decide was his famous discovery, made in March 1796, of the possibility of constructing a regular polygon with 17 sides. In 1807 Gauss was offered the position of director of the Göttingen observatory — a job he held until his death in 1855. Gauss' scientific interests were remarkably wide. He made major contributions not only to many areas of pure mathematics but also to the theory of planetary motions, to geodesy, and to electrodynamics. His two well-known contributions to the latter area are the CGS system of units and the notion of potential of an electric field. In pure mathematics Gauss published papers in number theory, algebra, analysis, and the geometry of surfaces. Regardless of the area, the depth of his penetration of the material, the boldness of his ideas, and the significance of his results are amazing. Gauss was called the prince of mathematicians. What produced special surprise was the study — in the second half of the 19th century — of Gauss' personal papers. It turned out that he made public only some of his ideas. Thus Gauss carried out profound and extensively elaborated investigations in the theory of elliptic and abelian functions barely hinted at in his published papers, mastered complex integration before Cauchy, elaborated non-Euclidean geometry, and published only some of his astronomical computations and observations. Sometimes Gauss mentioned his unpublished results in letters to some of his friends and, in time, this produced rumors in the mathematical world that provoked and directed the thoughts of younger mathematicians. We can hardly overestimate Gauss' scientific significance.

F. Klein's comments on Gauss' diary offer a glimpse of his personality as a young man (1796–1801).

> Here we see not the inaccessible, closed, cautious Gauss as he appears in his published papers. Here we see what Gauss was like when he experienced and conceived his great discoveries. He expresses his joy and pleasure in the liveliest manner, bestows laudatory epithets upon himself, and shows his mood in enthusiastic exclamations. A proud series of great discoveries in arithmetic, algebra and analysis (admittedly incomplete) passes before us and we experience the process of creation of the *Arithmetical Investigations*.[7]

Contemporaries describe Gauss as a cheerful man with a good sense of humor. He took a lively interest in literature, philosophy, politics and economics. The flourishing of Russian sciences attracted his special attention. He

---

7 Klein, F., *Vorlesungen über die Entwicklung der Mathematik im 19. Jahrhundert*, Bd. 1. Berlin 1926, p. 33.

C.F. Gauss

maintained scientific connections with the Petersburg Academy of Sciences which nominated him a corresponding member as early as 1801 and a foreign member in 1824. At 62 Gauss learned Russian and in letters addressed to the Petersburg Academy asked for Russian journals and books and, in particular, for a copy of Pushkin's "The captain's daughter". Gauss had few personal students but one can safely call him the teacher of all the world's mathematicians.

Solution of the Cyclotomic Equation

We now consider Gauss' contribution to the theory of the cyclotomic equation $x^n - 1 = 0$ dealt with in the seventh part of his *Disquisitiones* (*Disquisitiones arithmeticae*, Gottingae, 1801).

First we analyze, as an example, the case $n = 5$. This is a trivial case settled long before Gauss. Since

$$x^5 - 1 = (x - 1)(x^4 + x^3 + x^2 + x + 1),$$

the equation to be solved is really

$$x^4 + x^3 + x^2 + x + 1 = 0.$$

We put $z = x + 1/x$. Since $x^5 = 1$, we have $1/x = x^4$, $z = x^4 + x$. Further,

$$z^2 = x^2 + 2 + \frac{1}{x^2} = x^2 + 2 + x^3.$$

This yields the equation

$$z^2 + z - 1 = 0$$

for z and the equation

$$x^2 - zx + 1 = 0$$

for x. Solving these equations we obtain

$$z_{1,2} = \frac{-1 \pm \sqrt{5}}{2}, \qquad x_{1,2} = \frac{z \pm i\sqrt{z+3}}{2}.$$

Finally, we obtain four solutions

$$x_1 = \frac{-1 + \sqrt{5}}{2} + i\sqrt{\frac{5 + \sqrt{5}}{8}}, \qquad x_3 = \frac{-1 + \sqrt{5}}{4} - i\sqrt{\frac{5 + \sqrt{5}}{8}},$$

$$x_2 = \frac{-1 - \sqrt{5}}{4} + i\sqrt{\frac{5 - \sqrt{5}}{8}}, \qquad x_4 = \frac{-1 - \sqrt{5}}{4} - i\sqrt{\frac{5 - \sqrt{5}}{8}},$$

expressed in terms of radicals.

In modern terms, we say that the original equation

$$x^4 + x^3 + x^2 + x + 1 = 0$$

determines the field K of magnitudes $a + bx + cx^2 + dx^3$, a, b, c, d rational, and this field contains the subfield L of magnitudes of the form $\alpha + \beta z$. The field K is a radical extension of degree two of the field L and L is a radical extension of degree two of the field of rational numbers. Now we present a similar "translation" of Gauss' reasoning.

Gauss restricts himself, for simplicity, to prime values of n and considers the field of magnitudes

$$K = \{\alpha_0 + \alpha_1 x + \cdots + \alpha_{n-2} x^{n-2}, \alpha_0, \ldots, \alpha_{n-2} \in \mathbb{Q}\}.$$

To study the numbers in the field K Gauss considers the mappings $\varepsilon \to \varepsilon^k$, $0 < k < n$, on the set of roots of the equation $x^n = 1$, that is, the Galois group of K over Q. At one point he writes that the constructibility of a regular n-gon depends not on its geometric symmetry but on its more hidden algebraic symmetry.

Then Gauss selects in the field K the more convenient basis $x, x^2, \ldots, x^{n-1}$ and introduces for every decomposition $n - 1 = e \cdot f$ of the number $n - 1$ the subset K_e of magnitudes of the form

$$\alpha_1 \cdot (f, 1) + \alpha_2 \cdot (f, 2) + \cdots + \alpha_e \cdot (f, e),$$

where the symbols (f, i) denote Gauss' famous periods. To define a period Gauss considers not the numbers $1, \ldots, n - 1$ but the corresponding residue classes modulo n. Since n is prime, these classes form a multiplicative cyclic group, and for every divisor f of $n-1$ this group contains a unique subgroup of f elements. With G denoting the group, $H(f)$ the subgroup, and $H_{(f,i)}$ $(i = 1, 2, \ldots, e)$ the cosets of G with respect to $H(f)$, the period (f, i) is defined as the sum

$$(f, i) = \sum x^k, \quad k \in H_{(f,i)}.$$

Gauss knew the properties of cosets and could easily work with them. In particular, he shows that the product of periods for a given f is a linear combination of such periods,

$$(f, i) \cdot (f, j) = \sum a_{ijk}(f, k),$$

that is, that K_e is a field. He also shows that $K_{e_1} \subset K_{e_2}$ iff e_1 divides e_2. In the concrete situation he works with he shows that the generating element of the larger field satisfies an equation with coefficients from the smaller field and its degree is equal to the degree of the larger field over the smaller one. He also shows how to obtain these equations and proves that they are *solvable by radicals*. He concludes with the theorem: if n is prime and $n - 1 = a_1 \ldots a_k$ is its prime factorization, then the solution of the equation

$$x^{n-1} + \cdots + x + 1 = 0$$

reduces to the solution of K equations with respective degrees $a_1, \ldots, a_k$. In particular, since $16 = 2 \cdot 2 \cdot 2 \cdot 2$, the solution of the equation $x^{17} - 1 = 0$ reduces to the solution of four quadratic equations. This means that there is a ruler-and-compass construction of a 17gon. Gauss adds:

> If, in addition to 2, $n - 1$ involves other primes, then we obtain higher-degree equations ... and *we can prove with all necessary rigor that these higher-degree equations can in no way be eliminated or reduced*

to lower-degree equations [Gauss' italics (author)]. It would take us beyond the confines of this work to give the proof here. Nevertheless, we feel that it is our duty to point out that no one should expect to be able to reduce cyclotomies other than those given by our theorem — say, cyclotomies involving 7, 11, 13, 19, ... parts — to geometric constructions and waste his time in vain.[8]

Gauss combines the development of a general theory with the proof of an understandable and brilliant theorem. This makes for extremely interesting reading.

The significance of this part of Gauss' book is very great. The concepts that play an implicit role here are those of a field, group, basis of a field over a field, and even of a Galois group. Also, it is likely that but for Gauss' deeply and beautifully analyzed example it would have been far more difficult to think of these concepts and of their significance in the theory of equations.

Niels Henrik Abel

The further development of the theory of algebraic equations is linked to the name of the Norwegian mathematician Niels Henrik Abel (1802–1829). Abel was born in 1802, the son of a poor Norwegian pastor. In 1823, as a student at the University of Christiania, Abel thought that he had found a formula for the solution of the general quintic by radicals. He soon realized his error, and in 1824 published, in the form of a brochure, a very condensed proof of the impossibility of the solution of the general quintic by radicals. This research attracted attention and he was offered a scholarship for study abroad. From the fall of 1825 to the spring of 1826 Abel was in Berlin. There he befriended A. Crelle who was beginning the publication of the famous *Journal für die reine und angewandte Mathematik*. The first issue of the journal (1826) contained a number of Abel's papers. One of them was his "Proof of the impossibility of the solution by radicals of general equations of degree higher than the fourth" (*Démonstration de l'impossibilité de la résolution algébrique des équations générales qui passent le quatrième degré.* — J. für Math., 1826, 1).

After it was published in Crelle's Journal Abel's result became universally known. The importance of Abel's paper consisted in the result itself — a result that for a long time had defied the efforts of mathematicians.

In the summer of 1826 Abel visited Italy. He spent the last part of the year in Paris, where he presented to the Paris Academy a memoir containing his famous theorem from the theory of abelian functions. At first this paper produced no response and was almost lost. But it was for this very paper that the Paris Academy awarded Abel, posthumously, a major prize. After

---

8 Gauss, C.F., *Werke*, Bd. 1. Göttingen 1863, p. 462.

N.H. Abel

spending some time in Berlin at the beginning of 1827 Abel returned to Christiania in the summer of 1827.

His situation at home was disastrous. He had neither work nor means of sustenance. He earned some money by tutoring and continued to work strenuously on the theory of elliptic functions and algebraic equations. Two parts of his "Studies on elliptic functions" (*Recherches sur les fonctions elliptiques.* — J. für Math.) were published in 1827 and 1828, and his "Memoir on a particular class of solvable algebraic equations" (*Mémoire sur une classe particulière d'équations résoluble algébriquement.* — J. für Math.) was published in 1829. At the end of 1828 he contracted tuberculosis and died prematurely at the beginning of 1829, a short time before receiving the offer of a position in Berlin. Before his death he was at work on the problem of determination of all algebraically solvable equations.

We will give some details of the content of Abel's memoir of 1829. The first forward step taken by Abel in this memoir is the explicit introduction of the notion of a domain of rationality, the analogue of the modern concept of a field. Abel defines the domain of rationality with respect to magnitudes $a_1, \ldots, a_n$ as the set (of course, Abel did not use this term) of magnitudes obtained by means of the four arithmetical operations from $a_1, \ldots, a_n$ and all

56

real (or rational) numbers. The introduction of this concept was absolutely essential for all general investigations in the theory of equations. The next essential step was the proof of solvability of a remarkable class of equations. Abel determines the equations in this class by the following two conditions.

1. Every root x_i of the equation is a rational function of a fixed root

$$x_i = \theta_i(x_1).$$

2. The rational functions θ_i have the property

$$\theta_i(\theta_j(x_1)) = \theta_j(\theta_i(x_1)).$$

Nowadays we say of such an equation that it is normal and has an abelian Galois group. Abel's proof of their solvability is an extension of Gauss' work. The paper was inspired by Abel's investigation of the equation of the lemniscate, mentioned by Gauss in his *Disquisitiones*. Abel's paper supplements and develops Gauss' ideas in a fundamental manner and constitutes a remarkable addition to the theory of algebraic equations.

Evariste Galois

Great new discoveries that virtually transformed the theory of algebraic equations were made by the young French mathematician Evariste Galois.

In his short life Galois (1811–1832) managed to make remarkable discoveries and become one of the great mathematicians of the 19th century. He was born in 1811 near Paris into a well-to-do intellectual family. In 1823 his parents sent him to a lycée in Paris. He got interested in mathematics and, for a time, read the works of Legendre, Lagrange, and Gauss with ease and pleasure. Here is a report from one of his teachers.

> He has a passion for mathematics; I think that it would be better for him if his parents agreed that he should study just this science: here he merely wastes his time, exasperates his teachers and draws punishments.[9]

A number of misfortunes befell Galois between 1827 and 1829. His father committed suicide as a result of a great political intrigue. Galois himself did not finish his work at the lycée and twice failed the entrance examination to the Ecole Polytechnique. A paper that he submitted to the Paris Academy was lost. These misfortunes, and the tense situation due to political persecution and agitation in France at the time, conspired to make him tense and quick-tempered. His paper "Proof of a theorem on periodic continued fractions" (*Démonstration d'un théorème sur les fractions continues périodiques.*

---

9 Dupay, P., *La vie d'Evariste Galois*. Ann. sc. de l'Ecole Norm., s. 3, t. 13, 1896, p. 256.

— Ann. Math.) was published in 1829. By that time he had already made his most important discoveries in the theory of equations. In the fall of 1829 he entered the Ecole Normale whose level was, at the time, lower than that of the Ecole Polytechnique. His paper on his discoveries in the theory of algebraic equations was again lost. At the end of the year Galois was expelled from school for his republican speeches, and in June of 1831 he ended up in court for making provocative pronouncements about King Louis Philippe. The case was dismissed because of his young age. Within a month he was rearrested as one of the leaders of a youth demonstration. This time, without any reason, after a long trial, he was sentenced at the end of 1831 to six months in jail. For Galois, the tense young man with little experience of life, the stay in a common jail was particularly trying. Shortly after getting out of the jail, he was killed in a duel, in consequence of a dark love affair. The night before the duel Galois looked over and supplemented the manuscript that he was again preparing to present to the academy and sent it to his friend Chevalier. This fundamental paper of Galois was published in 1846 by Liouville.

The Algebraic Works of Evariste Galois

The most important of the notes and papers of Galois published during his lifetime is the remarkable paper "On number theory" (*Sur la théorie des nombres. —* Bull. sci. math., 1830). In it Galois considers polynomial congruences of the form

$$F(x) \equiv 0 (\operatorname{mod} p)$$

without integer roots. He writes that

> in that case, one must look at the roots of this congruence as a variety of imaginary symbols, for they do not satisfy the conditions required of integers; the role of such symbols and computations is often just as useful as the role of the imaginary $\sqrt{-1}$ in ordinary analysis.[10]

Galois goes on to consider a construction which is, essentially, that of adjunction to a field of a root of an irreducible equation (the requirement of irreducibility is clearly spelled out) and proves a number of theorems on finite fields.

We analyze in greater detail the content of Galois' fundamental paper "Memoir on the solvability conditions of equations by radicals" (*Mémoire sur les conditions de résolubilité des équations par radicaux. —* J. math. pures et appl., 1846). Galois begins the memoir with a definition of a domain of rationality.

> In addition, one can agree to consider as rationalities all rational functions of certain quantities regarded a priori as known. For example, one

---

10 Galois, E., Œuvres mathématiques. Paris 1951, p. 15.

58

Evariste Galois

can choose a root of a whole number and consider as rationalities all
rational functions of this radical.[11]

He notes that it is possible to change a domain of rationality by adjoining
new quantities as known. In this connection Galois writes:

> We will see that the properties and difficulties of an equation can vary
> drastically depending on the quantities adjoined to it.[12]

To determine the Galois group of an equation he proves the lemma of
the primitive element, to the effect that it is always possible (if there are no
repeated roots) to choose a function V of the roots of an irreducible equation
such that all the roots are rational functions of V. The proof is followed by
the following interesting remark:

---

11 Ibid., p. 34.

12 Ibid., p. 35.

It is remarkable that this proposition implies that every equation depends on an auxiliary equation all of whose roots are rational functions of one another.[13]

Such equations are now called normal. Galois does not embark on a study of normal equations. He merely notes that "such an equation is not easier to solve than any other."[14]

Now comes an extremely important lemma.

Suppose that we have formed an equation for V and have selected one of its irreducible factors such that V is a root of the irreducible equation. Let V, V', V'', ... be the roots of this irreducible equation. If $a = f(V)$ is one of the roots of the given equation, then $f(V')$ will also be a root of this equation.[15]

This lemma is the basis for the definition of the Galois group given in the theorem that follows it.

Theorem. Given an equation with m roots a, b, c, ... There is always a group of permutations of the letters a, b, c, ... with the following properties:

1. All functions of the roots invariant under the substitutions of the group are rationally determinable (belong to the domain of rationality).

2. Conversely, all rationally determinable functions of the roots are invariant under these substitutions.[16]

To understand this theorem we must bear in mind that by a permutation Galois means an ordering of the roots and by a substitution, a mapping of the set of roots into itself. A group of permutations is a set of permutations of which Galois says:

Since we always consider questions in which the original order of the letters has no effect whatever on the groups to be considered, we must have the same substitutions regardless of the initial permutation. Thus, if a group of this kind contains substitutions S and T then it certainly contains the substitution ST.[17]

It seems that this somewhat obscure phrase must be interpreted as follows: when the term *group of permutations* is used, one has in mind a set

13 Ibid., p. 37.

14 Ibid.

15 Ibid.

16 Ibid., p. 38.

17 Ibid., p. 35.

of permutations U with the additional property that for every permutation $u \in U$ the set $G(u, U)$ of all substitutions g such that $g(u) \in U$ is the same. Such a set of substitutions is called a *group of substitutions*, and in each case it must be clear from the context with respect to what set of permutations it was formed. It is easy to verify that in this way there actually arises a *group* of substitutions in our sense of the word, and Galois calls the reader's attention to this when he speaks of the substitutions S, T, and ST.

To prove this theorem Galois considers the primitive element V and the formulas

$$a = \varphi V, \; b = \varphi_1 V, \ldots, \; d = \varphi_{m-1} V$$

and introduces the group of substitutions of the roots

$$(\varphi V, \varphi_1 V, \ldots, \varphi_{m-1} V),$$

$$(\varphi V', \varphi_1 V', \ldots, \varphi_{m-1} V'),$$

$$(\varphi V'', \varphi_1 V'', \ldots, \varphi_{m-1} V''),$$

$$\ldots\ldots\ldots\ldots\ldots\ldots\ldots$$

where $V, V', V'', \ldots$ are all the roots of the irreducible equation for V. This is a perfectly clear definition of the action of the Galois group on the set of roots starting with its action on the primitive element.

That Galois understood more than he managed to write down is clear from his comments on this theorem. In these comments he tries to stress that what counts is not the permutation but the substitution. In the last comment he even says: "Substitutions don't even depend on the number of roots!"[18] The two subsequent propositions deal with the change of the group of an equation due to the adjunction of roots of the auxiliary equation to the domain of rationality. In his last letter to Chevalier, when listing his most important propositions, Galois has this to say about them:

Propositions II and III of the first memoir make clear the great difference between adjoining to the equation one of the roots of the auxiliary equation as against adjoining all of its roots.

In both cases the group of the equation is partitioned by the adjunction into groups such that one passes from one to the other by means of the same substitution; but it is only in the second case that it is certain that these groups have the same substitutions. This is called a *proper* decomposition.

In other words, when a group G contains a group H, then the group G can be divided into groups that are obtained by performing the same

18 Ibid., p. 40.

substitution on the permutations of H, so that

$$G = H + HS + HS' + \cdots.$$

It can also be divided into groups with the same substitution so that

$$G = H + TH + T'H + \cdots.$$

These two kinds of decomposition do not ordinarily coincide. When they do, the decomposition is said to be *proper*.[19]

If we consider Galois' exposition from the vantage point of modern knowledge, then it is clear that Galois is describing the decomposition of a group into right and left cosets with respect to a subgroup and singles out the case when the two decompositions coincide, that is, the case when the subgroup is normal.

The proof of proposition II is very beautiful. (There is no proof of proposition III). Then Galois states the criterion for solvability of an equation by radicals. This criterion takes up about two pages of text and describes the procedure for successive adjunction of radicals and the concomitant transformation of the group of the equation. This general theory is followed by the study of equations of prime degree.

"Thus for an irreducible equation of prime degree to be solvable by radicals it is *necessary and sufficient* that all the functions invariant under the substitutions

$$x_k \rightarrow x_{ak+l}$$

be rationally determinable".[20] In other words, the Galois group of such an equation must be metacyclic.

"For an irreducible equation of prime degree to be solvable by radicals it is necessary and sufficient that given any two of its roots the others are rationally derivable from them".[21]

This is the solution of the problem investigated already by Euler and Lagrange.

We quote the famous conclusion of Galois' letter to Chevalier:

You will publicly beg Jacobi or Gauss to give their opinion not of the truth but of the importance of these theorems. After this there will, I

---

19 Ibid., p. 25.

20 Ibid., p. 48.

21 Ibid., p. 49.

hope, be people who will find it to their advantage to decipher all this mess.[22]

Galois has in mind not only the theory of equations but also the deep results in the theory of abelian and modular functions.

The importance of Galois' works consists in that in them he fully revealed new and profound regularities of the theory of equations. After the assimilation of Galois' discoveries, the form and aims of algebra itself changed radically. The theory of equations vanished and its place was taken by the theory of fields, the theory of groups, and Galois theory. Galois' premature death was an irreplacable loss for science.

It took a few decades to fill the gaps in Galois' works, and to understand and improve them. The efforts of Cayley, Serret, Jordan and others transformed Galois' discoveries into Galois theory. Jordan's 1870 monograph "Treatise on substitutions and algebraic equations" presented the theory in a systematic manner, thus making it widely accessible. From then on Galois theory became a component of mathematical education and a foundation for new mathematical investigations.

The First Steps in the Evolution of Group Theory

The history of group theory proper begins in the middle of the 19th century, after the publication of the works of Galois. In the first half of the century we witness the process of its coming into being. Nevertheless, some group-theoretic considerations are found much earlier, namely, in the works of Euler and Fermat. The works of Lagrange and Vandermonde devoted to the theory of algebraic equations introduced into mathematics the first group-theoretic object, namely substitutions (see HM, v. 3, pp. 90–93).

Lagrange's memoir "Reflections on the algebraic solution of equations" (*Réflexions sur la résolution algébrique des équations*), published in 1771–1773, is of special significance. In addition to very important investigations in the theory of equations it contains a proof of the first group-theoretic theorem: the number of values taken on by a function of n variables under all permutations of these variables divides $n!$. This is a special case of the theorem that the order of a subgroup divides the order of the group. Among the followers of Lagrange and Vandermonde one must mention Paolo Ruffini. In his investigations on the theory of equations undertaken between 1808 and 1813 Ruffini studies not only the group of substitutions but also its subgroups, and introduces the notions of transitivity and primitivity (see HM, v. 3, p. 95).

An important step in the rise of group theory was the publication of Gauss' *Disquisitiones Arithmeticae*. This book is remarkable because of its constant

---

22 Ibid., p. 32.

and extensive use of general algebraic ideas. At the beginning of the book Gauss defines congruences. This is the first instance in history of the construction of a quotient ring. In the course of his systematic study of congruences with respect to a prime modulus Gauss proves the existence of a primitive root, that is, he proves that the multiplicative group of residue classes modulo p is cyclic. The proof is very general and carries over literally to the case of any finite field — a fact immediately noticed by Galois when the latter began to develop the theory of finite fields.

In connection with his study of the properties of periods Gauss finds it necessary to operate with cosets of the multiplicative group of the field of p elements with respect to its different subgroups and his arguments imply a clear understanding of the properties of these cosets.

What was most interesting and significant for group theory, however, was Gauss' construction of a whole series of groups, namely groups of classes of binary quadratic forms (of a given discriminant). These are the most abstract examples of groups constructed up to that time.

Gauss introduces composition of forms, a far from trivial operation, and shows that beginning with the composition of forms it is possible to define composition of classes of forms. He shows that composition of the principal class K yields K and that for each class there is an opposite one, that is, he verifies the elementary properties of a group operation. He does not check the associativity and commutativity of the composition of classes but these follow immediately from the associativity and commutativity of the composition of forms that he established earlier. We quote Gauss' proof of the existence and uniqueness of the solution of the equation $K + X = L$ in the group of classes.

> *It is convenient to denote composition of classes by the addition symbol "+" and the sameness of classes by the equality symbol.* Then the theorem just stated can be expressed as follows. If K' is the class opposite to the class K, then $K + K'$ is the principal class with the same determinant, and therefore $K + K' + L = L$. If we put $K' + L = M$, then it follows that $K + M = L$, as required. If there were another class M' with the same property, that is, if we had $K + M' = L$, then we would have $K + K' + M' = L + K' = M$. But then $M = M'$.[23]

In connection with his construction of the group of classes Gauss makes a very significant remark:

> We observe that since one base does not suffice in this case (singled out earlier), it is necessary to take two or more classes which when multiplied and composed yield all the remaining ones. Here we obtain

23 Gauss, C.F., *Werke*, Bd. 1. Göttingen 1863, p. 273.

double and multiple indices whose use is exactly the same as the use of simple indices ...[24]

What Gauss wants to say is that the group he is considering is not cyclic but is a direct sum of two or more cyclic groups.

There is no doubt about the great significance of Gauss' ideas for group theory.

When discussing the development of the theory of substitutions one must pay attention to Cauchy's papers. In the first paper dealing with this topic, namely his "Memoir on the number of values that can be taken on by a function if the magnitudes contained in it are permuted in all possible ways" (*Mémoire sur les nombres des valeurs qu'une fonction peut acquérir lorsqu'on y permute de toutes les manières possibles les quantités qu'elle renferme.* — J. Ec. Polyt. 1815), Cauchy investigates the number of values of an algebraic function under all permutations of its arguments. Let p be the number of values and n the number of arguments. Then his theorem asserts that p is 1 or 2 or greater than or equal to n. This means that the index of a subgroup H of the symmetric group S_n is 1, 2, or greater than or equal to n. Later, in 1844–1846, Cauchy publishes a whole series of papers and notes on substitutions. Many of them are again connected with the question of the possible number of values while some contain proofs of certain theorems on transitive groups of substitutions. The following is the best-known result: if the order of a group of substitutions is divisible by a prime p then the group contains a subgroup of order p.

The publication, in 1846, of the basic papers of Galois was a turning point in the rise of group theory. We have already discussed them in the section devoted to the theory of equations. Their special value for group theory is that they showed for the first time that the solution of an ancient and important problem can be reduced to the study of a new object — a group. For the first time groups appear not as an auxiliary reasoning tool but as a basic object of research. Notwithstanding their florid and imprecise definitions, Galois' skillful use of such complex concepts as a simple group, a normal subgroup, and a solvable group was of tremendous importance. For example, he gave rough formulations of the following assertions (we give their modern formulation):

a) the smallest nonabelian simple group is the group A_5 of order 60;

b) the group of fractional linear transformations with coefficients in the residue class field of order p is not solvable for $p > 3$;

c) for $p > 11$ this group has no subgroups of index p;

d) an irreducible equation of prime degree is solvable by radicals if and only if its group is metacyclic.

---

24 Ibid., pp. 374–375.

The systematic development of group theory began soon after the appearance of Galois' papers.

The first definition and the first investigations of abstract groups were published by Arthur Cayley in 1854. Cayley, an eminent English mathematician, was born in 1821, in a solid merchant family. He spent his childhood in Petersburg, where his father lived. From 1838 to 1841 Cayley studied at Cambridge and was one of the best students in mathematics. He began to publish mathematical papers in 1841. For 20 years, beginning in 1843, Cayley practiced law without breaking off his intensive mathematical studies. In 1863 he became a professor at Cambridge University and held this post until his death in 1895. He published some 200 papers in various areas of mathematics.

Cayley's most significant results belong to algebraic geometry, linear algebra, and group theory. In his lectures on the history of mathematics F. Klein says that Cayley was "the creator of modern algebraic geometry both in terms of the theory of invariants and in its geometric part."[25]

The two parts of Cayley's paper *On the theory of groups as depending on the symbolic equation $\theta^n = 1$* (Philos. Mag.) were published in 1854. In it Cayley defines a group as a set of symbols with a given law of composition such that the law is associative, there is an identity element, and the equations $ax = b$ and $ya = b$ are uniquely solvable for all a and b. It is true that at first Cayley writes that a group on a set of symbols is an associative law of composition, and only later does he add that the only laws that should be called groups are the laws whose multiplication tables contain all group elements in every row and column. In modern terms, the first requirement defines a semigroup whereas the second is an exact definition of a group. Cayley considers giving a group by means of its multiplication table and investigates giving it by means of generators and defining relations. Cayley notes that the group elements can be not only substitutions but also objects of a different kind, for example, quaternions. In 1859 Cayley published the third part of his paper and in it he shows that all groups of prime order are cyclic and lists all possible groups of order eight. Cayley used the name "group" to honor the memory of Galois. At first, Cayley's paper attracted little attention, but later it became a model for the definition of a group and appeared in practically all textbooks.

Speaking of the development of group theory in the middle of the 19th century we must mention, in addition to the works of Cayley, various investigations of groups of substitutions. Another aspect was the assimilation of Galois' *Nachlass*. In this area, a great deal of work was done at the time by J.A. Serret who included in his Sorbonne lectures on algebra ever larger portions of Galois theory.

Further great discoveries in group theory are connected with the name of Camille Jordan (1838–1922), an alumnus of the Ecole Polytechnique and a

---

25 Klein, F., *Vorlesungen* ..., Bd. 1. Berlin 1926, p. 148.

Arthur Cayley

professor at that school as well as at the Collège de France. In 1865 there appeared Jordan's first paper on Galois theory "Commentaries on the memoir of Galois" (*Commentaires sur le mémoire de Galois.* — C. r. Acad. sci. Paris), in 1869 its continuation "Commentaries on Galois" (*Commentaires sur Galois.* — Math. Ann.), and soon after his fundamental monograph "Treatise on substitutions and algebraic equations" (*Traité des substitutions et des équations algébriques.* Paris 1870). Jordan's treatise consists of parts devoted, respectively, to the study of groups of substitutions, to Galois theory proper, and to applications of Galois theory to equations arising in various areas of mathematics.

In his monograph Jordan explicitly singles out normal subgroups, introduces the notion of a simple group, presents the Jordan theorem (proved by him in 1869) and the first half of the famous Jordan-Hölder theorem, and investigates in detail multiply transitive groups. The notion of a homomorphism appears here for the first time (strictly speaking, it is a homomorphism onto, or an epimorphism). The term used by Jordan is *l'isomorphisme mériédrique.* It is interesting that when Jordan uses the phrase "the group Γ is isomorphic to the group G" he means that there is defined an epimorphism of G onto Γ. Another "first" in Jordan's treatise is his consideration of matrix groups with elements from a finite field. In the 20th century these groups became objects of detailed investigations. In his account of Galois theory Jordan uses the contemporary method of associating to an equation not a set of permutations of its roots but rather a group of substitutions, and his criterion for the solvability of an equation by radicals is formulated in terms of the solvability of its Galois group. For a time, Jordan's treatise served as a textbook of group theory as well as of Galois theory. Its publication symbolized the end of the period of birth of group theory. The last 30 years of the 19th century — a period not dealt with in our account — were marked by new great discoveries in group theory such as, above all, the discovery of continuous groups and of Lie groups.

The Evolution of Linear Algebra

There are two sides to linear algebra. One is formal algebraic formulas and computations and the other is their geometric interpretation. Linear algebra arose out of the merger of the theory of solution of systems of linear equations with analytic geometry. The first theory contributed the algebraic formulas and the second the geometric images. One could also call the resulting new discipline the (linear) geometry of n-dimensional space. The stress on geometric ideas in the whole theory is a 20th-century feature but the process of creation of the fundamental concepts and the basic construction of the theory took place in the middle of the 19th century. Until the middle of the 19th century, and sometimes even later, results are presented without involving their geometric content. At times it is difficult to decide to what extent the authors themselves perceived not only the algebraic but also the geometric sense of their works.

More concretely, in the first third of the 19th century topics begun earlier continue to be developed. They are the theory of determinants, the theory of quadratic forms (largely linked to number theory), of linear algebraic equations, and of linear differential equations.

In the fifth part of the *Disquisitiones* (1801) Gauss, following Lagrange, studies the problem of reduction of quadratic forms with integer coefficients to canonical form using invertible substitutions of the variables with integer coefficients. While his deep results do not belong to linear algebra proper, his methods and mode of thought influenced linear algebra.

One should mention Cauchy's important paper "Memoir on functions that can take on just two values, equal in magnitude and opposite in sign, under the permutations of the variables contained in them" (*Mémoire sur les fonctions qui ne peuvent obtenir que deux valeurs égales et des signes contraires par suite des transpositions opérées entre les variables qu'elles renferment.* — J. Ec. Polyt., 1815). In this paper Cauchy extends the investigations of Vandermonde and proves rigorously the properties of determinants established by Vandermonde for determinants of small order (HM, v. 3, p. 65). He views

Camille Jordan

a determinant as a function of n^2 variables which he arranges in a square table.

Some years later Jacobi published a number of important papers on the theory of determinants and quadratic forms. Carl Gustav Jacob Jacobi was born in 1804 in Potsdam into a wealthy family. After graduating from Berlin University he went to Königsberg. Soon after, in 1827, he became a professor, worked intensively and lectured until 1843. He then moved to Berlin. He died of smallpox in 1851. Jacobi had very wide mathematical interests. He is best known for his papers in the theory of abelian functions, in number theory, mechanics, the calculus of variations, and differential equations. He had many students and founded the Königsberg school.

Jacobi's first paper in linear algebra was "On the transformation of two arbitrary homogeneous functions of the second order by means of linear substitutions into two others containing only squares of the variables; together with many theorems on the transformation and computation of multiple integrals" (*De binis quibuslibet functionibus homogeneis secundi ordinis per substitutiones lineares in alias binas transformandis, quae solis quadratis variabilium constant; una cum variis theorematis de transformatione et determinatione integralium multiplicium.* — J. für Math., 1834, 12) devoted to the proof of the possibility of simultaneous diagonalization of two quadratic forms (if one of them is positive definite). At the same time Jacobi determined the conditions that must be satisfied by a transformation that preserves a sum of squares $x_1^2 + x_2^2 + \cdots + x_n^2$, that is, the orthogonality conditions of the matrix of the transformation.

Some of Jacobi's papers deal with determinants. In one of them he introduces his famous "jacobian" of a system of n functions in n variables. Incidentally, the manner of denoting a determinant by means of two vertical bars was introduced by Cayley in 1844.

In 1844 there appeared two works that marked a turning point in the evolution of linear algebra. They were Cayley's *Chapters in the analytical geometry of (n) dimensions* and Grassmann's "The science of linear extension" (*Die lineale Ausdehnungslehre*). An important role in the mastery of the concept of n-dimensional space was played by certain contemporary geometric investigations, in particular, Riemann's famous paper "On the hypotheses that lie at the foundations of geometry" delivered in Göttingen in 1854 and published by Dedekind in 1868. In it Riemann defines an n-dimensional manifold and introduces the notion of curvature of space, thus opening a boundless field of geometric investigations.

Another line of contemporary investigations that habituated mathematicians to multidimensional objects was that of hypercomplex numbers.

Thus the notion of n-dimensional space and the related geometric notions of linear algebra fell on fertile soil, and, as Klein puts it, "by 1870, the notion of the n-dimensional space R_n became the common property of the young generation that was forging ahead."[26]

An important step in the evolution of the algebraic side of linear algebra was the publication of Cayley's *A memoir on the theory of matrices* (Philos. Trans. 1858). In this paper Cayley introduces the notion of a matrix, defines addition of matrices, and defines multiplication of matrices by analogy with the composition of linear changes of variables. He introduces the identity and zero matrices, and notes that a determinant can be viewed as a function of a matrix rather than as a function of n^2 variables. This made it possible to state the multiplicative property of determinants in the transparent form

$$|A \cdot B| = |A| \cdot |B|.$$

By generalizing a theorem of Hamilton on quaternions, Cayley formulates the Hamilton-Cayley theorem that every matrix is a root of its characteristic polynomial and proves it for matrices of order two and three. The introduction

---

26 Klein, F., *Vorlesungen* ..., Bd. 1. Berlin 1926, p. 170.

Carl Gustav Jacobi

of the matrix concept made it clear that a linear change of variables — a linear transformation — must be the object of study rather than an auxiliary tool.

Somewhat earlier, in 1852, James Sylvester (1814–1897) proved the law of inertia of quadratic forms in the paper *The proof of the theorem that every homogeneous quadratic polynomial is reduced by real orthogonal substitutions to a form of sum of positive and negative squares* (Philos. Mag., London 1852).

The notion of rank of a matrix and the Kronecker-Capelli theorem were discovered independently by a number of investigators. The first published proof of this theorem is that of Charles L. Dodgson, author of the famous novels *Alice in Wonderland* and *Through the looking glass*. The theorem was published in his *An elementary treatise on determinants* in the following formulation:

For a system of n inhomogeneous equations in m unknowns to be consistent it is necessary and sufficient that the order of the largest nonzero minor be the same for the augmented and unaugmented matrices of the system.

The question of reduction of matrices of linear transformations to Jordan normal form was solved by Weierstrass and Jordan. In his paper "On the theory of bilinear and quadratic forms" (*Zur Theorie der bilinearen und quadratischen Formen.* — J. für Math. 1868) Weierstrass gave a necessary and sufficient condition for similarity of matrices in terms of elementary divisors. In his "Treatise on substitutions and algebraic equations" (1870) Jordan introduces the Jordan normal form of a matrix, and proves its uniqueness as well as the possibility of reducing a matrix to such form.

Thus, by 1870, all fundamental theorems of linear algebra had been proved and the fundamental notion of an n-dimensional space was known and in common use. However, the idea of combining all of these results into a single harmonious theory occurred, conceivably, to just a handful of the most eminent mathematicians. During the next 30 to 50 years the field of applications of linear algebra grew ever larger and the exposition and proofs continued to improve. The result of all this was that by the end of this period linear algebra was seen by the whole mathematical world as a remarkable and well-organized theory that must be a component of general education and an active tool in the hands of all mathematicians. It is safe to say that by now linear algebra has become simply a component of our thinking.

Hypercomplex Numbers

The search for hypercomplex numbers that led gradually to the rise of the theory of algebras coincides essentially with the course of English mathematics. Apart from isolated investigations, the uninterrupted development of the theory of algebras began on the continent in the 70s of the 19th century. This was already the second phase of its development — the period of creation of the structural theory of semisimple algebras and representation theory.

In order to understand the development of algebra in England it is important to describe the work of representatives of "the English school of symbolical algebra". This work lacked finished results that would form part of later mathematics, but it contained fundamental ideas, admittedly in an early and imperfect form, whose subsequent mastery constitutes an epoch in the history of mathematics that extends to our time.

The three most significant algebraists of the English school of symbolical algebra are George Peacock (1791–1858), Duncan F. Gregory (1813–1844), and Augustus De Morgan (1806–1871). Theirs was a close collaboration with Peacock, the apparent leader of the group. They published their ideas in a series of papers and handbooks on algebra in the 20-year period from 1830 to 1850. As professors at the universities in Cambridge and London, they were able to promote their ideas and spread them among scholars and students. They have undoubtedly exerted a marked influence on the subsequent development of English mathematics.

A clear exposition of these ideas is found in Peacock's *Report on the recent progress and present state of certain branches of analysis* (Rept. of the

British assoc. for the advanc. of sci. for 1833, London 1834) of 1834. Its name notwithstanding, this is basically not a report but a completely original work. Peacock considers the question of meaning and manner of use of symbols in mathematics. He mentions two different ways, namely, those of arithmetical and symbolical algebra. In the first, a symbol denotes a (known or unknown but definite) whole positive number. Operations with such symbols are, essentially, operations with numbers and their rules derive from our knowledge of arithmetic.

Symbolical algebra is

> the science of symbols and their combinations, constructed in accordance with their particular rules, which can be applied to arithmetic or to other sciences by means of interpretation.[27]

Peacock stresses that, like geometry, symbolical algebra must be constructed on the basis of certain "principles" that contain the rules of operation with symbols of completely undetermined nature. Thus Peacock states for the first time the idea of an axiomatic construction of algebra and notes that the axioms must not explain the "nature" of the employed objects. He insists that the symbols in symbolical algebra are completely general and their sense and mode of representation (realization) are completely unrestricted. In developing these ideas in his paper *On the nature of symbolical algebra* (Trans. Roy. Soc. Edinburgh, 1840, 14, 28–216) Gregory explains that there can be different classes of operations, and that, conversely, different theories can involve operations satisfying the same laws, so that their properties can be established at one time in the general case on the basis of these laws. Obviously, at the time these remarkable ideas could be illustrated only by the properties of the real and complex numbers and by rather artificial examples. Nevertheless, what we see here is an attempt to axiomatize algebra and the appearance of the notion of an algebraic structure. In his papers, and in particular in the paper *On the foundation of algebra* (Trans. of the Cambridge Philos. Soc., 1841), De Morgan tries to explain the concrete form of the rules of operations in symbolical algebra. He considers the symbols =, +, −, ×, :, 0, 1, singles out in the form of formulas — but without using the corresponding terms — such properties as associativity, commutativity, distributivity, and the algebraic characteristics of 1 and 0. He was the first to give the axiomatic properties of equality (=). We can say that this paper of De Morgan was the most concrete application of the general ideas of the school. Its ideas entered mathematics in the most direct manner. Returning to Peacock's previously mentioned paper we will show that he constructed one of the variants of symbolical algebra by means of a special principle of "permanence", or the principle of preservation of form. Specifically, he assumed that what must hold are the formulas that remain true if one replaces

---

27 Peacock, G., *Report on the recent progress ...*. London 1834, pp. 194–195.

the letters in them by arbitrary natural numbers, for example,

$$(a + x) - x = a.$$

This is, so to say, is a method of obtaining an algebraic structure from its realization. This method was later used far more skillfully by Dedekind to construct ideals, the irrational numbers, and points of a Riemann surface, and has been used ever since in mathematics and mathematical logic.

It should be noted that the mathematicians of the English school completely failed to notice certain difficulties connected with the axiomatic method. Thus, for example, they never posed the question of consistency of a system of axioms and sometimes obtained undefined and also meaningless assertions. But their ideas and ways of looking at algebra turned out to be very useful for the development of new abstract theories, namely group theory and the theory of systems of hypercomplex numbers.

In the 40s many English mathematicians tried to find various systems of hypercomplex numbers. The well-known ideas of the school of symbolical algebra offered faith and hope of progress. A system of hypercomplex numbers was a system of expressions

$$a \cdot 1 + b \cdot \xi + \cdots + d \cdot \zeta,$$

where $a, b, \ldots, d$ were real numbers, $1, \xi, \ldots, \zeta$ were "fundamental units", addition and subtraction were coordinatewise and multiplication required definition of the products of all possible pairs of fundamental units. It was not clear a priori what properties the resulting system would have, and it was only later that it was found that the systems obtained were either noncommutative, or nonassociative, or contained divisors of zero (that is, pairs of nonzero elements α, β with $\alpha\beta = 0$) and so could not be fields. Of course, one wanted to obtain systems whose properties were as close to those of the complex numbers as possible.

William Rowan Hamilton

A significant step forward in the search for systems of hypercomplex numbers was the discovery of the skew field of quaternions. This discovery was made in 1843 by the Irish mathematician William Hamilton.

William Hamilton (1805–1865) was born in Dublin. He was extremely talented. At the age of eight he knew five languages: French, Italian, Latin, Greek and Hebrew. At 12 he knew 12 languages, including Persian, Arabic, and Malaysian. In 1823 he entered Trinity College and was an excellent student. In 1827 he presented to the Royal Irish Academy his first paper, *Theory of systems of rays* (Trans. Roy. Irish Acad.) which is a profound investigation of geometrical optics. In the same year, before finishing college,

William Hamilton

he obtained the positions of professor of astronomy at Trinity College and of Royal Astronomer of Ireland. In 1830 and 1832 Hamilton publishes two supplements to *The theory of systems of rays* and in 1834 and 1835 there appear two of his papers on mechanics that brought him special fame. The latter papers dealt with so-called Hamiltonian mechanics — a new mathematical apparatus that later served as the basis for quantum mechanics. In 1837, in the 17th volume of the Transactions of the Royal Irish Academy for 1833 and 1835, there appears Hamilton's famous paper *Theory of conjugate functions or algebraic couples; with a preliminary and elementary essay on algebra as the science of pure time* (Trans. Roy. Irish Acad., 1837). This paper contains the now familiar definition of the complex numbers as pairs of reals with axiomatic definitions of addition and multiplication. In the next few years Hamilton, together with some other English mathematicians, embarked with great enthusiasm on a search for hypercomplex numbers. After large numbers of computations Hamilton concluded that systems with three fundamental units contained divisors of zero. He then began to study systems with four fundamental units and in 1843 discovered the quaternions. This discovery made a tremendous impression on him and he devoted the next 20 years of his life to the realization of the grand program of constructing for the quaternions analogues of all analytic and algebraic theories that hold for the complex numbers.

Hamilton thought — and we agree with him today — that the quaternions are a system of hypercomplex numbers closest to the complex numbers. But this point of view met with considerable criticism. The critics were some algebraists, including Peacock, who were greatly disturbed by the noncommutativity of quaternion multiplication. Hamilton's quaternion activity brought into being in Dublin a veritable movement of "quaternionists" but was in part a failure. The role of quaternions in mathematics is much more modest than that of the complex numbers. Nevertheless, the relevant results of Hamilton and his followers assisted the development of linear algebra and vector analysis. The last years of Hamilton's life were clouded over by mental derangement.

Hamilton was a remarkable scientist who invariably obtained deep and fundamental results in the areas of mathematics in which he worked. He was one of the best mathematicians of the 19th century.

We will now describe the quaternions in some detail. They can be defined as the set of elements of the form

$$\alpha = a + bi + cj + dk,$$

a, b, c, d real, satisfying the relations

$$i^2 = j^2 = k^2 = -1, \ ij = -ji = k, \ jk = -kj = i, \ ki = -ik = j.$$

The quaternions form a skew field. It was later shown that the quaternions are the only nontrivial (that is different from $\mathbb{R}$ and $\mathbb{C}$) skew field that is a finite-dimensional vector space over $\mathbb{R}$. They are usually denoted by $\mathbb{H}$ (in honor of Hamilton). Since quaternion multiplication is noncommutative, the quaternions yield examples of noncommutative groups. The norm $N(a + bi + cj + dk)$ of $a + bi + cj + dk$ is defined as $a^2 + b^2 + c^2 + d^2$. It turns out that $N(x_1 \cdot x_2) = N(x_1) \cdot N(x_2)$. Hamilton singled out the quaternion subspace V defined as

$$V = \{ai + bj + ck\}.$$

He called its elements vectors (this is the first appearance of the term). V can be regarded as a 3-dimensional Euclidean space. Any orthogonal transformation of this space with determinant 1 is of the form

$$x \to \alpha x \alpha^{-1},$$

where α is a quaternion with norm 1. This is a famous and remarkable formula whose analogue was found in Gauss' papers. Using quaternion multiplication Hamilton defined vector multiplication of vectors in 3-dimensional space. This was one of the earliest "nonarithmetical" operations in algebra.[28] Two especially important concepts introduced by Hamilton were those of a vector field, defined as a law that associated to each point in space a vector, and of the operation ∇ — the prototype of the d-operator in the theory of differential forms. Hamilton gave a systematic account of these and other of his discoveries in two works entitled *Lectures on quaternions* (Dublin 1853) and *Elements of quaternions* (Dublin 1866), respectively. The study of quaternions was a major stimulus for the development of the theory of algebras, linear algebra, and vector analysis.

Shortly after the discovery of quaternions Cayley published the paper *On Jacobi's elliptic functions, in reply to the Rev. B. Brouwin; and on quaternions* (Philos. Mag., 1845) in which he considers an 8-dimensional algebra over the reals known as the octonions, or Cayley numbers. In this algebra multiplication is not only noncommutative but also nonassociative (however, every pair of octonions generates an associative algebra — a property now known as "alternativity"), but every nonzero octonion has an inverse. In view of the nonassociativity of multiplication of octonions they can be used to yield examples of *loops*, that is, nonassociative analogues of groups.

At this time some mathematicians saw fit to study variants of hypercomplex systems containing divisors of zero. Of special interest is the paper *On algebraical triplets* (Dublin, 1847) of the Irish mathematician John. T. Graves (1806–1870), in which he considers "triplets" of the form

$$a + b\varepsilon + c\varepsilon^2,$$

with $\varepsilon^3 = 1$. Graves gave the following geometric interpretation of triplet multiplication. He associates to a triplet $a + b\varepsilon + c\varepsilon^2$ the point (a, b, c) in ordinary 3-space, draws in space the line (x, x, x) and the plane through the origin orthogonal to it and shows that when triplets are multiplied, their respective projections on the line and on the plane multiply respectively like the real and complex numbers. In this way Graves showed that his algebra is the direct sum of the fields $\mathbb{R}$ and $\mathbb{C}$; the zero divisors are the triplets whose projections on the line or the plane are zero.

Matrix Algebra

After these investigations of low-dimensional algebras, a notable step forward was Cayley's *A Memoir on the theory of matrices* (1858) in which he introduced matrices and addition and multiplication of matrices (cf. p. 70). He also realized the skew field of quaternions as a subalgebra of the algebra of

---

28 For details see Klein, F., *Elementarmathematik vom höheren Standpunkt aus*, Bd. 1. Berlin 1924.

complex matrices of order two by establishing the isomorphism

$$a + bi + cj + dk \cong \left(\begin{array}{cc} a + di & b + ci \\ -b + ci & a - di \end{array} \right).$$

Here the norm of a quaternion is equal to the determinant of its corresponding matrix. We can say that in this way Cayley is the first to consider a linear representation of an algebra. The matrix algebra introduced by Cayley soon obtained wide recognition. By this investigation Cayley revealed the connection between two apparently radically different branches of algebra, namely linear algebra and the theory of hypercomplex numbers. This discovery had a beneficial effect on the development of both.

The Algebras of Grassmann and Clifford

At this time, in 1862, there appeared the second edition of Hermann Grassmann's "Science of extension" (*Ausdehnungslehre*). The book deals with the geometry of n-dimensional space. In the first (1844) edition "everything is derived from the most general philosophical notions, virtually without formulas". The second edition, rewritten by the author, turned out to be more accessible and thus attracted some attention from the mathematical community. In particular, the book contains the construction of the famous Grassmann or exterior algebra, now well known in connection with exterior multiplication of differential forms.

After defining the n-dimensional linear space of vectors, Grassmann defines in it the skew-symmetric "exterior products"

$$[x_1 x_2] = -[x_2 x_1], [x_1 x_2 x_3] = -[x_2 x_1 x_3] = \ldots = -[x_3 x_2 x_1],$$

which he puts equal to zero if their factors are linearly dependent. If the vectors $x_1, \ldots, x_m, x_{m+1}, \ldots, x_p$ are linearly independent, then Grassmann also defines the product

$$[x_1 \ldots x_m] [x_{m+1} \ldots x_p] = [x_1 \ldots x_m x_{m+1} \ldots x_p].$$

If the vectors in question are linearly dependent, then their product is put equal to zero. In this way Grassmann defined an algebra of expressions of the form

$$a + \sum_i a_i e_i + \sum_i \sum_j a_{ij} [e_i e_j] + \cdots$$

$$+ \sum_{i_1} \sum_{i_2} \cdots \sum_{i_r} a_{i_1 i_2 \cdots i_r} [e_{i_1} e_{i_2} \cdots e_{i_r}] + \cdots + a_{12\ldots n} [e_1 e_2 \cdots e_n],$$

78

which he called "extensive magnitudes". These are now called Grassmann numbers of order n. These numbers have

$$1 + n + \binom{n}{2} + \cdots + \binom{n}{r} + \cdots + \binom{n}{2} + n + 1 = 2^n$$

coordinates.

The English mathematician William Kingdon Clifford (1845–1879), at one time a professor at University College, London, famous for his contributions to geometry, published *Application of Grassmann's extensive algebra* (Amer. J. Math. 1879) and in this paper proposed the following modification of Grassmann's algebra. Like Grassmann, he considered linear combinations of n vectors $e_1, e_2, \ldots, e_n$ and of products $e_{i_1} e_{i_2} \ldots e_{i_r}$ which, in the case of different factors, are defined just like Grassmann's "exterior product", but in the case of repeated factors are computed using the rule $e_i^2 = -1$ (for example, $e_1 e_2 e_1 = -e_1^2 e_2 = e_2$) rather than put equal to zero. Soon after the appearance of the Clifford numbers Rudolf Lipschitz (1832–1903) showed in his "Investigations on sums of squares" (*Untersuchungen über die Summen von Quadraten.* Bonn 1886) that orthogonal transformations of n-dimensional space can be represented by transformations of the form $x \to \alpha x \alpha^{-1}$, $x = \sum_i x^i e_i$ in these algebras. Later these representations came to be known as "spinor representations" of orthogonal transformations.

Associative Algebras

The field of complex numbers, the skew field of quaternions, the algebras of matrices, of dual and double numbers, as well as of Grassmann and Clifford numbers, are all instances of associative algebras. The general concept of an associative algebra was defined by the American algebraist Benjamin Peirce, professor at Harvard University, in the book *Linear associative algebras* (Harvard 1872). Peirce defined an algebra as an n-dimensional linear space in which there is defined an associative multiplication of vectors that is distributive with respect to vector addition and that commutes with multiplication of a vector by a number. Of the algebras considered by us the only ones that do not satisfy the associativity requirement are the Cayley numbers and the vectors in 3-dimensional space with vector multiplication.

If $e_1, e_2, \ldots, e_n$ are a basis, then the vector multiplication is determined as soon as we have prescribed multiplication for pairs of basis vectors by formulas

$$e_i e_j = \sum_k c_{ij}^k e_k,$$

known as the structure formulas of the algebra (the constants c_{ij}^k are called the structure constants of the algebra).

Peirce introduced the notions of nilpotent elements and idempotent elements. An element e is nilpotent if some power e^r of e is zero (such are, for example, the elements e_i of the Grassmann algebra), and idempotent if all of its powers are equal. By means of these notions Peirce classified the complex algebras of low dimension.

The general theory of algebras is found in Weierstrass' lectures of 1861 but his investigations were first published in the paper "On the theory of complex magnitudes formed of n principal units" (*Zur Theorie der aus n Hauptein-heiten gebildeten complexen Grössen. — Gött. Nachr.*, 1884). Weierstrass showed that every commutative algebra without nilpotent elements is the direct sum of a number of copies of the fields $\mathbb{R}$ and $\mathbb{C}$.

The Theory of Invariants

Classical invariant theory came into being in the middle of the 19th century in England and occupied one of the key positions in the mathematics of the second half of the 19th century. The three disciplines that gave rise to invariant theory are number theory (the Gaussian classification of binary quadratic forms), geometry (the projective properties of curves), and algebra (the theory of determinants). The initial period of its development is connected with the creative work of Cayley and Sylvester (who thought up almost all the terms of the theory, including the term "invariant") in England, Salmon in Ireland, Jacobi and Hesse in Germany, and Hermite in France. Later, German mathematicians begin to play a leading role in the development of invariant theory. Of these the most distinguished are Aronhold, Clebsch, and Gordan. The mathematician who at the end of the century (1884–1892) solved the fundamental problems of the theory and who, to quote H. Weyl, "almost killed the whole subject" was David Hilbert.

The fundamental problem of the theory that was the focus of all efforts was that of construction and study of the invariants of forms and of systems of forms. Let $x_1, \ldots, x_n$ be variables. A form of order k is a homogeneous polynomial

$$P = P(x_1, \ldots, x_n) = \sum_{i_1 + \cdots + i_n = k} a_{i_1 \ldots i_n} x_1^{i_1} \ldots x_n^{i_n}$$

of degree k. Let $GL(n)$ denote the full linear group, that is, the group of all nonsingular matrices of order n, and let $SL(n)$ denote the unimodular group of all matrices with determinant 1. Linear transformations on these groups act on the variables $x_1, \ldots, x_n$ of the form P and carry it into a new form

$$P^g(x_1, \ldots, x_n) = P(g(x_1), \ldots, g(x_n)), \quad g \in GL(n).$$

The coefficients of all possible forms for fixed values of n and k form a vector space E and the correspondence $P \to P^g$ determines a linear transformation of this space that corresponds to the element g of the group $GL(n)$. Later this correspondence was known as a representation of the group $GL(n)$ in the space E. An (integral) invariant is a function f on E which is 1) a polynomial in the coordinates of the space, that is, a polynomial in the coefficients of forms P and which 2) does not change under the action of the group $GL(n)$ or the group $SL(n)$, that is, $f(P^g) = f(P)$ for all $g \in GL(n)$ or $g \in SL(n)$.

In classical invariant theory one considered more general — say, rational — functions and one allowed other behavior, such as multiplication by a power of $\det(g)$ under the action of g supposed to belong to some subgroup of $GL(n)$.

The integral invariants f form a subring R^{inv} in the ring R of all polynomials in the coordinates of the space E. If one considers a system of forms $P_1, \ldots, P_s$ rather than a single form and the action

$$(P_1, \ldots, P_s) \to (P_1^g, \ldots, P_s^g)$$

of a linear transformation g on such a system, then an invariant is defined as a function of the coefficients of the forms $P_1, \ldots, P_s$ with the same properties as above.

The construction of invariants is closely related to the problem of projective classification of algebraic curves. In fact, the results of projective classification of algebraic curves served as the main starting point in the creation of invariant theory. The projective equivalence of algebraic systems X and Y determined by systems of forms $P_1, \ldots, P_s$ and $Q_1, \ldots, Q_s$, respectively, implies that the values of the invariants of these systems coincide. In general, the converse is false, so that the study of invariants was only the first step in the problem of classification of geometric objects.

The first ideas of algebraic invariant theory arise in number theory, more specifically, in Gauss' *Disquisitiones*, in connection with the study of binary quadratic forms

$$P = ax^2 + 2bxy + cy^2$$

under linear transformations of the unknowns x and y. Gauss considers a linear transformation

$$x = \alpha x' + \beta y', \quad y = \gamma x' + \delta y'$$

that carries the form P into a form

$$P' = a'x'^2 + 2b'x'y' + c'y'^2$$

and looks for magnitudes that remain unchanged or change in a particular manner, and hits first on the "determinant" (discriminant, in our terminology)

$$D' = b'^2 - a'c' = r^2 D; \quad r = \alpha\delta - \beta\gamma, \quad D = b^2 - ac.$$

The rise of invariant theory was also aided by the study of determinants, in particular, by two papers of Jacobi "On the formation and properties of determinants" (*De formatione et proprietatibus determinantium.* — J. für Math., 1841) and "On functional determinants" (*De determinantibus func-tionalibus.* — J. für Math., 1841) in which this line of research was developed to its fullest.

One of the important results in this area is that the determinant $f = \det(a_{ij})$ of a system of n linear forms in n variables

$$P_i = \sum a_{ij} x_j \quad (i = 1, \ldots, n)$$

is invariant under the group $SL(n)$. An important generalization of this result is the assertion that the value of the Jacobian

$$f = \det\left(\frac{\partial P_i}{\partial x_j}\right)$$

is an invariant at the null point.[29] This work of Jacobi was continued by L.O. Hesse (1811–1874), a professor at the universities of Königsberg, Heidelberg, and Munich, who was especially interested in its analytic-geometric ramifications — a departure that disrupted, for some time, the connection of invariant theory with number theory.

Hesse showed that for any form P in n variables the determinant

$$f = \det\left(\frac{\partial^2 P}{\partial x_i \partial x_j}\right)$$

made up of its second partial derivatives, later to be named a Hessian in his honor, is invariant under the group $SL(n)$.

One of the first to study algebraic invariants — in 1841 — was George Boole. His results attracted the attention of Cayley who appreciated their significance and, in turn, interested Sylvester in the new theory. They were joined by the Irish mathematician and theologian George Salmon. Cayley, Sylvester, and Salmon produced so many papers on invariant theory and their role in its development was so great that Hermite named them in one of his letters "the invariant trinity".

James Joesph Sylvester was born in London in 1814. In 1831 he entered St. John's College, Cambridge. Because of illness he completed his studies only in 1837. In 1838 Sylvester became a professor of natural philosophy at what is now University College in London. Then he went to America and

---

29 The Jacobian itself, like the Hessian considered below, is a form in n variables known as a so-called covariant.

James Sylvester

between 1841 and 1842 was a professor at the University of Virginia. Between 1845 and 1855 he worked as an actuary and a mathematician in an insurance company. Between 1855 and 1870 he held the post of a professor at the Royal Military Academy, Woolwich. In 1876 he became a professor at the newly established Johns Hopkins University in Baltimore. In addition to teaching he worked productively on problems in pure mathematics, essentially in the area of invariant theory. He founded the American Journal of Mathematics, one of the best known American mathematical journals to this day.

In 1883, at the age of 69, Sylvester returned to England and accepted a professorship at Oxford which he held for the rest of his life.

Sylvester had a brilliant and powerful mind. He studied with great interest a large variety of questions and was able to systematize and synthesize related phenomena from, apparently, widely different areas of knowledge. Sylvester attained his greatest successes in purely abstract combinatorial research of algebraic objects. During Sylvester's stay in London, Cayley introduced him to the new algebraic ideas of invariant theory. Soon Sylvester became one of the leading representatives of this branch of algebraic thought. In particular, he introduced all of the basic terms of the new theory such as invariant, covariant, discriminant, and so on.

An important part of Sylvester's creative mathematical work is the theory of elementary divisors of a pair of quadratic forms and the theory of canonical forms, that is, the theory of reduction of homogeneous quadratic polynomials to simplest form.

Boole's investigations suggested to Cayley the thought of computing invariants of homogeneous functions of degree n. Using, in addition, the ideas of Hesse and Eisenstein on determinants, he elaborated the technical means that enabled him to generalize the notion of invariants.

Also in 1841 Cayley begins to publish a series of papers in which he investigates the algebraic aspects of projective geometry. Between 1854 and 1878 Cayley publishes a string of papers devoted to the study of homogeneous polynomials in two, three, and more variables. He obtains a great many concrete results and discovers the symbolic method of computation of invariants. One of Cayley's results is that

$$g_2 = ae - 4bd + 3c^2$$

and

$$g_3 = \begin{vmatrix} a & b & c \\ b & c & d \\ c & d & e \end{vmatrix}$$

form a complete set of invariants with respect to the group $GL(2)$ for the binary form of degree four given by

$$P = ax_1^4 + 4bx_1^3x_2 + 6cx_1^2x_2^2 + 4dx_1x_2^3 + ex_2^4, \quad a, b, c, d \in \mathbb{C}.$$

In other words, the invariants g_2 and g_3 generate the ring of invariants

$$R^{\mathrm{inv}} = \mathbb{C}[g_2, g_3].$$

For example, the discriminant of the form P is given by

$$\Delta = g_2^3 - 27g_3^2.$$

For the binary cubic form

$$P = ax_1^3 + 3bx_1^2x_2 + 3cx_1x_2^2 + dx_2^3$$

Eisenstein obtained the following simplest invariant

$$f = 3b^2c^2 + 6abcd - 4b^3d - 4ac^3 - a^2d^2,$$

84

equal to within a multiplicative factor to the discriminant of the hessian

$$h = \frac{1}{36} \begin{vmatrix} \frac{\partial^2 P}{\partial^2 x_1} & \frac{\partial^2 P}{\partial x_1 \partial x_2} \\ \frac{\partial^2 P}{\partial x_2 \partial x_1} & \frac{\partial^2 P}{\partial x_2} \end{vmatrix}.$$

This result of Eisenstein was the starting point in the further study of binary invariants, in particular, the previously mentioned studies of Cayley.

Later, the Berlin mathematician S.H. Aronhold (1819–1884), who began his investigations in the theory of invariants in 1849, made a significant contribution to the study of the invariants of ternary cubic forms.

The examples just given show that the first fundamental problem facing the founders of invariant theory was finding particular invariants. This is a fair description of the work done between the 40s and 70s of the 19th century.

The second fundamental problem that arose after the computation of a large number of particular invariants was the problem of finding a complete system of invariants. It was noticed in many cases that it was possible to find among the invariants of a system of forms finitely many invariants $f_1, \ldots, f_n$ such that all other invariants of the system were polynomials in $f_1, \ldots, f_n$. Such invariants were first noticed by Cayley. Using the result of Eisenstein for binary cubic forms, Cayley found such invariants for binary forms of degree 3 and 4 and published them in the second of his famous *Memoirs on quantics* (1856). Such systems of invariants came to be known as complete or fundamental. From the modern viewpoint such a system is a finite system of generators of the relevant ring of invariants. Construction of such systems for higher-degree forms in many variables ran into great difficulties. Working along these lines, mathematicians (mostly) of the German school developed the so-called symbolic method. This method made possible the explicit computation of invariants of a prescribed degree for an arbitrary system of forms. Since the degrees of the invariants of a complete system are initially unknown, it is not possible, in principle, to construct complete systems of invariants by means of the symbolic method. Nevertheless, by combining the symbolic method with other considerations it was possible to obtain the required result in some special cases. The high point of this approach was the proof, obtained by the Erlangen professor Paul Gordan (1837–1912) in 1868, of the existence of a finite, complete system of invariants for binary forms of arbitrary degree. In 1870 he proved that any finite system of binary forms of arbitrary degree had a complete system of invariants. These achievements earned Gordan the name of "king of invariant theory". We note that Gordan's constructions, as well as other constructions connected with the symbolic method, were fully effective. The computations involved were very complex. For binary forms of degree n explicit complete systems of invariants were obtained only for $n \leq 6$. The case $n = 7$ involved computations that exceeded 19th-century possibilities; this in spite of the fact that a ten-page paper of wordless formulas was not considered unusual.

During a 12-year period researchers extended Gordan's result in various ways. Gordan himself found a complete system of invariants for ternary quadratic forms, for ternary cubic forms, and for systems of two and three ternary quadratic forms. However, the question of existence of a finite complete system of invariants in the general case remained one of the central unsolved problems of the theory.

This was the state of this area of mathematics in the mid 80s when Hilbert began to study invariant theory. His interest in this theory is quite natural if we bear in mind the traditions of the Königsberg mathematical school. The school was established in the first half of the century by Ernst Neumann (1798–1895) and Carl Jacobi, who lectured there from 1826 to 1843. Almost all German mathematicians involved in invariant theory came out of this school. Jacobi's student Hesse lectured there from 1840 to 1855. Clebsch and Aronhold also belonged to this school.

Hilbert's fundamental papers "On the theory of algebraic forms" (*Über die Theorie der algebraischen Formen.* — Math. Ann., 1890) and "On complete systems of invariants" (*Über die vollen Invariantensysteme.* — Math. Ann., 1893) contained a solution of the problem that defied for a long time the best efforts of many specialists in invariant theory. The problem is that of the finiteness of the number of generators of the ring of integral invariants. The proof, given by Hilbert in the first paper, relied on the basis theorem established by Hilbert with this application in mind. The proof is simple and brief. Judged in terms of the then common views, Hilbert's proof suffered from the fatal flaw of being noneffective. Hilbert's second paper completes his work on invariant theory. In it he presents an explicit and constructive solution of the problem based on his notion of a null form.

The contrast between the fates of these two papers is striking. The first paper stimulated the development of a number of chapters of 20th-century mathematics such as commutative algebra, algebraic geometry, representation theory, and homological algebra. By contrast, the second paper was forgotten soon after its publication and for 70 years had no effect on the evolution of mathematics.

3 The Theory of Algebraic Numbers and the Beginnings of Commutative Algebra

Disquisitiones Arithmeticae of C.F. Gauss

The works of Gauss, and especially his *Disquisitiones arithmeticae* (1801) and his "Theory of biquadratic residues" (*Theoria residuorum biquadraticorum.* Commentationes soc. reg. sci. Götting. recentiores. Gottingae, pt. 1, 1828; pt. 2, 1832), had a formative effect on the whole of number theory in the 19th century. To quote Gauss, the latter work resulted in what was, in a sense, an infinite extension of the domain of higher arithmetic. We will discuss it in the next subsection. Here we discuss his *Disquisitiones Arithmeticae*.

This book is remarkable in terms of content as well as form. In it Gauss proved very difficult theorems that had long resisted the best efforts of Euler, Lagrange, and Legendre, and laid the foundations for completely new theories that became models for 19th-century mathematicians. Above all, Gauss introduced the relation of *congruence* and expounded all of elementary number theory using the language of congruences. He called two numbers a and b congruent modulo a whole number c if the difference $a - b$ is divisible by c. He denoted congruence by the symbol "≡", analogous to the equality sign, and wrote

$$a \equiv b (\mathrm{mod}\, c).$$

Gauss' choice of symbol, stressing as it did the analogy between congruence and equality, was especially apt. Gauss proves that one can apply to congruences the rules applicable to equalities. Specifically, one can add and multiply congruences, and divide a congruence by a number coprime with the modulus.

Since the relation of congruence is symmetric, reflexive, and transitive, it splits the integers into disjoint classes of equivalent numbers. The classes are called residue classes modulo c. One can associate with each class one of its elements, say its least nonnegative element (called by Gauss the "least residue"). Then $0, 1, 2, \ldots, n - 1$ are a set of representatives modulo n of the classes. They form a so-called complete system of residues modulo n. One can define addition and multiplication of residue classes and obtain in this way a finite ring. If the modulus is a prime p then the residue classes form a finite field. This is so because each of the elements $1, 2, \ldots, p - 1$ has a (multiplicative) inverse. This was the first instance of a field in mathematics (and the first example of a "nonnatural" field, that is, a field not related to measurement). Gauss develops the theory of congruences in a manner analogous to the theory of equations. He first considers linear congruences

$$ax + b \equiv 0 (\mathrm{mod}\, c),$$

then systems of such congruences, and finally congruences of degree two and of higher degrees. He shows that an m-th degree congruence

$$Ax^m + Bx^{m-1} + \ldots + Mx + N \equiv 0 (\mathrm{mod}\, p),$$

p prime and $A \not\equiv 0 (\mathrm{mod}\, p)$, has at most m roots incongruent mod p. (Incidentally, this result was proved by Lagrange in the paper "A new method for solving indeterminate problems in integers" (*Nouvelle méthode pour résoudre les problèmes indéterminés en nombres entiers*, Mém. Acad. Berlin (1768), 1770) without the use of the congruence relation.)

In connection with higher-degree equations we note Gauss' rigorous proof of the existence of a primitive root with respect to a prime modulus, and his introduction and systematic use of the notion of index corresponding to that of a logarithm.

As for quadratic congruences, all of the basic results were earlier obtained by Euler. Gauss systematized the whole theory and gave the first rigorous proof of the law of quadratic reciprocity discovered by Euler (see HM, vol. 3, pp. 104–105).

In his book Gauss gave two proofs of this remarkable theorem (and later six more, for a total of eight). Using the Legendre symbol $(\frac{p}{q})$, p and q odd primes, and putting $(\frac{p}{q}) = +1$ when p is a quadratic residue $\bmod q$ (that is the equation $x^2 \equiv p \pmod{q}$ is solvable) and $(\frac{p}{q}) = -1$ otherwise, we can state the law of quadratic reciprocity as

$$\left(\frac{p}{q}\right)\left(\frac{q}{p}\right) = (-1)^{\frac{p-1}{2}\frac{q-1}{2}}.$$

Gauss called this theorem fundamental "because practically everything that can be alleged about quadratic residues is based on this theorem".[30] Studies devoted to this theorem and its generalizations provided one of the most important stimuli for the development of algebraic number theory.

Another fundamental direction developed by Gauss came to be known as the theory of quadratic forms. A quadratic form (or simply a form) in two variables is an expression

$$F(x, y) = ax^2 + 2bxy + cy^2, \tag{13}$$

$a, b, c \in \mathbb{Z}$. Gauss denotes such a form briefly by (a, b, c). A number N is said to be representable by the form (13) if there are numbers m and n such that

$$N = am^2 + 2bmn + cn^2.$$

The number $D = b^2 - ac$ is called the discriminant of the form.

Already Fermat posed the problem of determining all whole numbers representable by forms with a given discriminant. Fermat knew that the form $x^2 + y^2$ represents all primes $4n + 1$ and none of the primes $4n + 3$. He also knew which numbers are representable by the respective forms $x^2 \pm 2y^2$ and $x^2 + 3y^2$. Euler proved the facts discovered by Fermat and made substantial progress in the study of the properties of quadratic forms. He was the first to shift attention from the study of the numbers representable by a form to

---

30 Gauss, C.F., *Werke*, Bd. 1. Göttingen 1863, p. 99.

the study of their divisors. In this way he changed the additive problem to a multiplicative one (see HM, v. 3, p. 102).

Following this approach, Lagrange showed in his "Arithmetical investigations" (*Recherches arithmétiques*, Nouv. Mém. Acad. Berlin (1773), 1775) that the divisors p of a number N representable by a form

$$u^2 + dv^2 \tag{14}$$

are, in general, not representable by a form of that type but are representable by a form

$$p = ax^2 + 2bxy + cy^2 \tag{15}$$

with $ac - b^2 = d$, that is by a form whose discriminant is the same as that of the form (14). The latter came to be known as the principal form with given discriminant.[31]

Lagrange also establishes the converse proposition: if p is representable as in (15) then it divides a number representable by the principal form whose discriminant is equal to that of the form in (15). All this suggested to Lagrange the simultaneous study of all forms with the same discriminant. He divided the forms into classes by putting two forms (15) in the same class if they could be transformed into one another by means of a linear substitution

$$x = Lx' + My', \; y = \ell x' + my' \tag{16}$$

with determinant $Lm - \ell M = \pm 1$. Then a form $F(x, y)$ goes over into the form

$$F_1(x', y') = a_1 x'^2 + 2b_1 x'y' + c_1 y'^2 \tag{13'}$$

with $a_1 c_1 - b_1^2 = d(Lm - \ell M)^2 = d = ac - b^2$. This shows that the discriminant d is invariant under the transformation (16). Obviously, the form (13') can be transformed into the form (13) by means of a transformation like that in (16). Gauss later called two such forms "equivalent" (and "properly equivalent" if $Lm - \ell m = 1$). It is easy to see that if a number N is representable by a form (13) then it is also representable by all forms equivalent to it.

Like equality, this equivalence relation is reflexive ($F \sim F$), symmetric (if $F \sim F_1$ then $F_1 \sim F$), and transitive (if $F \sim F_1$ and $F_1 \sim F_2$ then $F \sim F_2$). Hence the forms with the same discriminant split into disjoint equivalence classes.

Lagrange proved the extremely important theorem that the number of equivalence classes of forms with given discriminant is finite. We reproduce

---

31 Lagrange defines the discriminant of a form as $ac - b^2$, that is $d = ac - b^2 = -D$, where D is the discriminant in Gauss' sense.

his argument, for all subsequent proofs of the finiteness of the number of classes of ideals are based on a similar principle.

First Lagrange shows that every form (15) can be changed by a transformation (16) to what he called a reduced form

$$\alpha x^2 + 2\beta xy + \gamma y^2$$

with $\alpha\gamma - \beta^2 = d$, $2|\beta| \leq |\alpha|$, $2|\beta| \leq |\gamma|$. By associating with each class of forms a reduced form (under Lagrange's reduction method more than one reduced form can correspond to a single class of forms) Lagrange shows that there are only finitely many inequivalent reduced forms with a given discriminant. Indeed, if $d > 0$, then $d = \alpha\gamma - \beta^2 \geq 4\beta^2 - \beta^2 = 3\beta^2$, so that $\beta \leq \sqrt{d/3}$, and, since β is a rational integer, it can take on only a finite number of different values. But for a given β

$$\alpha\gamma = d + \beta^2,$$

so that α and γ can take on just finitely many different values. This means that the number of reduced forms with given discriminant is indeed finite. A similar proof can be given for $d < 0$ (see HM, v. 3, pp. 114–116).

Gauss constructed in his book a complete theory of binary quadratic forms (that is, forms in two variables) and began similar constructions for ternary forms. He made precise the notions of equivalence and reduced form, splitting the forms with the same discriminant into equivalences classes of equivalent forms and the latter into genera and orders. He also introduced composition of forms, that is he defined on the set of forms with the same discriminant an operation analogous to addition (or multiplication): $F_1 \oplus F_2 = F_3$.

Gauss shows that if $F_1 \sim F_1'$ and $F_2 \sim F_2'$, then $F_3' = F_1' \oplus F_2'$ is equivalent to F_3. This means that the operation can be transferred to the set of classes of forms. In modern terms, we would say that he determines a quotient rule of composition. Gauss writes: "The notion of a class made up of two or a number of given classes becomes immediately clear."[32] And further: "It is convenient to denote composition of classes by the addition symbol $+$ and, similarly, the sameness of classes by the equality sign."[33]

Essentially, Gauss shows that the set of classes of forms under his law of composition forms a finite abelian group in which the role of zero is played by the principal class, that is the class containing the principal form $x^2 - Dy^2$. He shows that for every class there is an inverse and that the result of composing a class with its inverse is the principal class.

---

32 Ibid., p. 273.

33 Ibid.

In addition to the interest these investigations hold for the creation of group theory they are of utmost significance for the creation of the arithmetic of algebraic numbers. It is safe to say that Gauss created the arithmetic of quadratic fields not using the now current language of algebraic numbers but using quadratic forms.

Indeed, already Lagrange's results showed that the divisors of numbers of the form $N = u^2 - Dv^2$ are representable by forms $ax^2 + 2bxy + cy^2$ with discriminant D. But the representability of a number N by $u^2 - Dv^2$ shows that N is the norm of the number $\alpha = u + v\sqrt{D}$ in the field $K = \mathbb{Q}(\sqrt{D})$. As Lagrange showed, in general, a divisor p of the number N is not the norm of a number in K but is representable by a form with discriminant D. This is equivalent to p not being a prime in the field K. After Dedekind's construction of ideal theory it was shown that such a number p is the norm of an ideal K (see below). It was also discovered that the ideal classes of a quadratic extension of $\mathbb{Q}$ correspond to Gauss' classes of quadratic forms, and that the composition of classes of forms corresponds to the multiplication of the appropriate ideal classes.[34]

Using this correspondence it is possible to prove, among other things, that the number h_0 of ideal classes of a quadratic field is finite. The question of which prime numbers are representable by forms $F(x, y)$ with discriminant D — a question studied by Fermat, Euler, Lagrange and Gauss — is equivalent to the question of which primes are representable as norms of ideals in the field $\mathbb{Q}(\sqrt{D})$, that is, cease to be primes in the ring of integers of that field. For example, Gauss could immediately decide which rational primes cease to be primes in the ring $Z[i]$.

Conversely, the clumsy theory of composition of classes of forms can be expressed far more simply in terms of multiplication of the corresponding ideal classes.

At any rate, the foundations of the arithmetic of quadratic fields were contained in Gauss' theory of quadratic forms. His contemporaries found Gauss' book difficult to read. One of its first readers was P. Lejeune-Dirichlet. Dirichlet's understanding of Gauss' work was profound. To quote Felix Klein:

> That this work has exerted its deserved influence is due, above all, to Dirichlet's interpretative lectures. These lectures are a superb introduction to Gauss' formulation of problems and they explain his mode of thought.[35]

Dirichlet's lectures were published by R. Dedekind, one of his listeners (the first edition appeared in 1863; the book was subsequently reissued a number of times). After that, Gauss' ideas and methods acquired wide popularity.

---

34 For details see, for example, Hecke, E., *Vorlesungen über die Theorie der algebraischen Zahlen.* Leipzig 1923.

35 Klein, F., *Vorlesungen ...* , Bd. 1. Berlin 1926, p. 27.

Investigation of the Number of Classes of Quadratic Forms

A remarkable discovery in algebraic number theory is the derivation — by Gauss and Dirichlet — of exact formulas for the number of classes of quadratic forms with given discriminant. This discovery completed Gauss' theory of quadratic forms. Beyond that, it involves deep ideas and is the starting point of numerous investigations of Dirichlet's L-series and their applications in number theory.

Gauss did not publish his investigations in this direction during his life-time. His *Nachlass* contained an unfinished manuscript of a paper from the period of 1830–1835 on the determination of the number of classes of forms with negative determinant, including the correct answer and a basic outline of a proof. A complete published solution of the problem is due to Dirichlet and is found in the paper "Investigations on various applications of infinitesimal analysis to number theory" (*Recherches sur diverses applications de l'analyse infinitésimale à la théorie des nombres.* — J. für Math., 1838, 1840) published in volumes 19 and 21 of *Crelle's Journal.* This is one of a series of Dirichlet's remarkable papers on the application of analytic methods in number theory that are discussed in an essay in this book (see p. 171). While it is likely that as a close friend and virtually a student of Gauss' Dirichlet was familiar with his ideas on his topic, his paper is entirely original and independent of Gauss.

We consider Dirichlet's reasoning more closely. To begin with, he reduces the problem of determining the number of classes of binary forms with discriminant D to that of determining the number of classes of such forms

$$ax^2 + 2bxy + cy^2$$

with a, b and c pairwise coprime ("primitive forms"). Let S be a complete system of representatives of the classes of such forms. From Gauss' theory one can obtain the result that

> one and the same infinite set of numbers can be obtained in two ways: one is to combine the primes f for which D is a quadratic residue; and the other is to substitute all admissible pairs of numbers x and y in the forms of the system S. This result of the earlier findings on the equivalence of forms and the representability of numbers is the guiding principle of the following investigation.[36]

Here, if m is a number obtained by substituting an admissible pair (x, y) in one of the forms in S, then the number of such representations is $\kappa 2^\mu$, where

---

36 Lejeune-Dirichlet, P.G., *Vorlesungen über Zahlentheorie*, Braunschweig 1871. This work also contains a definition of the term "admissible pairs".

μ is the number of prime factors in m and κ is a constant that depends solely on the discriminant D. It follows that if we permute the terms in the series

$$\sum_{(a,b,d)\in S} \sum_{\substack{(x,y) \text{ varies over} \\ \text{all admissible pairs}}} (ax^2 + 2bxy + cy^2)^{-s},$$

where s is some real number, then we obtain the series

$$\kappa \sum \frac{2^\mu}{m^s}$$

in which the summation extends over all numbers m composed of prime factors with respect to which D is a quadratic residue. Both series converge absolutely for all $s > 1$, so that for such s we have the equality

$$\sum_{(a,b,c)\in S} \sum (ax^2 + 2bxy + cy^2)^{-s} = \kappa \sum \frac{2^\mu}{m^s},$$

where we sum as explained above.

Dirichlet investigates the function on the right and, in view of the above equality, obtains certain information about the classes of quadratic forms with given discriminant. To compute the number of classes Dirichlet investigates the behavior of both sides of the equality as $s \to 1$. If

the exponent s decreases and tends to 1, then each of these sums [Dirichlet has in mind the sums

$$\sum_{\substack{(x,y) \text{ varies over} \\ \text{all admissible pairs}}} (ax^2 + 2bxy + cy^2)^{-s}$$

for fixed $(a, b, c) \in S$] will increase without bounds and a closer investigation will show that the product of such a sum by $s - 1$ tends to a definite limit L that depends only on the discriminant D common to all the forms; hence the limit of the entire left side multiplied by $s - 1$ is hL, where h denotes the number of sums, that is the *number of forms* $(a, b, c), \ldots$ *in the system S of forms.* Since it is possible to compute directly the limit of the right side multiplied by $s - 1$, we obtain an expression for the number h of classes whose determination is the subject of our investigation.[37]

We have quoted Dirichlet's description of the general plan of his proof. The computation of the limit

$$\lim_{s \to 1} (s - 1) \sum (ax^2 + 2bxy + cy^2)^{-s}$$

_____

37 Ibid.

contains very refined arguments but is rather complicated and we won't reproduce it here.[38]

The general plan and the concrete ideas of the proof were later used in the proof of the formula for the number of classes of divisors of a field of algebraic numbers.[39] Series of the form $\sum_n \frac{a_n}{s^n}$, so skillfully used by Dirichlet in this and in other investigations, have become a standard mathematical tool and are known as Dirichlet series.

We wish to emphasize that this investigation, combining as it does deep algebraic results of the theory of forms and subtle analytic methods, is a persuasive demonstration of the unity of mathematics that invariably characterizes the works of the great masters.

Gaussian Integers and Their Arithmetic

The theory of algebraic numbers was created in the 19th century. Its development was furthered by two problems posed for rational integers. One was the law of reciprocity. The other was Fermat's Last Theorem, that is the assertion that for $n > 2$ the equation

$$x^n + y^n = z^n$$

has no integer solutions satisfying $xyz \neq 0$ (and thus no rational solutions).

When proving Fermat's Last Theorem for $n = 3$ Euler began to operate with expressions of the form

$$m + n\sqrt{-3}$$

$m, n \in \mathbb{Z}$, as with ordinary numbers. He used the concepts of "prime" number and "relatively prime" numbers, as well as the law of unique factorization into "prime" factors and its consequences, but without any justification (see HM, v. 3, p. 102).

The first rigorous introduction of algebraic integers is connected with the investigation of the law of biquadratic reciprocity. This was done by Gauss in his "Theory of biquadratic residues" (1828–1832). Contemplation of generalizations of the law of quadratic reciprocity led Gauss to the conclusion that it was possible and necessary to extend the notion of an integer that for over 2000 years seemed to be an inalienable and intrinsic property of the domain $\mathbb{Z}$ of rational integers. Gauss separated this notion from its natural carrier and transferred it to the ring of numbers of the form

$$a + bi, \tag{17}$$

38 For a modern account see Borevich, Z.I., and Shafarevich, I.R., *Number theory*. Ch. V.

39 Ibid.

where $a, b \in \mathbb{Z}$ and i is a root of the equation

$$x^2 + 1 = 0. \tag{18}$$

He showed that the numbers (17) not only form a domain $\mathfrak{O}$ closed under addition, subtraction and multiplication but also that in this domain it is possible to establish an arithmetic analogous to the usual one. Gauss designated 1, -1, i, $-i$ as the units of these new numbers and called numbers obtained from one another by multiplication by a unit "associates". He noted that one should not distinguish between factorizations that are the same up to associates. He associated with each number α of the form (17) a whole number, its norm $N\alpha = (a + bi)(a - bi) = a^2 + b^2$. The definition of norm implies that

$$N\alpha\beta = N\alpha N\beta.$$

Gauss called a number of the form (17) a prime if it is not a unit and cannot be written as a product of two numbers neither of which is a unit.

This definition implies that a composite rational integer remains composite in $\mathfrak{O}$. But primes in $\mathbb{Z}$ can be composite in $\mathfrak{O}$. Thus $2 = (1 + i)(1 - i)$, $5 = (1 + 2i)(1 - 2i)$ and, quite generally, every prime in $\mathbb{Z}$ of the form $4n + 1$ — which, as is well known, is representable as a sum of squares — ceases to be prime in $\mathfrak{O}$:

$$p = m^2 + n^2 = (m + ni)(m - ni).$$

Gauss showed that, on the other hand, primes in $\mathbb{Z}$ of the form $4n + 3$ remain prime in $\mathfrak{O}$ and the norm of a number q of this type is q^2. To find all primes in $\mathfrak{O}$ Gauss proves the following theorem:

An arbitrary integer $a + bi$, $a \neq 0$, $b \neq 0$, is prime in $\mathfrak{O}$ if and only if its norm is a prime or the square of a prime of the form $4n + 3$.

This implies that the primes in $\mathfrak{O}$ are

(1)   the number $1 + i$ (a divisor of 2);

(2)   the rational primes of the form $4n + 3$;

(3)   all complex numbers $a + bi$ whose norm is a prime of the form $4n + 1$.

It is of interest to note that Gauss uses here essentially *local methods*, namely he bases the primeness of a complex number on the nature of its norm.

In proving the uniqueness of prime factorization Gauss also relies on local methods in that he bases himself on the rules of divisibility of norms, that is rational integers.[40]

---

40 It is interesting to note that in his exposition of the divisibility theory of Gaussian integers in the appendices to Dirichlet's *Zahlentheorie* (published in 1879 and 1894) Dedekind relies on the Euclidean algorithm, that is he replaces local methods by global ones (for details on the local method see the chapter on the theories of Kummer and Zolotarev).

Here is Gauss' proof. Let $M = P_1^{\ell_1} \dots P_S^{\ell_S}$ be the prime decomposition of $M \in \mathfrak{O}$. Then M is not divisible by a prime Q different from $P_1, \dots, P_S$ (supposed distinct). Indeed, if M were divisible by Q then

$$N(M) = p_1^{\ell_1} \dots p_S^{\ell_S},$$

where $p_i = N(P_i)$, would be divisible by $q = N(Q)$. Since $p_1, \dots, p_s$ are either primes of the form $4n + 1$, or squares of primes of the form $4n + 3$, or one of them is 2, it follows that q must equal one of the numbers p_i, say $q = p_1$. But $Q \neq P_1$. Hence Q and P_1 must be complex conjugates, $P_1 = a + bi$, $Q = a - bi$. Note that q must be an odd prime. Now

$$P_1 \equiv 2a \pmod{Q}$$

and

$$M \equiv 2^{\ell_1} a^{\ell_1} P_2^{\ell_2} \dots P_S^{\ell_S} \pmod{Q}.$$

Hence the norm of the right side, that is the number

$$2^{2\ell_1} a^{2\ell_1} p_2^{\ell_2} \dots p_S^{\ell_S},$$

is divisible by q. Since q divides neither 2 nor a, it must coincide with one of the numbers $p_2, \dots, p_S$. Suppose that $q = p_2$. Then $P_2 = a + bi$ or $P_2 = a - bi$. In the first case $P_2 = P_1$ and in the second case $P_2 = Q$. Both of these possibilities contradict our assumption.

Gauss notes that now the uniqueness assertion can be proved just as in the case of the rational integers.

Next Gauss gives a geometric interpretation of the complex numbers as points in the plane and of the algebraic operations with these numbers. The integral complex numbers form a lattice which splits the complex plane into infinitely many squares. Gauss also considers residues with respect to a complex modulus and introduces the notion of a least residue and of a system of least residues with respect to a given modulus. To find the greatest common divisor of two complex numbers he introduces the Euclidean algorithm. He shows that if the remainders are chosen so as to be least residues with respect to the corresponding divisors then the algorithm is always finite; in symbols,

$$\alpha = \beta\gamma + \delta,$$

$$\beta = \delta\gamma_1 + \delta_1$$

$$\dots\dots\dots\dots$$

$$\delta_n = \delta_{n+1}\gamma_n,$$

where δ_{i+1} is a least residue mod δ_i.

We clarify this part of Gauss' argument. Suppose that we are to divide $\alpha = a + bi$ by $\beta = c + di$ and that

$$\alpha = \beta\gamma + \delta,$$

that is

$$N\delta = N(\alpha - \beta\gamma) = N(\alpha/\beta - \gamma) \cdot N\beta.$$

We must show that γ can be chosen so that $N(\alpha/\beta - \gamma) < 1$. Put $\alpha/\beta = \xi + i\eta$, $\gamma = x + iy$. Then

$$N(\alpha/\beta - \gamma) = (\xi - x)^2 + (\eta - y)^2.$$

If we choose integers x and y such that $|x - \xi| \leq 1/2$, $|\eta - y| \leq 1/2$, (which is always possible) then we have $N(\alpha/\beta - \gamma) \leq 1/2$.

Gauss develops for the numbers $m + ni$, $m, n \in \mathbb{Z}$, an arithmetic analogous to the usual one (he proves Fermat's Theorem, introduces indices, primitive roots, and so on), and uses it to formulate and partially prove the law of biquadratic reciprocity.

All this persuaded mathematicians everywhere that the complex integers are no less legitimate objects of higher arithmetic than the rational integers. So great was the impact of Gauss' paper that almost until the last third of the 19th century algebraic integers were called complex integers even if they were of the form $m + n\sqrt{D}$ with $D > 0$, that is even if they were real. It became clear to all that (1) the new numbers can be objects of arithmetic, that is are "really" integers, and (2) that by means of this extended arithmetic it is possible to obtain results about rational integers that can not be obtained by other means.

Gauss was perfectly aware that his paper opened boundless prospects to mathematicians. He wrote:

> This theory (that is, the law of biquadratic reciprocivity) requires what is, *in a sense, an infinite extension of higher arithmetic.*[41]

And further,

> *The natural source of the general theory must be sought in the extension of the domain of arithmetic.*[42]

Gauss noted that the theory of cubic residues (that is the cubic reciprocity law) must be based on the study of numbers of the form $a + b\rho$, where $\rho^3 - 1 = 0$ and $\rho \neq 1$, and added that

---

41 Gauss, C.F., *Werke*, Bd. 2, p. 67.

42 Ibid., p. 102.

in much the same way the theory of residues of higher degrees requires the introduction of other imaginary magnitudes.[43]

At the end of his memoir Gauss used the complex integers to formulate the general law of biquadratic reciprocity without, however, proving this fundamental result. He intended to give a proof in the third part of his memoir, but this was never written. His comment about the proof was that it "belongs to the most deeply hidden secrets of arithmetic".[44] It subsequently came to light that Gauss' *Nachlass* contains an outline of a proof of this theorem close to his sixth proof of the law of quadratic reciprocity (that is based on the theory of cyclotomic extensions).

Stimulated by Gauss' work, C.G. Jacobi and G. Eisenstein (1823–1852) continued his investigations. Already in his lectures on number theory of 1836–1837 Jacobi gave a proof of the law of biquadratic reciprocity based on the theory of cyclotomic extensions. Eisenstein was the first to publish a proof of this law (in 1844), basing it on the theory of complex multiplication of special elliptic functions. He also proved the law of cubic reciprocity and in this connection constructed arithmetic in the ring $\mathbb{Z}[\rho]$. Then he proved the reciprocity law for residues of eighth degree. These researches of Eisenstein are found in " The proof of reciprocity laws for cubic residues in the theory of numbers composed of cubic roots of unity" (*Beweis der Reciprocitätsgesetze für die cubischen Reste in der Theorie der aus dritten Wurzeln der Einheit zusammengesetzten Zahlen.* — J. für Math., 1844). In the 27th and subsequent volumes of *Crelle's Journal* there are 25 notes by Eisenstein, then still a student at Berlin University. Gauss had a very high opinion of the gifts of the young mathematician and the lectures he gave at Berlin University in the last years of his short life attracted many listeners. In addition to the results just mentioned, Eisenstein made a number of discoveries in the theory of binary cubic forms and in other questions of number theory as well as in the theory of elliptic functions. Some of his unproved but published results in number theory were proved by the Oxford professor H.J.S. Smith, one of the few English mathematicians who studied number theory in the 19th century (see pp. 159–160).

In the 40s, generalizing Gauss' idea, Eisenstein, Dirichlet and Hermite independently defined an algebraic integer as a root of an equation of the form

$$F(x) = x^n + a_1 x^{n-1} + \ldots + a_{n-1}x + a_n = 0, \tag{19}$$

where $a_1, \ldots, a_n \in \mathbb{Z}$. Only Eisenstein found it necessary to prove that sums and products of integers defined in this manner are likewise integers.

---

43 Ibid., p. 102.

44 Ibid., p. 139.

The next step in the evolution of the arithmetic of algebraic integers was the theory of units. Eisenstein studied the group of units of a cubic extension field, L. Kronecker that of cyclotomic extension fields. Hermite came very close to the general case, but it was Dirichlet who created the theory in finished form. In the memoir "On the theory of complex units" (*Zur Theorie der complexen Einheiten.* — Bericht über Verhandl. Königl. Preuss. Akad. Wiss., 1846) Dirichlet investigates expressions of the form

$$\varphi(\theta) = b_0 + b_1\theta + \ldots + b_{n-1}\theta^{n-1}, \; b_i \in \mathbb{Z}, \tag{20}$$

where θ is a root of an equation of the form (19), that he calls integers. He calls $\varphi(\theta)$ a unit if

$$\varphi(\theta_1)\varphi(\theta_2)\ldots\varphi(\theta_n) = 1,$$

where $\theta_1, \theta_2, \ldots, \theta_n$ are the roots of (19). If h is the number of real roots and of pairs of conjugate-complex roots of the equation (19), then in the number ring (20) there are $h-1$ fundamental units such that all units of this ring are uniquely representable as products of powers of the fundamental units times a root of unity. In other words, the general form of a unit of the ring under consideration is

$$\varepsilon_i e_1^{m_1} e_2^{m_2} \ldots e_{h-1}^{m_{h-1}}, \; i = 1, 2, \ldots, h.$$

Here ε_i is a root of unity in the ring and $e_1, \ldots, e_{h-1}$ are a set of fundamental units.

Fermat's Last Theorem. The Discovery of E. Kummer

In the meantime, attempts to prove Fermat's Last Theorem continued at the beginning of the 19th century. Proofs of special cases were proposed by Legendre, Sophie Germain, Gauss, Dirichlet and Lamé. Since

$$x^\lambda = z^\lambda - y^\lambda = (z - y)(z - \zeta y)\ldots(z - \zeta^{\lambda-1}y),$$

where λ is a prime, $\zeta^\lambda = 1$ and $\zeta \neq 1$, the proof involved working with numbers of the form

$$b_0 + b_1\zeta + \ldots + b_{\lambda-1}\zeta^{\lambda-1}, \tag{21}$$

where

$$b_0, b_1 \ldots, b_{\lambda-1} \in \mathbb{Z}.$$

Finally, in 1847, Lamé published his general "proof" of Fermat's theorem on the unsolvability in integers of the equation $x^n + y^n = z^n$ (*Démonstration générale du théorème de Fermat sur l'impossibilité, en nombres entiers, de l'equation $x^n + y^n = z^n$.* — C.r. Aca. Sci. Paris, 1847). He assumed that in questions of divisibility numbers of the form (21) behave just like rational integers.

Lamé read his memoir at the meeting of the Paris Academy of Sciences on March 1, 1847. On this occasion Liouville made the following observation:

> The idea of introducing C.N. [complex numbers] into the theory of the equation $x^n + y^n = z^n$ would, quite naturally, occur to mathematicians owing to the binary form $x^n + y^n$.
>
> From this I have not derived a satisfactory proof. At any rate, my attempts showed me that it was first necessary to establish for C.N. a theorem similar to the elementary proposition for integers that a product can be factored into primes in just one way. Lamé's analysis confirms me in this opinion. Is there no gap here that must be filled?[45]

The result of Liouville's remark was that, for a short time, the arithmetic of the field of algebraic numbers was at the center of attention of French mathematicians. It was studied by Lamé, Vantzel, and Cauchy. During the meeting of the Academy on March 22, 1847 Cauchy presented a memoir in which he tried to introduce Euclid's algorithm for numbers of the form (21). The memoir was published in parts in Comptes rendus for 1847. In the last part Cauchy concluded that this could not be done.

Similar events took place somewhat earlier in Germany except that Kummer's paper containing a "proof" of Fermat's Last Theorem was not published, and it was Dirichlet rather than Liouville who observed that it was necessary to investigate the uniqueness of factorization.[45a]

Ernst-Eduard Kummer was born in Sorau (now Zary), in Germany, in 1810. He lost his father, a physician, at an early age. After that his family was left without means of livelihood. It was only because of his mother's exceptional energy (she bravely tackled every job, including the sewing of soldiers' underclothing) that Kummer acquired an education.

In 1828 Kummer entered the university of Halle. First he studied theology but he soon got interested in mathematics, which he at first regarded as a kind of "preparatory discipline" for the subsequent study of philosophy. He retained a life long interest in philosophical matters. After completing work at the university in 1831 Kummer taught physics and mathematics in a gymnasium in Lignitz (now Lignica) until 1842. One of his students was L. Kronecker, who later became his friend. In addition to teaching, Kummer worked on problems in analysis.

---

45 Quoted after: Nogúes, R., *Théorème de Fermat*. Paris 1966, p. 28.

45a See H.M. Edwards' paper "The background of Kummer's proof of Fermat's Last Theorem for regular primes", Archive for history of exact sciences, vol. 14, 1975, pp. 219–236. (Transl.)

In 1842, following recommendations by Dirichlet and Jacobi, Kummer obtained a position as professor in Breslau (now Wrocław). In 1855 he took over Dirichlet's position in Berlin. (Dirichlet took over Gauss' position in Göttingen.) Beginning in 1861 Kummer and Weierstrass organized the first seminar on pure mathematics in Germany.

Kummer's students include L. Kronecker, P. du Bois-Reymond, P. Gordan, H.A. Schwarz, and G. Cantor.

By his own testimony, Kummer's work was greatly influenced by Gauss and Dirichlet. His researches were at first directed towards the theory of the hypergeometric series and then towards number theory.

Here Kummer's greatest achievement was the introduction of ideal numbers in connection with the proof of Fermat's Last Theorem. A third direction of his work pertains to geometry; more specifically, the theory of general congruences of lines.

Kummer began to work on Fermat's Last Theorem in 1837. He noted the following curious fact about the ring of integers (21): the product $\alpha\beta$ of two irreducible numbers of the form (21) can be divisible by a third irreducible number of the same form without either α or β being so divisible.

We will explain Kummer's discovery in greater detail.

It is well known that the primes in $\mathbb{Z}$ can be characterized by either one of the following properties:

1) they cannot be written as products of two factors different from a unit;
2) if a product ab is divisible by a prime p then at least one of the factors is divisible by p.

One usually takes the first property as the defining property of a prime and proves the second. But one can do the opposite. Kummer discovered that numbers of the form (21) with property 1) need not have property 2). This fact put in question the possibility of developing arithmetic for numbers of the form (21). Kummer saved the situation by introducing new objects called by him "ideal factors". In connection with Lamé's proof Kummer wrote to Liouville:

> As for your justified complaint that [Lamé's] proof, also flawed in some other respects, does not include, for these complex numbers, the elementary theorem that a composite complex number can be factored into primes in just one way, I can assure you that it fails in general for complex numbers of the form $\alpha_0 + \alpha_1 r + \ldots + \alpha_{n-1}r^{n-1}$ but can be saved by the introduction of a new kind of complex numbers that I have called ideal complex numbers.[46]

Kummer also states that the results of his new theory were presented in 1846 at a meeting of the Berlin Academy of Sciences and published in its

---

46 J. math. pures et appl., 1847, **12**, 136.

Ernst-Eduard Kummer

proceedings. This paper, dated 1844 and entitled "On complex numbers made up of roots of unity and real integers" (*De numeris complexis qui radicibus unitatis et numeris integris realibus constant.* — Gratulationsschrift der Univ. Breslau) was published in Liouville's journal in the same year — 1847 — as Kummer's letter. Also in 1847 *Crelle's Journal* published Kummer's two other papers containing a more complete exposition of his theory: "On the theory of complex numbers" (*Zur Theorie der complexen Zahlen.* — J. für Math., 1847) and "On the decomposition of complex numbers formed of roots of unity into their prime factors" (*Über die Zerlegung der aus Wurzeln der Einheit gebildeten complexen Zahlen in ihre Primfaktoren.* — J. für Math., 1847).

What follows is a brief account of Kummer's basic ideas and methods.

Kummer's Theory

Kummer's method was local. His theory was based on two ideas each of which would have sufficed for the construction of the arithmetic of cyclotomic fields.

The first idea was that the factorization of a rational prime p in $\mathbb{Z}[\zeta]$ is determined by the factorization of the polynomial $\frac{x^\lambda - 1}{x - 1} = x^{\lambda - 1} + x^{\lambda - 2} +$

$\ldots + x + 1 \bmod p$. The germ of his idea is present already in the works of Lagrange. It was subsequently developed by E.I. Zolotarev and K. Hensel.

The second idea had to do with the definition of local uniformizing elements in $\mathbb{Z}[\zeta]$ and, using the latter, of "valuations". This idea was fully grasped and developed by Zolotarev.

Finally, when developing his theory, Kummer essentially introduced the notion of a splitting field that later played an important role in Hilbert's work.

Such an abundance of ideas and methods, redundant for the realization of some immediate objective, usually characterizes works that reveal a new theory.

We describe briefly the basic stages of Kummer's construction.

Let ζ be a root of the equation

$$\Phi(x) = x^{\lambda-1} + x^{\lambda-2} + \ldots + x + 1 = 0, \tag{22}$$

where λ is a prime. Kummer investigates the ring of integers in $\mathbb{Z}[\zeta]$ whose elements are of the form

$$\varphi(\zeta) = b_0 + b_1\zeta + \ldots + b_{\lambda-2}\zeta^{\lambda-2}, \quad b_i \in \mathbb{Z}. \tag{23}$$

He notes that in order to find all the primes in $\mathbb{Z}[\zeta]$ it suffices to consider the factorizations in $\mathbb{Z}[\zeta]$ of the primes of $\mathbb{Z}$. To this end he considers the behavior of $\Phi(x)$ in the field of residues $\bmod p$. If $\Phi(x)$ is irreducible $\bmod p$ then Kummer regards p as prime in $\mathbb{Z}[\zeta]$. If $p = m\lambda + 1$, then Kummer shows that $\Phi(x)$ splits $\bmod p$ into linear factors

$$\Phi(x) = \prod_{k=1}^{\lambda-1}(x - u_k).$$

In this case Kummer supposes that p also splits into a product of $\lambda - 1$ different prime factors, real or ideal,

$$p = \mathfrak{p}_1\mathfrak{p}_2\ldots\mathfrak{p}_{\lambda-1}.$$

Thus there is a complete parallelism between the factorization of $\Phi(x)$ $\bmod p$ and the factorization of p itself in $\mathbb{Z}[\zeta]$. Kummer notes that this parallelism served as a guiding thread in the introduction of ideal divisors. Here the factor $\mathfrak{p}_i$ is regarded as associated with $x - u_i$. If $\Phi(u_i) \equiv 0 \pmod p$ then the number $\Phi(\zeta)$ is divisible by the ideal prime factor of p that belongs to $x - u_i$.

Now suppose that p belongs to an exponent f mod λ (that is $p^f \equiv 1 (\text{mod } \lambda)$ and $p^k \not\equiv 1 (\text{mod } \lambda)$ if $k < f$). In order to factor it Kummer uses Gaussian periods to construct a subfield κ of the field $\mathbb{Q}(\zeta)$, which later came to be known as a splitting field of the number p. If $\eta_0, \eta_1, \ldots, \eta_{e-1}$ are the f-term Gaussian periods, where $ef = \lambda - 1$, then the defining equation of this field is

$$\Phi_e(y) = (y - \eta_0)(y - \eta_1) \ldots (y - \eta_{e-1}). \tag{25}$$

Kummer shows that

$$\Phi_e(y) \equiv \prod_{k=0}^{e-1} (y - v_k) (\text{mod } p). \tag{26}$$

With each $y - v_k$ Kummer associates the ideal factor $\mathfrak{p}_k$ of p, so that in the field κ

$$p = \mathfrak{p}_0 \mathfrak{p}_1 \ldots \mathfrak{p}_{e-1}.$$

In this way Kummer obtains all the prime ideal factors of the ring $\mathbb{Z}[\zeta]$.

Next Kummer turns to the question of local uniformizing elements for the prime factors of p, that is numbers in $\mathbb{Z}[\zeta]$ that contain exactly one ideal factor of p. If $\pi(\eta)$ is one such local uniformizing element then Kummer shows that $N_\kappa \pi(\eta)$ is divisible by p but not by p^2. Kummer uses this important property to construct local uniformizing elements.

Let $\varphi(\zeta) \in \mathbb{Z}[\zeta]$. Let $\mathfrak{p}_i$ be a factor of the number p that belongs to the exponent f. To decide what power of $\mathfrak{p}_i$ enters $\varphi(\zeta)$ Kummer computes the norm of the uniformizing element $\pi(\eta)$ in the field κ (η is an f-term Gaussian period):

$$N_\kappa(\pi(\eta)) = \pi(\eta_0)\pi(\eta_1) \ldots \pi(\eta_{e-1}) = \pi(\eta)\psi(\eta).$$

$\mathfrak{p}_i$ enters $\varphi(\zeta)$ m times if

$$\varphi(\zeta)\psi^m(\eta_i) \equiv 0 (\text{mod } p^m), \quad \varphi(\zeta)\psi^{m+1}(\eta_i) \not\equiv 0 (\text{mod } p^{m+1}). \tag{27}$$

Denote the exponent with which $\mathfrak{p}_i$ enters $\varphi(\zeta)$ by $\nu_{\mathfrak{p}_i}(\varphi)$. This is a positive integer-valued function on $\mathbb{Z}[\zeta]$. Kummer shows that its fundamental property is

$$\nu_{\mathfrak{p}}(\varphi\chi) = \nu_{\mathfrak{p}}(\varphi) + \nu_{\mathfrak{p}}(\chi).$$

Now he can easily develop the fundamentals of the divisibility theory.

To study the structure of the set of ideal factors in $\mathbb{Z}[\zeta]$ Kummer defines a notion of equivalence of [ideal] factors and splits the factors into equivalence classes. In this he follows the road built by Lagrange and Gauss in connection with their study of quadratic forms. Kummer proves the finiteness of the number of classes of ideal factors using the method once employed by Lagrange to prove the finiteness of the number of classes of forms with a given discriminant. Specifically, Kummer associates with each class of ideal numbers an element of that class whose norm does not exceed a certain number (in the case of quadratic forms such an element is a reduced form). Like his predecessors, Kummer computes the number of classes of ideal factors in the ring $\mathbb{Z}[\zeta]$.

While the term "ideal factor" turns up constantly in Kummer's theory it is never formally defined. Kummer considered the congruences (27) as a means of defining the exponent with which a particular ideal factor (to which he granted independent existence) enters into a given number. At the present time the situation has changed. Now the congruences (27) are regarded as a definition of an ideal factor, for they enable us to find the exponent with which this factor enters into an arbitrary integer of the field. And that is all we need to know to construct a divisibility theory. We can thus say that for us an ideal factor is determined by its exponents. Similar situations have frequently arisen in the history of mathematics. Thus, until the 19th century the integral was not regarded as defining the mathematical notion of the area of a figure and the derivative was not regarded as defining the mathematical notion of the slope of the tangent to a curve. Both were viewed as means for computing these magnitudes to which, on the basis of intuition, one granted some kind of independent existence. The 19th century brought with it a new viewpoint according to which the way a magnitude is computed is the way it is defined. This way of viewing ideal factors arose only in our own time.

We note that Kummer called ideal factors "numbers". What confirmed him in this view was the theorem he had proved that raising such a factor to a suitable finite power yields an existing complex number. It was only the work of Dedekind and Weber that made clear the profound difference between the numbers in the field and the ideal factors — the divisors. Owing to the absence of general concepts, Kummer's theory was so intertwined with Gauss' work on cyclotomic equations that a major effort was needed to separate the general ideas from the specific apparatus of cyclotomic fields.

Beyond this, construction of the arithmetic of general fields of algebraic numbers ran into great difficulties.

As for Kummer's further investigations on the arithmetic of cyclotomic fields we note the following.

1. In papers published in Crelle's Journal for 1847 and 1850 Kummer proved Fermat's Last Theorem for all exponents λ that are odd primes which do not occur in the numerators of the first $(\lambda - 3)/2$ Bernoulli numbers.

2. Kummer formulated and proved the reciprocity law for power residues with prime exponents (in a number of papers between 1858 and 1887).

For a long time it seemed impossible to improve on Kummer's result. Only D. Hilbert (1862–1943) managed to advance in the area of reciprocity laws.

Using his class field theory Hilbert showed that the first part of the equality in the expression for reciprocity laws can be represented as a product of a finite number of symbols (so called norm symbols) whose properties he described. Hilbert failed to give a direct construction for these symbols. It was the Soviet mathematician I.R. Shafarevich who formulated and proved the general reciprocity law.[47]

At first sight, the cyclotomic fields studied by Kummer seem to be rather special cases of algebraic number fields. That this is not so was shown by Kronecker who proved that every abelian number field over the rationals is a subfield of some cyclotomic field (see his "On algebraically solvable equations" (*Über die algebraisch auflösbaren Gleichungen.* — Ber. Königl. Akad. Wiss. zu Berlin, 1853)). A rigorous proof of this fact was supplied in 1886 by H. Weber. A generalized version of this theorem became part of the foundation of Hilbert's class field theory.

Difficulties. The Notion of an Integer

From the time of Dirichlet the term "algebraic integer" denoted a root of a monic polynomial equation

$$x^n + a_1 x^{n-1} + \ldots + a_{n-1} x + a_n = 0, \tag{28}$$

where $a_1, a_2, \ldots, a_n$ are rational integers.

Let θ be a root of such an equation and consider the number field $\mathbb{Q}(\theta)$. How does one characterize the integers of $\mathbb{Q}(\theta)$?

In all cases considered by Dirichlet, Eisenstein, Cauchy and Kummer the integers in $\mathbb{Q}(\theta)$ turned out to be of the form $b_0 + b_1 \theta + \ldots + b_{n-1}\theta^{n-1}$, $b_i \in \mathbb{Z}$, that is they coincided with the ring $\mathbb{Z}[\theta]$. Is this always so? In other words, is it always the case that the ring $\mathbb{Z}[\theta]$ contains all the integers in $\mathbb{Q}(\theta)$, with "integer" as just defined?

It is difficult to say which 19th-century mathematician first realized that this is not so. The first complete investigation of this problem was presented by Dedekind in his famous Supplement X to the second edition of Dirichlet's "Lectures on number theory" (1871).[48]

---

47 Shafarevich, I.R., *Collected Mathematical Papers*. Berlin etc. 1989, p. 20.

48 We note that this problem was completely solved for real quadratic fields $\mathbb{Q}(\sqrt{D})$ by Newton in his *Universal arithmetic* (1707). Newton showed that if D is squarefree and $D\equiv 2$ or $D\equiv 3 \pmod 4$ then the integers are $m+n\sqrt{D}$, $m,n\in\mathbb{Z}$. On the other hand, if $D\equiv 1 \pmod 4$ then the integers are

$$(m + n\sqrt{D})/2, \quad m \equiv n \pmod 2.$$

But these investigations of Newton were ignored in the 19th century.

We note that in 1877 E.I. Zolotarev gave a description of the integers in the field $\mathbb{Q}(\theta)$ different from that of Dedekind. The two descriptions provide a striking demonstration of the deep difference between the two scientists in their respective approaches to questions of arithmetic.

Dedekind's approach was global. He showed that the "power" basis

$$1, \theta, \ldots, \theta^{n-1}$$

is not always minimal and that the ring $\mathfrak{O}$ of the integers in $\mathbb{Q}(\theta)$ always contains a basis (later called minimal)

$$\omega_1, \omega_2, \ldots, \omega_n$$

of integers such that all numbers in $\mathfrak{O}$ are linear forms in the ω's whose coefficients are rational integers.

Zolotarev solves the problem locally. He constructs a minimal basis in the semilocal ring $\mathfrak{O}_p$ of p-integral numbers rather than in $\mathfrak{O}$ and obtains $\mathfrak{O}$ by subsequent globalization as the intersection of all $\mathfrak{O}_p$.

Another difficulty that arose in the transition from cyclotomic to general fields is clearly described in Zolotarev's paper "On the theory of complex numbers" (*Sur la théorie des nombres complexes. —* J. für Math., 1880) as well as in Dedekind's memoir "On the connection between ideal theory and the theory of congruences of higher degree" (*Über den Zusammenhang zwischen der Theorie der Ideale und der Theorie der höheren Congruenzen. —* Abh. Ges. d. Wiss., Göttingen, 1878). Kronecker seems to have run into the same difficulty. We describe its Zolotarev version.

Let

$$F(x) = 0 \tag{29}$$

be an irreducible equation of the form (28) and let θ be one of its roots. By Kummer's method, to find the prime ideal factors of a prime p we must consider $F(x)$ over the field of residues $\bmod\, p$. If $F(x)$ stays irreducible over that field then p is prime in the field $\mathbb{Q}(\theta)$. On the other hand, if

$$F(x) = V_1^{\ell_1}(x) \ldots V_s^{\ell_s}(x) - pF_1(x),$$

then there arise two cases:

1) $F_1(x)$ is not divisible $\bmod\, p$ by any of the V_i whose exponent ℓ_i is > 1. In this case Zolotarev calls p "nonsingular" and puts $p = \mathfrak{p}_1^{\ell_1} \ldots \mathfrak{p}_s^{\ell_s}$, where $\mathfrak{p}_i$ is associated with $V_i(\theta)$;

2) $F_1(x)$ is divisible $\bmod\, p$ by one of the V_i whose exponent ℓ_i is > 1. Zolotarev calls such numbers "singular" and shows that they are necessarily divisors of the discriminant Δ of the equation $F(x) = 0$.

The latter case caused difficulties which seemed to rule out the possibility of using Kummer's method to construct a divisibility theory in an arbitrary ring of algebraic integers.

When they encountered the singular case, Dedekind and Kronecker abandoned local methods and went along completely different paths, described in detail below. At this point we discuss the Zolotarev approach. This remains in the shadows to this day,[49] and yet it is precisely in Zolotarev's works that the ideas and methods that are now an integral part of commutative algebra have been developed. He followed the path later chosen by Hensel. Whereas Hensel introduced p-adic numbers and local rings, Zolotarev operated with p-integral numbers and investigated the first important instance of a semilocal ring.

The Zolotarev Theory. Integral and p-Integral Numbers

Egor Ivanovich Zolotarev, one of the most talented representatives of the Chebyshev mathematical school, was born in Petersburg in 1847 in the family of an owner of a watch store. Already as a high school student the boy showed mathematical talent. At 17 he entered the physical-mathematical faculty of Petersburg University where he attended lectures by P.L. Chebyshev and A.N. Korkin, with whom he later collaborated on a number of investigations. In 1867 he completed university work and earned the degree of a candidate. One year later he defended a dissertation for *venia legendi* on the theory of best approximation of functions by polynomials, and after successful presentations of two test lectures began to lecture at Petersburg University as a *Privatdozent*. A year later he defended his Master's dissertation "On the solution of the third-degree indeterminate equation $x^3 + Ay^3 + A^2z^3 - 3Axyz = 1$" (Petersburg 1869), and in 1874 his doctoral dissertation "The theory of whole complex numbers with application to the integral calculus" (Petersburg 1874). This remarkable paper made him one of the most prominent number theorists of his time. In 1876 Zolotarev was appointed professor at Petersburg University. He lectured on the theory of elliptic functions, on mathematical analysis, on mechanics and algebra, and was the first to give an introductory course in analysis. In 1876 he was elected member of the Petersburg Academy of Sciences.

In the summers of 1872 and 1876 the university sent Zolotarev abroad for four months. He visited Berlin, Heidelberg, and Paris, attended lectures by Weierstrass, Kummer, and Kirchhoff, and talked with Hermite, who had a high opinion of his joint research with Korkin on the arithmetic theory of quadratic forms. Zolotarev wrote to Korkin, who was a close friend, about his impressions from the trip abroad and about German and French mathematicians. Their correspondence has come down to us virtually complete (64

---

49 See, for example, N. Bourbaki's "Historical note" in the books dealing with commutative algebra.

Egor Ivanovič Zolotarev

letters in all) and is of major interest, for it deals with various questions pertaining to concrete problems as well as to the evaluation of various theories and trends in contemporary mathematics.

Zolotarev's life came to an abrupt and tragic end in the summer of 1878. On July 2, when he was preparing to visit relatives in their summer home, he fell under a vehicle and died of blood poisoning on July 19.

To judge by the opinions of his contemporaries, Zolotarev was an extremely kind, direct, and friendly person. He was liked by his colleagues and students who describe him as an excellent teacher. While his scientific activity lasted little more than ten years, he managed to enrich mathematics with many important discoveries. His most important papers pertain to number theory, the theory of algebraic functions, and the theory of best approximation of functions by polynomials.

Already in his Master's dissertation we discern two important directions of his creative work. In the course of obtaining the integer solutions of the indeterminate cubic equation given above, in which A is an integer that is not a cube, Zolotarev finds the minima of a quadratic form in three variables whose coefficients depend on parameters, and on this basis constructs a clever algorithm for computing the fundamental unit of the pure cubic field $\mathbb{Q}(\sqrt[3]{A})$. Not surprisingly, Zolotarev's later papers on number theory dealt with the arithmetic theory of quadratic forms and the arithmetic of algebraic numbers.

Zolotarev turned for the first time to the arithmetic of algebraic numbers in his doctoral dissertation of 1874. He was led to this issue by a certain problem in the theory of algebraic functions. Abel had shown that if an integral of the form

$$\int \frac{\rho(x)dx}{\sqrt{R(x)}}, \tag{30}$$

where $R(x)$ and $\rho(x)$ are polynomials whose respective degrees are $2n$ and $n-1$, is expressible in terms of elementary functions, then $\sqrt{R(x)}$ can be expanded in a periodic continued fraction. Conversely, if $R(x)$ can be so expanded, then it is always possible to find a polynomial $\rho(x)$ such that the integral (30) can be expressed in terms of elementary functions. In spite of its importance, Abel's criterion is not effective. This is so because the aperiodicity of a partial quotient does not rule out the eventual periodicity of the expansion.

In the special case when $R(x)$ is a fourth-degree polynomial with rational coefficients, Chebyshev found an algorithm for deciding in a finite number of steps whether or not the integral (30) is expressible in terms of elementary functions. He published his algorithm without a proof. In 1872 Zolotarev justified Chebyshev's algorithm and in his doctoral dissertation he posed the problem of extending it to the case when the coefficients of $R(x)$ are real numbers. To solve this problem he found it necessary to construct the arithmetic of fields of algebraic numbers. He produced such a construction in the nonsingular case, in the sense given in the previous subsection. This case sufficed for the solution of the problem involving the integral (30).

Later Zolotarev turned to the arithmetic of fields and constructed it in the most general case. (He sent his relevant paper *On the theory of complex numbers* to Liouville's journal in 1876. The paper was published in 1880, that is, two years after his death.)

The second sequence of number-theoretic papers, coauthored with Korkin, dealt with the problem of minima of quadratic forms posed by Hermite; they will be considered in Chapter III.

We note that in 1872 Zolotarev gave a very original proof of the law of quadratic reciprocity based on group-theoretic considerations.

On Chebyshev's advice, Zolotarev worked on questions of best approximation of functions by means of polynomials when he was still a student. He returned to these problems in his dissertation and in two papers of 1877 and 1878 respectively.

A cursory analysis of Zolotarev's creative work shows that he was not only an excellent solver of difficult concrete mathematical problems — a charac-

110

teristic of the Chebyshev school — but also a creator of new methods and theories. He was one of the founders of the arithmetic of algebraic numbers and of local methods that turned out to be very fruitful. We now turn to an analysis of these methods.

In addition to his doctoral dissertation, Zolotarev investigates questions of the theory of divisibility in two papers: "On complex numbers" (*Sur les nombres complexes*, Bull. Acad. sci. St.-Pétersbourg, 1878) and "On the theory of complex numbers" (*Sur la théorie des nombres complexes*, J. math. pures et appl., 1880).

The first of these papers contains the proof of a fundamental lemma that is the basis for a divisibility theory applicable to the nonsingular and singular cases. The second paper contains a completely general construction of a divisibility theory.

We have already mentioned that Kummer obtained the ideal divisors of a prime p by considering the fundamental equation over the field of residues mod p and then also the local uniformizing elements for the prime divisors of p and used them to determine the multiplicity with which a prime $\mathfrak{p}$ divides a given integer α.

Zolotarev's idea was to construct a divisibility theory by finding the local uniformizing elements for every prime p without the use of functional congruences.

To this end he constructed in effect the semilocal ring of p-integral numbers in $\mathbb{Q}(\theta)$ and showed that this ring contains a finite number of prime elements (determined up to p-units)

$$\pi_1, \pi_2, \ldots, \pi_s \tag{31}$$

which are the local uniformizing elements. Finally, he showed that every number in $\mathfrak{O}_p$ is uniquely representable (to within p-units) as a product of the factors (31).

Zolotarev begins his second paper with a more precise definition of the ring of integers $\mathfrak{O}$ of the field $\mathbb{Q}(\theta)$. In his doctoral dissertation he considered only fields with $\mathfrak{O} = \mathbb{Z}[\theta]$ and defined integers as polynomials

$$b_0 + b_1\theta + \ldots + b_{n-1}\theta, \; b_i \in \mathbb{Z}. \tag{32}$$

He shows that in general fields such a definition is inadequate, that there exist numbers

$$\alpha = \varphi(\theta)/M, \; \varphi(\theta) \in \mathbb{Z}[\theta], \; M \in \mathbb{Z},$$

which can be roots of equations of the form (28). He notes that, firstly, M can contain only singular primes and that, secondly, the powers with which they enter are bounded, so that if $\alpha = \varphi(\theta)/M$ is an integer, then the discriminant of the defining equation is divisible by M^2.

After this explanation Zolotarev gives the following definition of an integer:

We will call a complex number

$$y = a + bx + \ldots + \ell x^{n-1},$$

where $a, b, \ldots, \ell$ are rational, an integer if it satisfies an equation of the form (α).[50] (In our numbering scheme, of the form (28), I.B.)

Then Zolotarev considers, essentially, p-integral numbers. True, he gives no explicit definition of p-integral numbers. Such a definition was given 30 years later by Hensel. We give Hensel's definition:

A rational number $\alpha = m/n$ i called "an integer mod p" if its denominator n is not divisible by p.[51]

And:

An algebraic number β is called "an integer mod p" if it satisfies at least one equation

$$\beta^m + B_1\beta^{m-1} + \ldots + B_m = 0$$

whose coefficients are rational integers mod p.[52]

It is easy to see that if α is an integer mod p, or, as we will say in the sequel, a p-integer in $\mathbb{Q}(\theta)$, then it can always be represented in the form β/M with $\beta \in \mathfrak{D}$ and M not divisible by p. Conversely, if a number $\alpha = \beta/M$ satisfies these conditions, then it is p-integral in Hensel's sense. Zolotarev uses the latter definition. Thus, for example, if he has to show that β/α is p-integral, he shows that β can be multiplied by an M, $M \in \mathbb{Z}$, $(M, p) = 1$, such that $M\beta/\alpha$ is an integer. This type of argument is used many times.

It is easy to see that the p-integral numbers in $\mathbb{Q}$ form a local ring $\mathbb{Z}_p$. We recall that a ring A is said to be local if it has a unique maximal ideal other than A.

$\mathbb{Z}_p$ consists of all fractions m/n whose denominators are not divisible by the fixed prime p. The unique maximal ideal of this ring is (p).

A ring A is called semilocal if it has a finite number of maximal ideals $\mathfrak{p}_1, \mathfrak{p}_2, \ldots, \mathfrak{p}_k$. Zolotarev was the first to investigate such a ring, namely the ring $\mathfrak{D}_p$ of numbers in $\mathbb{Q}(\theta)$ "integral mod p". He proved (see below) that in such a ring there are only finitely many ideal prime factors.

Zolotarev uses the locutions given previously. He does not introduce the concept of "ring A" and certainly not the concept of "semilocal ring A_p".

---

50 Zolotarev, E.I., *Collected works*, vol. 2, Leningrad 1931, pp. 105–106 (in Russian).

51 Hensel, K., *Theorie der algebraischen Zahlen*. Leipzig-Berlin 1908, p. 76.

52 Ibid.

Nevertheless, in describing his train of thought, we will use these terms for the sake of brevity. In particular, we will speak of numbers in the ring $\mathfrak{O}_p$ and of divisibility in this ring.

Zolotarev shows that α is an integer in $\mathbb{Q}(\theta)$ if and only if it is p-integral for all p, that is, if and only if

$$\mathfrak{O} = \cap_p \mathfrak{O}_p.$$

To further study the arithmetic of p-integral numbers Zolotarev constructs in $\mathfrak{O}_p$ a complete system of residues $\bmod\, p$:

$$\alpha_1, \alpha_2, \ldots, \alpha_\sigma, \tag{33}$$

where $\sigma = p^n - 1$ and $\alpha_i \in \mathfrak{O}$. He chooses as a representative of each residue class an α_i whose norm is divisible by the least possible power of p.

Zolotarev calls numbers β in $\mathfrak{O}_p$ relatively prime to p (we would call them p-units) if none of the products $\alpha_i\beta$, with α_i in the sequence (33), is divisible in $\mathfrak{O}_p$ by p. He proves that β is relatively prime to p if and only if $N\beta$ is not divisible by p. He calls two numbers β and γ in $\mathfrak{O}_p$ relatively prime if there is no number α_i in the sequence (33) such that $\alpha_i\beta$ and $\alpha_i\gamma$ are both divisible by p.

Then he proves the following theorems:

1. If α is relatively prime to β and γ, where $\alpha, \beta, \gamma \in \mathfrak{O}_p$, then α is relatively prime to $\beta\gamma$.

2. If α and β are relatively prime in $\mathfrak{O}_p$, then there exist γ, δ such that

$$\alpha\gamma + \beta\delta \equiv 1 (\bmod\, p).$$

Also, if m is any natural number, then it is possible to find γ' and δ' in $\mathfrak{O}$ such that

$$\gamma'\alpha + \delta'\beta \equiv 1 (\bmod\, p^m).$$

If all numbers in the sequence (31) are relatively prime to p (that is, are units in $\mathfrak{O}_p$), then Zolotarev regards p as a prime in $\mathfrak{O}$.

To factor other rational primes Zolotarev proves the following fundamental lemma.

Let $\alpha \in \mathfrak{O}_p$ satisfy the equation

$$\alpha^n + b_1\alpha^{n-1} + \ldots + b_{n-1}\alpha + b_n = 0, \tag{34}$$

where $b_1, b_2, \ldots, b_n$ are p-integral. Then if b_n is divisible by p^μ, b_{n-1} by p^{μ_1}, b_{n-2} by p^{μ_2}, and so on, then p^λ/α will be p-integral if

$$\lambda = r/s = \max\left(\mu - \mu_1, \frac{\mu - \mu_2}{2}, \frac{\mu - \mu_3}{3}, \ldots\right).$$

This is the first assertion of the lemma.

If we now choose α so that its norm is divisible by the least possible power of p compared with all numbers in the same residue class as α, then

$$\lambda \le 1,$$

that is, $\mu_k = \mu - k$.

We note that while

$$p^\lambda/\alpha = p^{r/s}/\alpha$$

does not belong to $\mathfrak{O}_p$,

$$(p^{r/s}/\alpha)^s = p^r/\alpha^s \in \mathfrak{O}_p.$$

From this lemma Zolotarev deduces two important consequences:

1) the number p is divisible in $\mathfrak{O}_p$ by all α_i in the sequence (33);
2) if $\beta \in \mathfrak{O}_p$, then there is an α_i in the sequence (33) such that β is divisible by α_i.

These two consequences enable Zolotarev to prove the fundamental theorem of divisibility theory.

If the product of two numbers $\beta\gamma$ in $\mathfrak{O}_p$ is divisible in $\mathfrak{O}_p$ by α and β is relatively prime to α, then γ is divisible by α.

After this Zolotarev proves a proposition that plays the same role in $\mathfrak{O}$ as the Euclidean algorithm in $\mathbb{Z}$.

If α and β are not p-units, are not relatively prime and neither of them is divisible mod p by the other, then there is a number γ such α and β are both p-divisible by γ. Also, p is a factor of $N\gamma$ fewer times than of $N\alpha$ and $N\beta$.

This proposition enables him to select in (33) the prime elements, that is, to find the local uniformizing elements for the ring $\mathfrak{O}_p$.

We will call a number α in (33) prime if every α_i that is not relatively prime to α is p-divisible by α.

Consequence 2) of the fundamental lemma shows that in order to select the prime elements it suffices to compare α with all terms of the sequence (33).

114

We make the selection in stages: if α_1 is a p-unit then we reject it. If α_1 is not a p-unit, then we compare it with $\alpha_2, \alpha_3, \ldots, \alpha_\sigma$. Suppose there is an α_k such that α_1 and α_k are not p-relatively prime and neither is p-divisible by the other. Then, by the established proposition, there is a γ in (33) such that α_1 and α_k are both p-divisible by γ and p goes into $N\gamma$ fewer times than into $N\alpha_1$ and $N\alpha_k$. If γ is not p-prime, then we repeat the process. In this way, after a finite number of steps we obtain a prime element π_1.

Continuing the selection we obtain

$$\pi_1, \pi_2, \ldots, \pi_s, \tag{35}$$

and we can always assume that the numbers (35) have been selected so that they are pairwise relatively prime. In this way we obtain a complete selection of local uniformizing elements each of which contains just one prime ideal factor of p:

$$p = \varepsilon \pi_1 \ldots \pi_s,$$

where ε is a unit.

This result shows that the ring $\mathfrak{O}_p$ contains only finitely many prime ideal factors, that is, in modern terms, it is a semiring.

After this Zolotarev considers the global theory. He does not define an ideal prime factor of the number p but from his paper it is clear that he associates to each π_p in (35) a symbol $\mathfrak{p}_i$ so that to each π_i there corresponds just one such symbol and to different elements in (35) there correspond different symbols. It remains to determine how a number $\alpha \in \mathfrak{O}$ is factored into ideal factors.

Zolotarev introduces the following definitions: we will say that α contains m times the factor of the number p belonging to π_i if α is p-divisible by π_i^m and is not p-divisible by π_i^{m+1}.

From this it is clear that the ideal factors are determined, essentially, as exponents (valuations).

Actually, what is constructed is a function $\nu_{\mathfrak{p}}(\alpha)$ which takes on nonnegative integral values.

It is clear that this function must have the property

$$\nu_{\mathfrak{p}}(\alpha\beta) = \nu_{\mathfrak{p}}(\alpha) + \nu_{\mathfrak{p}}(\beta).$$

Zolotarev proves this in the last theorem of his paper. Its statement is that

The product $\beta\gamma$ of two complex numbers contains the ideal divisors $\mathfrak{p}$ belonging to π as many times as both numbers β and γ together.[53]

---

53 Zolotarev, E.I., *Collected works*, vol. 2. Leningrad 1931, p. 129 (in Russian).

Now it is easy to carry over the usual arithmetic to the ring $\mathfrak{O}$.

It is striking that both Kummer and Zolotarev talked of "ideal factors" but limited themselves, essentially, to the determination of the exponent with which an ideal factor enters an integer. That is all we need to know about the ideal factor.

We see that, in effect, Zolotarev developed Kummer's local methods and used them to construct a completely rigorous arithmetic of algebraic numbers. He also began the study of local and semilocal rings (without explicitly introducing these concepts) and applied the methods of localization and globalization. This very road was later followed by Hensel, and it led him to introduce p-adic numbers. The papers of Zolotarev and Hensel were the beginning of the circle of ideas and methods which constitute the core of local algebra.

It is also remarkable that in his construction of algebraic numbers Zolotarev relied on a lemma that holds for algebraic numbers and algebraic functions, so that his theory can be immediately carried over to rings of algebraic functions.

Dedekind's Ideal Theory

Richard Julius Wilhelm Dedekind was born in 1831 in Brunswick (Germany) where his father was a professor and corporation lawyer. Dedekind attended Gymnasium and then the Collegium Carolinum in his native town. His interest turned first to chemistry and physics and he regarded mathematics as a tool of these sciences. Soon, however, purely mathematical questions began to interest him more and more. He mastered the elements of analytic geometry, algebraic analysis and the differential and integral calculus. In 1850 he matriculated at the University of Göttingen, which was one of the finest centers of thought in the natural sciences. There, at a seminar on mathematical physics, he was introduced to the elements of number theory, which was to become the main subject of his investigations. It was in Göttingen that he befriended Riemann. In 1852, with Gauss as his supervisor, Dedekind wrote his doctoral dissertation on the theory of Eulerian integrals.

In the summer of 1854, almost at the same time as Riemann, Dedekind qualified as university lecturer and then began his teaching activities as *Privatdozent*. After Dirichlet succeeded Gauss, Dedekind became his friend. Their discussions stimulated Dedekind to embark on investigations in entirely new directions. Dedekind later remembered that Dirichlet had made a "new man" of him by expanding his scholarly horizons.

In 1858 Dedekind was called to the Polytechnikum in Zurich as the successor to J.L. Raabe (1801–1859), whose name is associated with a convergence test for number series. From Dedekind's time it became an established tradition that Zurich was the first step toward professorial position in Germany;

this route was taken by E. Christoffel, H. Schwarz, G. Frobenius, A. Hurwitz, H. Weber and H. Minkowski, to mention a few.

In September 1859 Dedekind traveled to Berlin with Riemann and met the leaders of the Berlin mathematical school including K. Weierstrass, E. Kummer, K. Borchardt and L. Kronecker. In 1862 he was appointed successor to A.W. Uhde at the Polytechnikum in Brunswick, and he held this position for the rest of his life. He died in 1916.

We have already mentioned that in his work on the arithmetic of algebraic numbers Dedekind encountered difficulties that made it impossible to apply Kummer's methods to the general case without essential modifications. As Dedekind himself put it, he tried to base arithmetic on the theory of higher congruences but finally chose a completely different path that led to the goal. He first described this new foundation, later known as ideal theory, in Supplement X to the second edition of Dirichlet's "Number theory" (1871) and then (in final form) in Supplement XI in the later editions of the same work (1879, 1894).

Dedekind's new approach can be briefly characterized as set-theoretical and axiomatic. Analyzing Kummer's papers, Dedekind noted that Kummer had defined divisibility by ideal numbers without defining ideal numbers and concluded that, rather than operate with ideal numbers, one should work with sets of integers in $K = \mathbb{Q}(\theta)$ that are divisible by some ideal number. It was necessary to define these sets, later called "ideals" (F. Klein was probably right when he noted that a better name would have been "reals"), and define number-theoretic notions, such as divisibility, congruence with respect to a modulus, and so on, in terms of set theory. Dedekind did all this in Supplement X. This is the first work in which fundamental objects of algebra are introduced axiomatically. To quote Bourbaki, this supplement was written "in a general manner and in a completely new style". "In practically one rapid move the theory of algebraic numbers accomplished the transition from first attempts and diffident steps to a mature discipline with its own fundamental apparatus".[54]

In view of the tremendous importance of Dedekind's new conception for algebra and number theory we shall try to describe Supplement X.

First Dedekind introduces the notion of field (Körper), defined as follows:

By a field we will mean every infinite system of real or complex numbers so closed in itself and perfect that addition, subtraction, multiplication, and division of any two of these numbers again yields a number of the system.[55]

---

54 Bourbaki, N., *Eléments d'histoire des mathématiques*. Paris 1974, p. 130.

55 Dedekind, R., *Gesammelte mathematische Werke*, Bd. 3. Braunschweig 1932, p. 224.

R. Dedekind

We see that Dedekind's definition differs from the modern one by the requirement that the field elements are real or complex numbers. But most of the consequences established by him are quite general and can be applied to an arbitrary field.

In Supplement XI of the fourth edition of Dirichlet's "Number theory" (1894) Dedekind presents his theory anew. Now the definition of a field is simpler.

"A system A of real or complex numbers is called a field if the sum, difference, product, and quotient of any two numbers of A belong to A."

Dedekind adds that "the smallest field is formed by the rational numbers and the largest by all the numbers". He introduces the notions of subfield, basis, and degree of a field as well as the notions of conjugate number, norm, and discriminant of n numbers $\alpha_1, \alpha_2, \ldots, \alpha_n$ defined as the square of the determinant formed by these numbers and all their conjugates. He considers the ring of all algebraic numbers and shows that there is no reasonable way of transferring to it the theory of divisibility. However, this can be done for the integers in a finite extension $\Omega = \mathbb{Q}(\theta)$, where θ is a root of a monic polynomial equation

$$x^n + a_1 x^{n-1} + \ldots + a_{n-1} x + a_n = 0,$$

$a_i \in \mathbb{Z}$, which is irreducible over $\mathbb{Q}$.

Before investigating the ring $\mathfrak{O}$ of integers in the field Ω Dedekind introduces the important new concept of a module, which was destined to play such an important role in algebraic number theory as well as in the algebra of our time. We quote its definition:

A system $\mathfrak{a}$ of real or complex numbers whose sums and differences belong to the same system $\mathfrak{a}$ is called a module.[56]

If $\alpha - \beta \in \mathfrak{a}$ then Dedekind writes

$$\alpha \equiv \beta \pmod{\mathfrak{a}}.$$

If $\mathfrak{a} \in \mathfrak{b}$ then Dedekind calls $\mathfrak{a}$ a multiple of $\mathfrak{b}$ and $\mathfrak{b}$ a divisor of $\mathfrak{a}$. He calls the intersection $\mathfrak{a} \cap \mathfrak{b}$ of two modules $\mathfrak{a}$ and $\mathfrak{b}$ their least common multiple.

We see that Dedekind's modules are over the ring $\mathbb{Z}$. That is why he does not state the requirement that if ω is a number in $\mathfrak{O}$ and $\alpha \in \mathfrak{a}$, then $\omega \alpha \in \mathfrak{a}$. This requirement turns up later in connection with his definition of an ideal. We see also that Dedekind systematically replaces number-theoretic notions ("is divisible by", "is a multiple of",) by set-theoretic ones ("is contained in", "is a subset of", and so on). In the course of his study of modules Dedekind introduces the notion of a basis of a module and splits a module into classes incongruent with respect to another module. He uses his knowledge of modules to study the ring $\mathfrak{O}$ of integers in the field Ω. He notes — and this was "a first" in mathematical literature — that if $\omega_1, \ldots, \omega_n$ is a basis of the field Ω and the ω_i are integers then, in general, the set of numbers

$$h_1 \omega_1 + \ldots + h_n \omega_n \tag{36}$$

with $h_i \in \mathbb{Z}$ does not exhaust all integers in Ω. More specifically, some integers in Ω may be expressible in the form (36) with fractional multipliers $h_1, \ldots, h_n$.

Dedekind's fundamental result is to prove the existence of a basis

$$\bar{\omega}_1, \bar{\omega}_2, \ldots, \bar{\omega}_n$$

(Dedekind calls it a fundamental series) such that all integers in the field Ω are expressible in the form (36) with rational integers as coefficients. Dedekind

---

56 Ibid., p. 242.

calls the discriminant of this basis the discriminant of the field or the "fundamental number" (Grundzahl).[57]

As an example he considers the quadratic extension $\Omega = \mathbb{Q}(\sqrt{D})$ and states that if $D \equiv 1 \pmod 4$ then the numbers 1 and $(1 + \sqrt{D})/2$ form a "fundamental series" and the "fundamental number" is D. On the other hand, if $D \equiv 2$ or $3 \pmod 4$ then 1 and $\sqrt{D}$ form a "fundamental series" and the "fundamental number" is $4D$. In essence, this fact was already known to Newton.[58]

Dedekind recalls that in the domain $\mathfrak{D}$ of integers in the field Ω — which he calls an "order" (Ordnung) — the notions of irreducibility and primeness need not coincide[59] (this was noticed already by Kummer) and in order to construct the arithmetic of such numbers he introduces the new notion of an ideal defined axiomatically:

A system $\mathfrak{a}$ of infinitely many numbers in $\mathfrak{D}$ is called an ideal if it satisfies the following two conditions:

(I)   the sum and difference of any two numbers in $\mathfrak{a}$ is again a number in $\mathfrak{a}$;

(II)   the product of a number in $\mathfrak{a}$ by a number in $\mathfrak{D}$ is again a number in $\mathfrak{a}$.[60]

We see that Dedekind's ideals are nonzero modules over the order $\mathfrak{D}$.

Divisibility of numbers by ideals is also defined in set-theoretic terms. Dedekind states that if $\alpha \in \mathfrak{a}$ then α is divisible by $\mathfrak{a}$. Also, if $\alpha - \beta \in \mathfrak{a}$ then $\alpha \equiv \beta \pmod{\mathfrak{a}}$. He notes that, like equations, such congruences can be added, subtracted and multiplied by numbers. Since the relation of congruence is transitive, congruence $\mathrm{mod}\,\mathfrak{a}$ splits the numbers in $\mathfrak{D}$ into disjoint classes of congruent numbers. Dedekind calls the number $N(\mathfrak{a})$ of such classes the norm of $\mathfrak{a}$.

If $\eta \in \mathfrak{D}$ and $\eta \neq 0$ then, clearly the set $I(\eta)$ of all numbers divisible by η satifies (I) and (II) and is thus an ideal. Dedekind calls such an ideal principal. Next Dedekind defines divisibility of ideals and prime ideals.

1) If $\mathfrak{a} \subset \mathfrak{b}$ then we say that $\mathfrak{a}$ is divisible by $\mathfrak{b}$, or is a multiple of $\mathfrak{b}$, and $\mathfrak{b}$ is said to be a divisor of $\mathfrak{a}$.

It is clear that the numbers in $\mathfrak{b}$ form one or more classes mod $\mathfrak{a}$. If the number of classes is r then

$$N\mathfrak{a} = rN\mathfrak{b}.$$

---

57 This term is absent from the fourth edition of Dirichlet's "Lectures".

58 See the paper of I.G. Bashmakova, "On a question in the theory of algebraic numbers in the works of I. Newton and E. Waring". IMI, issue 12. M., Fizmatgiz, 1959, pp. 431–456 (in Russian). See also footnote 48.

59 In subsequent presentations of his theory Dedekind gives a simple example of numbers that are indecomposable but not prime. Thus in the field $\Omega = \mathbb{Q}(\sqrt{-5})$ we have $6 = 2.3 = (1+\sqrt{-5})(1.\sqrt{-5})$ and $2,3,1+\sqrt{-5}$ and $1-\sqrt{-5}$ are indecomposable integers.

60 Dedekind, R., *Gesammelte mathematische Werke*, Bd. 3. Braunschweig 1932, p. 251.

Dedekind shows that every ideal has only finitely many divisors. The principal ideal $\mathfrak{O}$ divides all ideals and thus plays the role of a unit. This ideal is uniquely characterized by either of the following conditions: 1) it contains 1 (or any other field unit), 2) $N\mathfrak{O} = 1$.

Dedekind calls the intersection $m = \mathfrak{a} \cap \mathfrak{b}$ of two ideals $\mathfrak{a}$ and $\mathfrak{b}$ their least common multiple and the system D of sums $\alpha + \beta$, $\alpha \in \mathfrak{a}$, $\beta \in \mathfrak{b}$, their greatest common divisior. He shows that $N\mathfrak{a}N\mathfrak{b} = NmND$. Finally, he calls an ideal $\mathfrak{p}$ prime if its only divisors are $\mathfrak{O}$ and $\mathfrak{p}$.

The following theorem holds:

If $\eta\rho \equiv 0 (\mathrm{mod}\,\mathfrak{p})$ then at least one of the numbers η, ρ is divisible by $\mathfrak{p}$.

We give Dedekind's proof of this result.

Suppose η is not divisible by $\mathfrak{p}$. Then the roots of the congruence

$$\eta x \equiv 0 (\mathrm{mod}\,\mathfrak{p})$$

form an ideal X containing the ideal $\mathfrak{p}$. But this ideal does not contain 1, that is $X \neq \mathfrak{O}$. Hence $X = \mathfrak{p}$, that is $\rho \in \mathfrak{p}$.

We see that this proposition characterizes a prime number as well as a prime ideal.

Before defining multiplication of ideals Dedekind defines the power $\mathfrak{p}^r$ of a prime ideal $\mathfrak{p}$. He does this as follows:

1) he shows that if η is not divisible by an ideal $\mathfrak{a}$, then there is always a number ν divisible by η such that the roots π of the congruence

$$\nu x \equiv 0 (\mathrm{mod}\,\mathfrak{a})$$

form a prime ideal;[61]

2) he claims that if $\mu \in \mathfrak{O}$, $\mu \neq 0$, $\mu \neq 1$, then, by 1), there is a number ν such that the roots π of the congruence

$$\nu x \equiv 0 (\mathrm{mod}\,\mu)$$

form a prime ideal $\mathfrak{p}$.

Dedekind calls such an ideal "simple" (einfach). If r is a natural number then the roots ρ of the congruence

$$\nu^r x \equiv 0 (\mathrm{mod}\,\mu^r)$$

form an ideal that Dedekind calls the r-th power of the ideal $\mathfrak{p}$ and denotes by $\mathfrak{p}^r$.

---

61 In the proof Dedekind uses the method of descent which he applies to the norms of ideals.

We note that the notion of a power of a prime ideal $\mathfrak{p}$ as defined by Dedekind coincides with the notion of a primary ideal.

Dedekind proves that $\mathfrak{p}$ is the only prime divisor of $\mathfrak{p}^r$. Then he establishes for ideals fundamental theorems analogous to the divisibility properties of rational integers:

1) every ideal is the least common multiple of the powers of the prime ideals contained in it;

2) an ideal m is divisible by an ideal $\mathfrak{a}$ if and only if all powers of the prime ideals contained in $\mathfrak{a}$ are also contained in m.

In his proofs Dedekind relies on properties of congruences $\bmod \mathfrak{a}$, $\mathfrak{a}$ an ideal, and on the method of infinite descent.

But to construct in the order $\mathfrak{O}$ an arithmetic fully analogous to the usual arithmetic it was necessary to define multiplication of ideals and to establish the connection between this operation and the operation of division introduced earlier. Here is Dedekind's definition of multiplication of ideals:

> If we multiply each number in an ideal $\mathfrak{a}$ by each number in an ideal $\mathfrak{b}$ then these products and their sums form an ideal $\mathfrak{ab}$ divisible by $\mathfrak{a}$ and $\mathfrak{b}$.[62]

He notes that this definition implies $\mathfrak{ab} = \mathfrak{ba}$, $\mathfrak{aO} = \mathfrak{a}$ and $(\mathfrak{ab})\mathfrak{c} = \mathfrak{a}(\mathfrak{bc})$, that is that the ideal $\mathfrak{O}$ plays the role of a unit and the operation of multiplication is associative and commutative. But in his 1871 work Dedekind was unable to relate his operations of multiplication and division of ideals. All he was able to show was that if $\mathfrak{p}^a$ and $\mathfrak{p}^b$ are the highest powers of the prime ideal $\mathfrak{p}$ contained in the ideals $\mathfrak{a}$ and $\mathfrak{b}$ then $\mathfrak{p}^{a+b}$ is the highest power of $\mathfrak{p}$ contained in $\mathfrak{ab}$, and that $N\mathfrak{ab} = N\mathfrak{a} \cdot N\mathfrak{b}$.

To establish the full analogy with ordinary arithmetic it was necessary to prove that if an ideal $\mathfrak{a}$ is divisible by an ideal $\mathfrak{b}$, that is if $\mathfrak{a} \subset \mathfrak{b}$, then there exists a unique ideal $\mathfrak{c}$ such that $\mathfrak{a} = \mathfrak{bc}$. It was only some years later that Dedekind was able to overcome the fundamental difficulties connected with this theorem. He presented his new ideal theory in Supplement XI to the third edition of Dirichlet's "Number theory" (1879). There he wrote:

> Whereas our previous investigation of ideals relied essentially on just the application of divisibility properties of modules, now we go over to a new construction of ideals, namely to multiplication of ideals that form the essential core of ideal theory.[63]

Here Dedekind proves the following theorem (see above):

---

62 Dedekind, R., *Gesammelte mathematische Werke*, Bd. 3. Braunschweig 1932, p. 259.

63 Ibid., p. 297.

If an ideal $\mathfrak{a}$ is divisible by a prime ideal $\mathfrak{p}$ then there exists just one ideal $\mathfrak{q}$ such that

$$\mathfrak{a} = \mathfrak{p}\mathfrak{q}.$$

The ideal $\mathfrak{q}$ is a proper divisor of $\mathfrak{a}$ and therefore $N\mathfrak{q} < N\mathfrak{a}$.[64]

After that he proves the theorem on the unique decomposition into prime ideals in the form familiar to us:

> Every ideal different from $\mathfrak{O}$ is prime or is uniquely representable as a product of primes.[65]

We note that in the process of proving the theorem Dedekind investigates the order $\mathfrak{O}$ and singles out the following characteristic properties:

1) $\mathfrak{O}$ is a finite module whose basis $[\omega_1, \dots, \omega_n]$ is also a basis of the field Ω;

2) every product of two numbers in $\mathfrak{O}$ is itself in $\mathfrak{O}$;

3) 1 is in $\mathfrak{O}$.

These properties are, essentially, the first definition of the notion of ring.

The last (third) exposition of ideal theory is found in Supplement XI to the fourth edition of Dirichlet's book. It is closest to the modern one. We note that here Dedekind introduces fractional ideals and shows that they form a group. Even now Dedekind's exposition is a model of logical clarity, transparency, and rigor.

We add that in the paper "On discriminants of finite fields" (*Über die Diskriminanten endlicher Körper*) Dedekind introduces the notion of a different which he uses to give a new definition of the discriminant and determines the powers of the prime ideals contained in the prime divisors of the discriminant.

On Dedekind's Method. Ideals and Cuts

We saw that in his ideal theory Dedekind consistently employs the set-theoretic and axiomatic methods. We are used to these investigative tools but they were completely new to his contemporaries. This being so, Dedekind found it necessary to explain his method in the paper "On the theory of algebraic integers" (*Sur la théorie des nombres entiers algébriques.* — Bull. sci. math. et astron., sér. 1, 1876, 9; sér. 2, 1877, 1), published also in book form in Paris in 1877. He resorts to a comparison with the method of constructing the theory of cuts which he used to introduce the irrational numbers. A definition must:

---

64 Ibid., p. 309.

65 Ibid.

1) be based on the facts and relations that obtain in the domain $\mathbb{Q}$ of rational numbers (for example, he rejects a definition based on the ratio of homogeneous magnitudes as foreign to arithmetic); 2) generate all irrationals at once; 3) allow a clear introduction of arithmetic operations.

Dedekind noted that every rational number splits $\mathbb{Q}$ into two classes: A containing all rationals $q < r$, and B containing all rationals in $\mathbb{Q}$ greater than r; r itself can be assigned to either A or B. Dedekind called such a split a "cut". The next step was to single out the basic properties of a cut that would enable one to define a cut axiomatically, without relying on the number r that generates it. Therefore Dedekind gave a new definition (α) of a cut: it is a split of the rational numbers $\mathbb{Q}$ into two classes A and B such that every number in A is less than every number in B.

With this definition it becomes clear that there is seldom a rational number that generates a given cut:

> There are infinitely many cuts which cannot be generated in the indicated manner by rational numbers. For each such cut one creates or introduces into arithmetic an irrational number corresponding to this cut.[66]

On the basis of his definition Dedekind introduces in the new set of number-cuts an order relation ("larger" and "smaller") and four arithmetic operations. Dedekind shows that the new domain $\mathbb{R}$ of all number-cuts has the property of completeness in the sense that to each split satisfying the definition (α) there is a number in $\mathbb{R}$ that generates it.

A similar situation arose in connection with the definition of ideals. If $\mu \in \mathfrak{O}$ then, clearly, the set $\mathfrak{O}$ of numbers divisble by μ satisfies properties (I) and (II) that is, if α and β are divisible by μ then $\alpha \pm \beta$ are also divisible by μ, that is $\alpha \pm \beta \in \mathfrak{a}$ (property I), and if α is divisible by μ and $\omega \in \mathfrak{O}$ then $\omega\alpha$ is divisible by μ, that is $\omega\alpha \in \mathfrak{a}$ (property II). Now, says Dedekind, since we want to define the rules of divisibility in $\mathfrak{O}$ so that they are analogous to the usual ones,

> we must so define ideal numbers and divisibility by ideal numbers that the two elementary theorems (I) and (II) formulated above continue to hold even if μ is not an existing number but an ideal one.[67]

This is why Dedekind takes properties (I) and (II) as axioms that define not just principal ideals but all ideals.

If $\mathfrak{O}$ is $\mathbb{Z}$ then we get nothing new. If $\mathfrak{a}$ is a set of rational integers closed under sums and differences then (as is easily shown) $\mathfrak{a}$ contains a number of

---

66 Dedekind, R., *Gesammelte mathematische Werke*, Bd. 3. Braunschweig 1932, p. 269.

67 Ibid., p. 271.

least absolute value that divides all the other numbers in $\mathfrak{a}$. In Dedekind's terminology this means that all ideals in $\mathbb{Z}$ are principal. If $\mathfrak{O}$ is the ring of integers in a field $\Omega = \mathbb{Q}(\theta)$ then it will have ideals $\mathcal{I}$ containing no number $\mu \in \mathfrak{O}$ of which all other numbers in $\mathcal{I}$ are integral multiples.

We saw that by providing a direct definition of the notions of divisibility and primeness Dedekind was able to construct in the ring $\mathfrak{O}$ of algebraic integers in $\mathbb{Q}(\theta)$ an arithmetic analogous to the usual one.

Thus in the theory of cuts and in ideal theory Dedekind followed the same path: he found basic determining properties of certain objects (rational cuts and sets of integers divisible by a certain number, respectively), took these properties as axioms, and obtained a new domain (the set of all cuts of the original domain and the set of all ideals in a field, respectively) larger than the "old" one and containing the "old" one as a special case. Also, the new domain had the required properties: completeness in the case of the cuts and unique factorization into primes in the case of ideal theory.

We add that Dedekind used the same approach a third time, namely when constructing the arithmetic of fields of algebraic functions; a point of a Riemann surface was defined in this way. We consider this question in some detail below.

Construction of Ideal Theory in Algebraic Function Fields

We saw that, already when founding the theory of divisibility of algebraic numbers, Dedekind singled out such extremely important concepts as field, module and ideal. He introduced them axiomatically and on a set-theoretic basis. This was the first important step that led to the formation of commutative algebra.

The second, no less important step, consisted in the transfer of the whole theory to fields of algebraic functions. This Dedekind did in collaboration with H. Weber.

Heinrich Weber was born in 1843 in Heidelberg. He studied at Heidelberg University and attended the lectures of Helmholtz and Kirchhoff. Between 1873 and 1883 he worked in Königsberg and between 1892 and 1895 he was a full professor in Göttingen. He died in Strassburg in 1913. Weber left a significant mark on virtually all areas of 19th-century mathematics. F. Klein notes that he has "a remarkable ability to master concepts new to him, such as Riemann's function theory and Dedekind's number theory".[68]

The development of the theory of algebraic functions in the 19th century paralleled the investigations of fields of algebraic numbers induced by the discovery of the nonuniqueness of factorization into "primes". In the case

---

68 Klein, F., *Vorlesungen über die Entwicklung der Mathematik im 19. Jahrhundert*, Bd. 1. Berlin 1926, p. 275.

Heinrich Weber

of algebraic functions the central problem was the treatment of multivalued functions of a complex variable. This problem was studied by mathematicians such as Abel and Jacobi, and its final solution was due to B. Riemann, who constructed the so-called Riemann surfaces. To each point of such a surface there corresponded just one value of the corresponding function. However, Riemann's construction did not satisfy Dedekind and Weber since it relied on continuity and representability in terms of series — concepts based on geometric intuition of one kind or another. In their memoir "Theory of algebraic functions of a single variable" (*Theorie der algebraischen Funktionen einer Veränderlichen. — J. für reine und angew. Math., 1882, 92*) Dedekind and Weber proposed to "provide a basis for the theory of algebraic functions, the major achievement of Riemann's researches, in the simplest and at the same time rigorous and most general manner".[69]

It is remarkable that the exposition of Dedekind and Weber applies to any ground field of characteristic zero (it seems that the only assumption is that the field is algebraically closed): "There would be no gap anywhere if one

---

69 Dedekind, R., *Gesammelte mathematische Werke*, Bd. 1. Braunschweig 1930, p. 238.

wished to restrict the domain of variable numbers to the system of algebraic numbers".[70]

When constructing their theory, Dedekind and Weber were guided by the analogy between algebraic functions and algebraic numbers noted long before. Already Stevin observed that polynomials in one variable behave like integers, with irreducible polynomials playing the role of primes. Stevin introduced for polynomials the Euclidean algorithm that can be used to prove that every polynomial is uniquely representable as a product of irreducibles. Later, the method of congruences was also transferred to polynomials. But the full depth of this analogy was disclosed by the Dedekind and Weber memoir under consideration.

True to his general principle described in the previous subsection, Dedekind turned Riemann's presentation "upside down". Riemann begins with the construction of a "Riemann surface" to which there corresponds a class of algebraic functions. Our authors begin with the construction of a field of functions corresponding to Riemann's class, develop for this field a theory fully analogous to the theory of algebraic numbers, define in the field modules and ideals, and only then do they construct the points of the Riemann surface with the help of their theory.

The whole first part of the memoir is devoted to the formal theory of algebraic functions. It begins with the definition of an algebraic function. Consider an irreducible equation

$$F(\theta, z) = 0, \tag{37}$$

with

$$F(\theta, z) = a_0 \theta^n + a_1 \theta^{n-1} + \ldots + a_{n-1} \theta + a_n,$$

where $a_0, \ldots, a_n$ are integral rational functions of z. The authors call a function θ, defined by an equation (37), an algebraic function.

Next they use the "Kronecker construction" (see pp. 47–49) to construct the field of algebraic functions Ω containing all rational functions of z and θ (of degree $\leq n$). The elements of this field are of the form

$$\zeta = b_0 + b_1 \theta + \ldots + b_{n-1} \theta^{n-1},$$

where $b_0, \ldots, b_{n-1}$ are rational functions of z, and its degree over the field $\mathbb{Q}(z)$ is n. Then the authors transfer almost literally Dedekind's theory to fields of algebraic functions: they define the norm and trace of a function and the discriminant of a system of n functions, introduce the notion of an integral function in the field and study the field of such functions, define the notion of a

---

70 Ibid., p. 240.

module (*Funktionenmodul*), introduce congruences with respect to a modulus, and, finally, define the notion of an ideal and prove the fundamental theorem of divisibility theory.

By way of example we quote their definition of a module:

A system of functions (in Ω) is called a module if it is closed under addition, subtraction, and multiplication by an integral rational function of z.[71]

Of greatest interest is the definition of a point of the Riemann surface corresponding to the field of functions Ω. This is the problem with which the second half of the memoir begins. As already indicated, we encounter here again Dedekind's characteristic mode of reasoning. If we already had a point $\mathfrak{p}$, then by considering the values of all functions in the field Ω at this point we would obtain a mapping of Ω into the field of constants C:

$$F \to F(\mathfrak{p}) = F_0 \in C.$$

If $F \to F_0$ and $G \to G_0$, then it is clear that

$$F \pm G \to F_0 \pm G_0, \ FG \to F_0 G_0, \ F/G \to F_0/G_0.$$

The authors note that for the sake of generality C must be supplemented by the number ∞ for which there are defined the usual arithmetic operations with the exception of $\infty \pm \infty$, $0 \cdot \infty$, $0/0$, ∞/∞, to which no numerical values are assigned. If we now take this extended domain $\bar{C}$ and consider all homomorphisms of Ω into $\bar{C}$, then to each such homomorphism we can associate a point P. This is the definition used by Dedekind and Weber:

If all individuals $\alpha, \beta, \gamma, \ldots$ of the field Ω are associated to definite numerical values $\alpha_0, \beta_0, \gamma_0, \ldots$ in such a way that (I) $(\alpha)_0 = \alpha$ if α is a constant, and in general (II) $(\alpha+\beta)_0 = \alpha_0 + \beta_0$, (III) $(\alpha-\beta)_0 = \alpha_0 - \beta_0$, (IV) $(\alpha\beta)_0 = \alpha_0\beta_0$, (V) $(\alpha/\beta)_0 = \alpha_0/\beta_0$, then one should associate to all these values a point $P \ldots$ and we say that $\alpha = \alpha_0$ at P or α has at P the value α_0. Two points are said to be different if and only if there is a function α in Ω that takes on different values at the two points.[72]

The authors note that this definition is an invariant of the field Ω since it does not depend on the choice of the independent variable used to represent a function in this field.

This method of introduction of a point P has become very popular in modern mathematics. N. Bourbaki[73] notes that I.M. Gel'fand used this method

---

71 Ibid., pp. 251–252.

72 Ibid., p. 294.

73 Bourbaki, N., *Eléments d'histoire des mathématiques*. Paris 1974, p. 134.

128

to find the theory of normed algebras (1940), and that the same method has been used afterwards on a number of occasions.

In order to construct a Riemann surface out of points the authors prove the following theorems:

1. If $z \in \Omega$ has a finite value at P, then the set of all integral functions $\pi \in \Omega$ (with respect to z) that vanish at P form a prime ideal $\mathfrak{p}$ (with respect to z).

The authors say that the point P generates the prime ideal $\mathfrak{p}$. If ω is an integral function in Ω that takes on a finite value ω_0 at P, then they say that

$$\omega \equiv \omega_0 (\mathrm{mod}\, \mathfrak{p}).$$

2. The same prime ideal cannot be generated by two different points.

3. If $z \in \Omega$ and $\mathfrak{p}$ is a prime ideal (relative to z), then there exists one (and only one (by 2.)) point P that generates this ideal; it is called the null point of the ideal $\mathfrak{p}$.

These theorems imply the following method of constructing a Riemann surface T: one takes an arbitrary function $z \in \Omega$, forms the ring of integral functions of z, considers all its prime ideals $\mathfrak{p}$ and their corresponding null points P. In this way one obtains all points of the Riemann surface T at which the function z is finite. To include the remaining points P' (at which $z = \infty$) one takes the function $z' = 1/z$ which vanishes at these points, forms the ring of integral functions of z', and constructs its prime ideals $\mathfrak{p}'$ that contain z'. If we add to the points of T the new points P' corresponding to the ideals $\mathfrak{p}'$, then we obtain all points of the Riemann surface T.

We see that the transfer of the methods of number theory to function theory turned out to be very productive. It became clear that these methods, as well as the new concepts introduced with their help, carry over almost literally.

It is well known that the axiomatic method in geometry came into being in antiquity. In the third century BC it formed the foundation of Euclid's *Elements*. It took more than 2000 years for its analogue to arise in algebra.[74] We saw that this was connected with a deep transformation of algebra. Algebra changed from the science of equations (essentially of the first four degrees) and of elementary transformations to the science of algebraic structures defined on sets of objects of arbitrary nature.

Dedekind's works devoted to the arithmetic of algebraic numbers enable us to retrace the first, and thus the most important, steps of the axiomatic method in algebra. We see that Dedekind 1) based his axiomatics on set theory; 2) introduced axiomatically the notions of field, module, ring, and ideal

---

74 See the paper of Yanovskaya, S.A., "From the history of axiomatics" in the book *The methodological problems of science*. Moscow 1972, pp. 150–180 (in Russian).

for algebraic numbers; all these new objects were defined as sets of numbers satisfying various explicitly formulated properties; 3) finally, he transferred the whole theory into the domain of algebraic functions, and not only gave the first rigorous construction of the whole theory but also defined an abstract Riemann surface.

It turned out that his axioms determined the new notions of field, module, ring, and ideal for two radically different sets of mathematical objects. In fact, these notions can be applied to sets of objects of arbitrary nature. Many of Dedekind's theorems and arguments had a very general character. But it took a few decades before mathematicians understood this fact.

We note that Dedekind himself took this last step in the construction of abstract algebra. He did this while investigating "dual groups" (*Dualgruppen*), subsequently known as Dedekind structures or lattices. In the papers "On the factorization of numbers in terms of their greatest common divisor" (*Über Zerlegungen von Zahlen durch die größten gemeinsamen Teiler*. Festschrift der Techn. Hochschule zu Braunschweig, 1897, S. 1–40)[75] and "On the dual group generated by three modules" (*Über die von drei Modulen erzeugte Dualgruppe*. — Math. Ann., 1900, 53, 371–403)[76] he defined the "dual group" for arbitrary objects and investigated its properties relying solely on explicitly formulated axioms. In the beginning of the second of these papers he states that if one denotes the greatest common divisor of two modules $\mathfrak{a}$ and $\mathfrak{b}$ by $\mathfrak{a} + \mathfrak{b}$ and their least common multiple by $\mathfrak{a} - \mathfrak{b}$, then these two operations satisfy the following conditions

(1) $$\mathfrak{a} + \mathfrak{b} = \mathfrak{b} + \mathfrak{a}; \qquad \mathfrak{a} - \mathfrak{b} = \mathfrak{b} - \mathfrak{a};$$

(2) $$(\mathfrak{a} + \mathfrak{b}) + \mathfrak{c} = \mathfrak{a} + (\mathfrak{b} + \mathfrak{c}); \quad (\mathfrak{a} - \mathfrak{b}) - \mathfrak{c} = \mathfrak{a} - (\mathfrak{b} - \mathfrak{c});$$

(3) $$\mathfrak{a} + (\mathfrak{a} - \mathfrak{b}) = \mathfrak{a}; \qquad \mathfrak{a} - (\mathfrak{a} + \mathfrak{b}) = \mathfrak{a},$$

which imply that

(4) $$\mathfrak{a} + \mathfrak{a} = \mathfrak{a}, \quad \mathfrak{a} - \mathfrak{a} = \mathfrak{a}.$$

Then Dedekind introduces the definition:

If two operations $\pm$ on two arbitrary elements $\mathfrak{a}$, $\mathfrak{b}$ of a (finite or infinite) system G generate two elements $\mathfrak{a} \pm \mathfrak{b}$ of the same system G that satisfy the conditions (1), (2), (3), then, regardless of the nature of these elements, G is called a dual group with respect to the operations $\pm$.[77]

---

75 Dedekind, R., *Gesammelte mathematische Werke*, Bd. 2. Braunschweig 1932, pp. 103–147.

76 Ibid., pp. 236–271.

77 Ibid., p. 237.

As one example of such a group Dedekind cites the previously considered system of modules. Other examples are found in his first paper. There he notes, among other things, that in a logical system $a+b$ can be interpreted as logical sum and $a-b$ as logical product. This was actually done by Schröder in his "Algebra of logic" (see Chapter I). Then Dedekind deduces the properties that must be satisfied by any dual group, and only then does he study groups generated by two and three modules.

We cannot analyze these very interesting papers in greater detail. What is important for us here is that both were written fully in the style of the abstract algebra of the 20s of this century.

Concerning general axiomatic definitions of other algebraic concepts we note that the general definition of a field was formulated at the end of the 19th and at the beginning of the 20th century, and the abstract definition of a ring was given somewhat later (1910–1914; Fraenkel, Steinitz). Finally, we owe to Emmy Noether and her school the definitive formulation of the part of mathematics that unified the ideas and methods of algebra and number theory of the last century and, due to the widely known book of B.L. van der Waerden, came to be known as modern algebra.

We emphasize once more that number theory — the workshop of mathematical methods — played a fundamental role in the creation of modern algebra, set theory, and the axiomatic method.

L. Kronecker's Divisor Theory

Leopold Kronecker was one of those concerned with the problem of constructing a general divisor theory. He was born in Germany, in Liegnitz (now Legnica, Poland), in 1823. After graduating from the local gymnasium, where his teacher was E. Kummer, later a friend, he attended lectures at the universities of Berlin, Bonn, and Breslau. In addition to Kummer, a major influence on Kronecker as a mathematician was Dirichlet, whose lectures he attended in Berlin. In 1861 Kronecker was elected to the Berlin Academy of Science and in 1883 became professor at the university of Berlin. Kronecker's fundamental works deal with algebra and number theory, where he continued Kummer's investigations on quadratic forms and group theory. Kronecker's views on the "arithmetization" of mathematics, according to which mathematics must be based on the arithmetic of the integers, are widely known. These views prompted him to actively oppose the function-theoretic principles of the Weierstrass school and Cantor's set-theoretic approach. Kronecker died in Berlin in 1891. He was one of the most eminent mathematicians of the 19th century. According to F. Klein, what distinguished Kronecker was the following remarkable talent: "In various areas of his work he anticipated the true sense of a number of relations of fundamental significance before being in a position to elaborate them with full clarity".[78]

---

78 Klein, F., *Vorlesungen* ... , Bd. 1. Berlin 1926, p. 275.

Leopold Kronecker

In particular, Kronecker formulated and partially proved the Kronecker-Weber theorem, the first theorem of class field theory (see. p. 106). We will now describe his divisor theory in algebraic number fields.

While rumours and fragmentary information about the theory created by Kronecker circulated from the middle of the century, a full exposition of the "Kroneckerian" construction of divisor theory appeared only in 1882 in the paper "Foundations of an arithmetic theory of algebraic magnitudes". (*Grundzüge einer arithmetischen Theorie der algebraischen Grössen.* — J. für Math., 1882) published in Crelle's Journal in the *Kummer Festschrift* volume. Kronecker's paper is characterized, above all, by the greater generality of formulation of the problem, for he considers the divisibility theory in a field obtained by adjoining independent variables to the field of rational numbers and magnitudes algebraic over the resulting field of rational functions. Kronecker notes that the adjunction of independent variables is equivalent to the adjunction of independent transcendental numbers.

Kronecker resolves the resulting algebraic difficulties by developing a special technique, that of adjunction of auxiliary unknowns. We will not go into the details of Kronecker's arguments. Such an exposition is found in H. Weyl's book *Algebraic Number Theory*.

The problem considered by Kronecker is more complex than the problem of constructing a divisor theory in fields of algebraic numbers, a problem solved at that time by Dedekind and Zolotarev. From a modern viewpoint, this can be explained by noting that the ring of integers in a field of algebraic numbers has dimension 1 and is a Dedekind domain, whereas the ring of integers in a field obtained by adjoining k independent variables and some algebraic magnitudes has dimension $k + 1$. While we can consider ideals in the second case as well, their dimensions can vary (they can be $0, 1, \ldots, k$), whereas in the first case all ideals have dimension 0. In the second case, the only ideals that correspond to the divisors are ideals of dimension k rather than all ideals.

Comparing the papers of Kronecker and Dedekind, Hermann Weyl writes:

> In sum, we can say that K (the Kronecker theory) is more fundamental whereas D (the Dedekind theory) is a more finished theory.[79]

His contemporaries failed to appreciate the merits of Kronecker's paper. The significance of the more general formulation of the problem was not understood. After the clear Dedekind papers, with their attractive, set-theoretic language, the new approach to the construction of the arithmetic of algebraic numbers seemed superfluous. In his book Weyl makes a great effort to convince the reader of the merits of Kronecker's theory. Of course, today we can say that Kronecker was able to take only the first step in the construction of a divisibility theory in rings of algebraic functions. Nevertheless, this was a very important step towards the creation of a divisor theory in an algebraic manifold (or in the ring of functions on such a manifold). Such a theory is an indispensable tool for the investigation of many problems in algebraic geometry.

Conclusion

We have already said that the 70s of the last century were, in a sense, a watershed in the development of algebra and algebraic number theory. It was at that time that the ideas and methods that came into being in the first third of the century became current in mathematics. The main ones were: the idea of a group and its invariants, the notions of field, module, ring and ideal, and the notions and apparatus of linear algebra. At that time, too, arithmetic was transplanted into the new domain of algebraic numbers.

In the subsequent 50 years all concepts we have discussed are treated more abstractly, and at the same time there begins the penetration of algebraic ideas and methods into various areas of mathematics.

---

79 Weyl, H., *Algebraic number theory*, Ch. II, § 11.

In group theory, the study of infinite and topological (continous) groups is pursued side by side with the study of finite groups and occupies the key position. This was due to the applications of group theory in geometry (beginning with Klein's Erlangen program), in the theory of functions of a complex variable (especially in the work of Poincaré), and in differential equations. The creation of the theory of continuous groups, and especially the investigations of Lie, Killing, and E. Cartan, had a major impact on the subsequent development of algebra and topology.

In the 80s of last century there begins the systematic construction of the theory of group representations. Here the fundamental results during the first stage of development are due to Frobenius and Molien. Of special importance was the discovery of the deep inner connections between the theory of group representations and the theory of associative algebras. It demonstrated the unity of different algebraic ideas and the fruitfulness of their interactions, and played a major role in the development of the new algebra.

We saw that in the last century mathematicians noticed the deep analogy between the fields of algebraic numbers and algebraic functions. This analogy guided us in our analysis of the memoir of Dedekind and Weber in which the methods of number theory developed by Dedekind were transferred to the theory of functions. On the other hand, when constructing his divisibility theory, Zolotarev used the methods of the theory of algebraic functions. The same approach was used at the end of the last century by Hensel, who managed to introduce into number theory an analogue of Puiseux series. He did this by defining p-adic numbers. Somewhat later, topology was introduced into the field of p-adic numbers. This made possible the creation of p-adic analysis which enabled the extensive use of local examination — a development that simplified considerably the investigation of many difficult problems.

At the end of the century (1899) Hilbert gave the first systematic account of the theory of algebraic numbers. Before that, this theory was scattered in papers. Hilbert's famous *Zahlbericht* became the basis for further development of the theory. In this book Hilbert deepened the analogy between the fields of algebraic numbers and functions by giving a new form to the norm residue symbol and formulated the general reciprocity law by analogy with Cauchy's residue theorem. He also introduced into number theory the term 'ramification divisor', originally associated with functions.

Finally, as indicated, Kronecker was the first to sketch in his memoir a unified theory of algebraic numbers and algebraic functions of many variables. However, he lacked sufficient technical means to fully realize his design.

At the end of the last century and at the beginning of this century the development of the theory of algebraic functions came to a halt. Mathematicians seemd to retreat somewhat in order to fortify their positions. At the time the axiomatic method based on set-theoretic notions, evolved in algebra. What came into being were constructions of theories of abstract groups,

134

A. Cauchy

fields, rings, ideals, local and semilocal rings, up to the theory of schemes; that is, the apparatus of modern abstract algebra was developed. The first variant of this algebra (without the theory of schemes) was created by the school of Emmy Noether and presented by B.L. van der Waerden in a book known to every modern mathematician. Several generations of scientists learned their algebra out of this book.

At the same time, algebraic methods have been ever more extensively applied in geometry, in mathematical analysis, and later in physics. This being so, we may well speak of the algebraization of mathematics.

Chapter Three
Problems of Number Theory

1 The Arithmetic Theory of Quadratic Forms

The General Theory of Forms; Ch. Hermite

The preceding chapter contains an exposition of Gauss' investigations pertaining to binary quadratic forms

$$ax^2 + 2bxy + cy^2,$$

$a, b, c \in \mathbb{Z}$. Gauss began to study ternary forms

$$\sum_{i,k=1}^{3} a_{ik} x_i x_k,$$

$a_{ik} = a_{ki}$, in Part V of his *Disquisitiones* entitled "Digression containing an investigation of ternary forms". He introduced the notion of a discriminant (he called it a determinant) for such forms and showed that the number of classes of ternary forms with given discriminant is finite. Gauss sketched a program for the further development of a theory of ternary forms, considered their applications to the problem of representation of numbers by means of a sum of three squares, and to the proof of the theorem that every positive integer can be represented as a sum of three triangular numbers or four squares.

The problem of equivalence of positive definite ternary quadratic forms was studied by L.A. Seeber (see p. 154). These investigations were continued by Eisenstein.

Following Gauss, P.G. Lejeune-Dirichlet elaborated a general theory of quadratic forms. He devoted to it many papers and a major part of his "Lectures on number theory" (1863) discussed previously.

The first of these papers was "Investigations on the theory of quadratic forms" (*Untersuchungen über die Theorie der quadratischen Formen*. Abhandl. Preuss. Akad. Wiss., 1833) in which he considered questions of representation of numbers by means of quadratic forms, of prime divisors of quadratic forms, and some other issues. He summarized what had been done in this theory up to Gauss' time. In the 1838 paper "On the use of infinite series in number theory" (*Sur l'usage des séries infinies dans la théorie des nombres*. J. für Math., 1838, 18) Dirichlet used the finiteness of the number of classes of quadratic forms with given discriminant to prove the infinitude of primes in arithmetic progressions (see p. 174). In this connection he compared the classification of forms due to Gauss and Lagrange and found various expressions for the number of classes of positive forms.

Dirichlet devoted the paper "On various applications of infinitesimal analysis to number theory" (*Recherches sur diverses applications de l'analyse infinitésimale à la théorie des nombres*. J. für Math., 1839, 19; 1840, 20)[1] to the problem of determination of the number of classes of quadratic forms of given positive or negative discriminant D. He noted that his approach could also be used to prove other assertions of Gauss in the second half of the fifth part of *Disquisitiones*. Dirichlet presented this part of Gauss' work in a simpler and more accessible form by using a geometric interpretation. Here he also established analogues of Euler's identity for Dirichlet series of various kinds and deduced formulas for the number of classes of forms with given discriminant.[2] Some of the many results in this paper are due to Jacobi. One such result is that "the number of solutions of the equation $x^2 + y^2 = n$ is equal to four times the difference between the number of factors of n of the form $4\nu + 1$ and of the form $4\nu + 3$". Another is that "the number of solutions of the equation $x^2 + 2y^2 = n$ is equal to twice the difference between the number of factors of n of the form $8\nu + 1$ or $8\nu + 3$ and the number of factors of the form $8\nu + 5$ or $8\nu + 7$".[3]

This paper of Dirichlet is remarkably rich in new results, methods, and connections among different branches of mathematics such as the integral calculus, series, number theory, the theory of quadratic forms, indeterminate equations, and trigonometric sums. It served as the starting point of many investigations including the works of P.L. Chebyshev on analytic number theory.

---

1 Lejeune-Dirichlet, P.G., *Werke*, Bd. 1. Berlin 1889, pp. 411–496.

2 Already Gauss tried to compute the number of different forms of a given discriminant. See C.F. Gauss, *Werke*, Bd. 1. Göttingen 1863, pp. 278–290, 365–379; Bd. 2, pp. 269–304.

3 Lejeune-Dirichlet, P.G., *Werke*, Bd. 1. Berlin 1899, pp. 462–463.

Dirichlet was also the author of a geometric exposition of the theory of reduction of positive quadratic forms.

Another eminent scientist who worked in the area of quadratic forms was Charles Hermite of whom F. Klein wrote that "as a result of the attractive force of his charming personality . . . he was for many decades one of the most important centers of the mathematical world".[4] The vast number of his students, his extensive correspondence, his involvement in almost all scientific societies in the world, his friendships with mathematicians in various countries — all these attest Hermite's tremendous influence on mathematicians in the second half of the 19th century.

Charles Hermite (1822-1901) attended the Collège of Louis-le-Grand in Paris, and was admitted to the Ecole Polytechnique in 1842. His first papers were published in 1842. One of them, "Remarks on the algebraic solution of a 5th degree equation" (*Considérations sur la résolution algébrique de l'équation du 5-me degré* — Nouv. Ann. Math., 1842, 12) contained an original proof of the impossibility of an algebraic solution of the general quintic.

In January 1843 Liouville advised Hermite to write to Jacobi about his investigations on the division of abelian transcendentals: he had extended to abelian functions theorems of Abel and Jacobi on the division of the argument of elliptic functions. Jacobi's response was enthusiastic. In the same year Hermite was forced to leave the Ecole Polytechnique and later studied on his own.[5]

In 1848 Hermite was appointed examiner and *répétiteur* at the Ecole Polytechnique. He carried out these modest functions for many years, even after being elected in 1856 to the Paris Academy of Science. Only in 1862, through Pasteur's influence, was Hermite able to lecture at the Ecole Normale. In 1869 he was appointed professor at the Ecole Polytechnique, where he worked until 1876, and at the Sorbonne Faculty of Science, where he lectured until 1897. His lectures were truly memorable. Hermite would use the simplest issue to open boundless horizons to his listeners and to chart the future paths of science. Almost all French mathematicians in the last third of the 19th century were his immediate students. Some of the most famous are H. Poincaré, P. Appell, E. Picard, G. Darboux, P. Painlévé, and P. Tannery.

Hermite worked in algebra, number theory, and the theories of elliptic, abelian, and modular functions, and contributed new ideas and important results to each of theses areas. The most widely known is his proof of the transcendence of e.

---

4 Klein, F., *Vorlesungen über die Entwicklung der Mathematik im 19. Jahrhundert*, Bd. 1. Berlin 1926, p. 292.

5 He was told that he would not be appointed to a public service position upon graduation because of a congenital defect of a foot.

Charles Hermite

Hermite was deeply interested in the paths of the evolution of science, the mechanism of mathematical creativity, and the connections between mathematics and the real world. In a letter to T. Stieltjes, dated October 28, 1882, he wrote:

> I am convinced that to the most abstract speculations of analysis there correspond real relations that exist outside of us and will one day become the property of our consciousness...It seems to me that the history of science proves that an analytical discovery occurs at a moment suitable for making possible each new advance in the study of the phenomena of the real world that can be investigated mathematically.[6]

Hermite regarded the method of observation as one of the component parts of mathematical research. He attached a great deal of importance to the connection between different areas of mathematics and valued above all works that established such connections.

---

6 *Correspondance d'Hermite et de Stieltjes*, publiée par les soins de B. Baillaud et H. Bourget avec une préface d'Emile Picard, T. 1. Paris 1905, p. 8.

Hermite introduced the notion of a bilinear form with conjugate complex variables (hermitian form) and gave a complete theory of reduction of definite forms of this type. From this work he deduced numerous consequences, including results on the approximation of complex magnitudes by means of Gaussian fractions.

Hermite studied forms in complex variables whose coefficients are Gaussian integers and obtained new results. He used hermitian forms to obtain new proofs of the theorems of Sturm and Cauchy on isolating roots of algebraic equations.

In 1844 Hermite renewed his correspondence with Jacobi.[7] This correspondence contains the germ of the fundamental ideas of subsequent investigations of Hermite and his students. One of the first objectives outlined by Hermite was to study the new method of approximation of irrational magnitudes introduced by Jacobi in connection with his attempt to prove the nonexistence of functions with more than two periods. Hermite extended Jacobi's method to a larger class of irrationalities. In his letters to Jacobi Hermite presented for the first time his idea of the method of continuous parameters, used it to obtain upper bounds for the minima of quadratic forms, and pointed to other possible applications.

Hermite was one of the first to pose the problem of determination of bounds for the minima of quadratic forms. Consider a positive definite quadratic form in n variables

$$f = \sum_{i,k=1}^{n} a_{ik} x_i x_k, \quad a_{ik} = a_{ki},$$

where the a_{ik} are some real numbers. The discriminant of the form is the determinant

$$D = \begin{vmatrix} a_{11} & a_{12} & \cdots & a_{1n} \\ a_{21} & a_{22} & \cdots & a_{2n} \\ \cdots & \cdots & \cdots & \cdots \\ a_{n1} & a_{n2} & \cdots & a_{nn} \end{vmatrix}.$$

Assign rational integral values (other than the n-tuple $0, 0, \ldots, 0$) to the variables $x_1, x_2, \ldots, x_n$ and call the least of the resulting values of the form f its minimum. This minimum is a function of the coefficients of the form. Consider the set of forms of given discriminant D obtained from one such form by continuously varying its coefficients. The minimum will also vary continuously and will be the same for all forms equivalent to the initial one. As it varies, the minimum can take on one or more maximal values for inequivalent forms. The problem is to determine an upper bound for the minima of all

---

7 Jacobi, C.G.J., *Opuscula mathematica*, Bd. 1, 1846; Bd. 2, 1851; Hermite, Ch., *Œuvres*, T. 1. Paris 1905, pp. 100–163.

positive quadratic forms in n variables with given discriminant D. Hermite investigated the minima of the binary quadratic form

$$(y - ax)^2 + x^2/\Delta,$$

under the assumption that a is some real number and Δ varies continuously from 0 to ∞ and found that the corresponding ratios y/x form a set of convergents of the continued fraction representing the number a. Then Hermite considers the same problem for the ternary form

$$A(x - az)^2 + B(y - bz) + z^2/\Delta,$$

with A and B positive, a and b real, and Δ varying continuously from 0 to ∞. He shows that we can always arrange things so that

$$A(x - az)^2 + B(y - bz)^2 < \sqrt[3]{2AB/\Delta}.$$

Then, letting Δ tend to ∞ we can establish that, as Δ increases continuously, the fractions x/z and y/z tend to the respective limits a and b, and that at each approximation the sum of the squares of the errors $x - az$ and $y - bz$ multiplied by the constants A and B respectively is minimal (a formulation reminiscent of the Legendre-Gauss method of least squares).

Next Hermite considers the problem of the minima of quadratic forms in any number of variables, determines an upper bound for the minima of a quadratic form in n variables, and formulates the following hypothesis about the least upper bound:

> My initial investigations in the case of a form in n variables with given discriminant D yield the bound (least upper bound for the minima)
>
> $$(4/3)^{(n-1)/2} \sqrt[n]{D};$$
>
> I think, but cannot prove, that the numerical coefficient $(4/3)^{(n-1)/2}$ should be replaced by $2/\sqrt[n]{n+1}$.[8]

Hermite tells Jacobi of a number of other problems whose solution depends on finding minima of quadratic forms. He writes:

> We again arrive ... at the curious (problem) of finding all minima of a quadratic form corresponding to different systems of a number of parameters of which we must assume that they take on all possible values. This is the road to the solution of numerous problems that the

---

8 Hermite, Ch., *Œuvres*, T. 1. Paris 1905, p. 142.

preceding analysis opens up for us. One such problem is the following: Let $\varphi(\alpha)$ be a complex integer that depends on a root α of an equation $F(x) = 0$ with integer coefficients and leading coefficient 1. Find all solutions of the equation Norm $\varphi(\alpha) = 1$.[9]

The problem posed by Hermite is equivalent to finding the units of the field $\mathbb{Q}(\alpha)$. In his 1869 Master's dissertation E.I. Zolotarev used a method based on an idea of Hermite to solve this problem for the equation

$$x^3 + Ay^3 + A^2 z^3 - 3Axyz = 1, \tag{1}$$

with A a given integer that is not a cube. Incidentally, this equation was investigated by many mathematicians, including Gauss and Eisenstein.

The method of continuous parameters and the results obtained by its use enabled Hermite to prove a number of theorems in number theory such as the theorem on the representation of a number as a sum of four squares, on the divisors of numbers of the form $a^2 + 1$ and $x^2 + Ay^2$, and so on. He also applied the theory of quadratic and hermitian forms to prove theorems in algebra. The method of continuous parameters was subsequently developed by G.F. Voronoĭ, Ya.V. Uspenskiĭ, and other mathematicians at the end of the 19th and at the beginning of the 20th century.

Many Russian mathematicians worked on the theory of quadratic forms. V.Ya. Bunyakovskiĭ (1804–1889) wrote a number of entries devoted to this topic in the *Dictionary of pure and applied mathematics* (St.-P. 1839) as well as related materials for the continuation of the dictionary now kept in the Leningrad division of the archive of the AS USSR. In the paper "Contiguous forms" (*Formes contiguës*) he quotes definitions from Gauss' *Disquisitiones* and recommends reading its Fifth Part. In the unpublished paper "The theory of forms" (*Théorie des formes*) he gives a concise formulation of the contents of the theory of forms. But Bunyakovskiĭ went beyond mere propaganda in favor of Gauss' theory. In the paper "Researches on various new laws pertaining to the sum of divisors of numbers" (*Recherches sur différentes lois nouvelles relatives à la somme des diviseurs des nombres*. Mém. Acad. sci. St.-Pétersbourg (6), sci. math. et phys., 1850, 4) he gives a number of relations for the representation of numbers by means of quadratic forms. A related paper, "New method in investigations bearing on the representation of numbers by means of quadratic forms" (*Nouvelle méthode dans les recherches relatives aux formes quadratiques des nombres*. Mém. Acad. sci. St.-Pétersbourg (6), sci. math. et phys., 1853, 5), contains representations of numbers by means of special quadratic forms.

P.L. Chebyshev worked on quadratic forms at almost the same time as Bunyakovskiĭ. In this area he followed Euler, Lagrange, and Legendre rather

---

9 Hermite, Ch., *Œuvres*, T. 1. Paris 1905, p. 146.

than Gauss. Chapters VII and VIII of his "Theory of congruences" (St.-P. 1849) deal with quadratic forms. Here Chebyshev applies the theory of divisors of quadratic forms to the factorization of integers. He shows how to find linear and quadratic divisors of a quadratic form by linking this question with the possibility of solving binomial quadratic congruences with respect to a prime modulus. Then Chebyshev determines the linear factors of $x^2 + ay^2$ and $x^2 - ay^2$ and applies these investigations to the prime factorization of numbers.

In the paper "On quadratic forms" (*Sur les formes quadratiques*. J. math. pures et appl. (1), 1851, 16) Chebyshev showed that to verify the primeness of numbers one can use not only forms with negative discriminant but also forms with positive discriminant. Let $x^2 - Dy^2$ have a positive discriminant $D > 0$ ($D = b^2 - ac$). Its quadratic divisors are of the form $\lambda x^2 - \mu y^2$. Let N be a number relatively prime to D whose linear factors are contained in the quadratic forms $f = \pm(x^2 - Dy^2)$. Let α be the least value of $x > 1$ satisfying the equation $x^2 - Dy^2 = 1$ and let x and y belong to the respective intervals

$$0 \leq x \leq \sqrt{\frac{(\alpha \pm 1)N}{2}}, \quad 0 \leq y \leq \sqrt{\frac{(\alpha \mp 1)N}{2D}}.$$

N is prime if it has a unique representation by means of the forms f with x and y relatively prime. In all other cases N is composite. Chebyshev gave tables of bounds for x and y in linear representation of N for all squarefree $D \geq 33$. Chebyshev's method was later used by A.A. Markov (see p. 152).

Chebyshev's students A.N. Korkin, E.I. Zolotarev and A.A. Markov also worked on quadratic forms. For Chebyshev himself this was a relatively minor one of his many interests.

Korkin's and Zolotarev's Works on the Theory of Quadratic Forms

The first work in this area by A.N. Korkin and E.I. Zolotarev was the above-cited Master's thesis of Zolotarev "On a certain indeterminate equation of the third degree" (St.-P. 1869).

The thesis consisted of two parts. In the first part Zolotarev continued Hermite's investigation concerning upper bounds for the minima of quadratic forms. This problem was solved for binary forms by Lagrange, and for ternary forms by Gauss. Hermite generalized Gauss' investigations to the case of forms in n variables and showed that the minimum of an n-ary quadratic form is less than $(4/3)^{(n-1)/2} \sqrt[n]{D}$, where D is the absolute value of the discriminant of the form. Hermite conjectured that his upper bound is not least and needs to be replaced by

$$2\sqrt[n]{\frac{D}{n+1}};$$

Zolotarev proved this assertion for the case $n = 2$ and showed that it is always possible to choose a quadratic form whose minimum is equal to this bound. One such form is

$$2\sqrt[n]{\frac{D}{n+1}}(x_0^2 + x_1^2 + \ldots + x_{n-1}^2 + x_0 x_1 + x_0 x_2 + \ldots).$$

He gave a simple proof of Hermite's theorem mentioned above, suggested a method of reduction of definite ternary forms different from Gauss', and examined solutions of other problems which are also reducible to the determination of the minima of quadratic forms. Zolotarev employed Hermite's idea on the application of the method of continuous parameters in number theory and suggested his own method for finding all minima of the quadratic form

$$(x - az)^2 + (y - bz)^2 + z^2/\Delta,$$

where a and b are given real numbers, and Δ is a variable parameter. In the second part of his dissertation Zolotarev applied his new method to solving the indeterminate equation (1).

It is reasonable to date the beginning of Korkin's and Zolotarev's joint investigations from the moment when Zolotarev defended his Master's thesis and Korkin made his comments on it. Their meetings and discussions led to their close scientific collaboration.

Before we proceed to the analysis of their work, we remark briefly on Korkin's life and activity.

Aleksandr Nikolaevich Korkin (1837–1908), son of a peasant, graduated from the gymnasium of Vologda, and later from the Department of Physics and Mathematics at the University of St.-Petersburg. Upon Chebyshev's recommendation, Korkin began teaching at the university in October of 1860. In 1868 he received a professorship which he held until the end of his life. Korkin also taught at the Naval Academy, where his student, A.N. Krylov succeeded him in 1900.

Korkin's scientific work covered three main areas: integration of ordinary differential equations, integration of partial differential equations, and the theory of quadratic forms. His early work was devoted to partial differential equations and mathematical physics; for example, his doctoral dissertation was titled "On simultaneous partial differential equations of the first order and some problems of mechanics" (St.-P. 1864). Korkin's investigations in the mathematical theory of geographic maps belongs to the same field of his research.

From 1871 to 1877 Korkin collaborated with Zolotarev on the theory of quadratic forms.

In their first joint article "On positive quaternary quadratic forms" (*Sur les formes quadratiques positives quaternaires*. Math. Ann., 1871) Korkin and

A.N. Korkin

Zolotarev discussed the problem of finding the least upper bound for the minima of positive quadratic forms in four variables. They obtained their result proceeding from the conclusions made by Zolotarev in his Master's dissertation for forms in three variables. Using a unimodular substitution they passed from a form in four variables to a form in three variables: the minimum of the new form remains the same as that of the initial form and is equal to the first coefficient (in the initial form this coefficient was assumed to be equal to the minimum of the form). Korkin and Zolotarev introduced the notion of "adjoint form" and used it to derive the following inequality for the first coefficient of the initial form:

$$a_{11} \leq \sqrt[4]{4D}.$$

They noted that this bound is least, since there exists a positive form having the minimum $\sqrt[4]{4D}$, for example, the form

$$V_4 = \sqrt[4]{4D}(x_1^2 + x_2^2 + x_3^2 + x_4^2 + x_1x_2 + x_1x_3 + x_1x_4).$$

146

They summarize their investigation as a theorem: "The variables of any positive quaternary form of discriminant D can be assigned integer values such that the value of the form does not exceed $\sqrt[4]{4D}$, and there exist forms whose minima are $\sqrt[4]{4D}$."[10]

Their next joint paper, "On quadratic forms" (*Sur les formes quadratiques.* Math. Ann., 1873, 6), deals with the question of minima of positive quadratic forms in n variables with discriminant D and arbitrary real coefficients, in which they consider the set of all positive quadratic forms in n variables with discriminant D and arbitrary real coefficients. All these forms can be obtained from one of the forms by continuously varying its coefficients. The minimum of this form varies continuously and assumes the same value for all equivalent forms. They introduce the concept of a limit or extremal form, namely, a form whose minima can only decrease for any infinitesimal variations of the coefficients which leave the discriminant unchanged.

Korkin and Zolotarev realized that the quantity $2\sqrt[n]{D/(n+1)}$ given by Hermite is the minimum of the extremal form, that is, it is the least upper bound for a certain set of forms. But there exist minima which exceed $2\sqrt[n]{D/(n+1)}$, that is, it is not the least upper bound for all forms with a given discriminant. The least upper bound of the minima of the entire set of these forms is the largest minimum of the extremal forms belonging to this set.

Next Korkin and Zolotarev study in their paper properties of extremal forms in n variables. One of the properties is the following: an extremal form in n variables has at least $n(n+1)/2$ representations of its minimum (that is, systems of values of the variables for which the form assumes the minimum value). They make a remark à propos indefinite forms:

> The limits (bounds) which we have considered above are also useful in the theory of indefinite forms ... Thus, it is possible to obtain several limits for values of indefinite forms, but here, as well as in the theory of positive forms, we speak of finding least upper bounds. Thus, for binary indefinite forms with determinant D the bound is given by $\sqrt{4D/5}$. However, when it comes to these limits, there is a basic difference between indefinite and definite forms. To demonstrate this distinction for binary forms and the bound $\sqrt{4D/5}$ we add that if we exclude the form
>
> $$f_0 = \sqrt{4D/5}(x^2 + xy - y^2)$$
>
> and equivalent forms, then the least upper bound for other forms with this determinant is $\sqrt{D/2}$.[11]

---

10 Zolotarev, E.I., *Collected works*, vol. 1. Leningrad 1931, p. 68 (in Russian).

11 Zolotarev, E.I., *Collected works*, vol. 1. Leningrad 1931, pp. 111–112 (in Russian).

Korkin and Zolotarev suggested a special method for positive forms which they called "decomposition of forms by minima", and thus obtained for a quadratic form in n variables with discriminant D the inequality

$$A = \min f \leq (4/3)^{(n-1)/2} \sqrt[n]{D},$$

that is, the upper bound for the minima obtained earlier by Hermite. Employing their decomposition by minima, Korkin and Zolotarev obtained another bound closer to the least upper bound than the one obtained by Hermite. They established that this bound is least for $n = 2, 3, 4$ but not for $n = 5$. They also found upper bounds for the minima for $n = 2m$ and $n = 2m + 1$, respectively.

In their last paper on the theory of quadratic forms, "On positive quadratic forms" (*Sur les formes quadratiques positives*. Math. Ann., 1877, 11), Korkin and Zolotarev again dealt with upper bounds for the minima of positive quadratic forms. They studied basic properties of extremal forms and found all such forms for $n = 2, 3, 4, 5$. In Chapter One they consider forms in n variables. If a form is specified, it is possible to know whether it is extremal or not using the method which Korkin and Zolotarev apply to three concrete forms, two of which, U_n and V_n, are in n variables and the third, Z, is in five variables. But other methods must be used to find all extremal forms for a given number of variables. Korkin and Zolotarev prove the theorem that they formulated in the article "On Quadratic Forms": the number of representations of the minimum of an extremal form cannot be less than $n(n + 1)/2$. If the minimum is assumed known, then these representations determine the form completely. The determination of the least upper bound for the minima is reduced to the determination of extremal forms with the greatest minimum. The problem will be solved if one can obtain all extremal forms for a given discriminant and a given number of variables.

Next Korkin and Zolotarev proceed to the investigation of determinants made up of numbers representing the minimum of a form (made up of "representations of the minimum"). They call these determinants "characteristic" and denote them by Δ. For a form in n variables Δ is an nth-order determinant. For each extremal form there exists at least one characteristic nonzero determinant. For any positive form in n variables for which the number of representations of the minimum is $\geq n$, the absolute values of the characteristic determinants do not exceed a bound depending only on the number n of variables. For $n = 2$ this follows from the obvious identity

$$(u^2 + v^2)(u'^2 + v'^2) = (uv' - u'v)^2 + (uu' + vv')^2.$$

In fact, let $f(x, y)$ be a binary form with discriminant D, represented as a sum of squares:

$$f(x, y) = (ax + by)^2 + (a'x + b'y)^2.$$

The double substitution

$$u = ap_1 + bp_2, \quad u' = aq_1 + bq_2, \quad v = a'p_1 + b'p_2, \quad v' = a'q_1 + b'q_2,$$

yields

$$f(p_1, p_2) = u^2 + v^2, \quad f(q_1, q_2) = u'^2 + v'^2.$$

The product $f(p_1, p_2) \cdot f(q_1, q_2)$ is

$$f(p_1, p_2) \cdot f(q_1, q_2) = (ab' - ba')^2 (p_1 q_2 - p_2 q_1)^2 + (uu' + vv')^2.$$

Hence

$$f(p_1, p_2) f(q_1, q_2) \geq D\Delta^2,$$

where

$$D = (ab' - ba')^2$$

and

$$p_1 q_2 - p_2 q_1 = \Delta,$$

and if the values $f(p_1, p_2)$ and $f(q_1, q_2)$ equal the minimum M, then $M^2 \geq D\Delta^2$. But $M \leq \sqrt{\frac{4}{3}D}$. Therefore, the characteristic determinant $\Delta \leq \sqrt{4/3}$ and, since p_1, p_2, q_1, q_2 are integers, $\Delta = 0$ or $|\Delta| = 1$. But $\Delta = 0$ is impossible (in this case we would have $p_1 q_2 - p_2 q_1 = 0$, in other words, $p_1/p_2 = q_1/q_2$, and because $(p_1, p_2) = 1$, $(q_1, q_2) = 1$, we would have $p_1 = \pm q_1$, $p_2 = \pm q_2$ and the representations $(p_1, p_2) = 1$, $(q_1, q_2) = 1$ would not be distinct). Hence $|\Delta| = 1$.

Korkin and Zolotarev apply similar considerations to ternary forms using Euler's well-known formula

$$(p^2 + q^2 + r^2 + s^2)(p'^2 + q'^2 + r'^2 + s'^2)$$
$$= (pp' - qq' + rr' - ss')^2 + (pq' + qp' - rs' - sr')^2$$
$$+ (pr' + sq' - qs' - rp')^2 + (ps' + qr' + rq' + sp')^2.$$

They thus obtain $|\Delta| \leq \sqrt{2}$, which in turn implies that $\Delta = 0$ or $|\Delta| = 1$. Next they develop a general proof for positive quadratic forms in n variables. From the inequality $|\Delta| \leq \mu$ thus obtained it follows that 1) for binary forms $|\Delta| = 1$; 2) for ternary forms $|\Delta| = 1$ or $\Delta = 0$; 3) for quaternary forms $|\Delta| = 0, 1, 2$. One knows from "On Quadratic Forms" that $|\Delta|$ can be equal to 2 only for the form V_4. The minimum of this form is equal to $\sqrt[4]{4D}$. For a form in five variables, $|\Delta| = 0, 1, 2$.

Korkin and Zolotarev proceed to find extremal forms using these results. To do this, they seek the representations of the minimum of each form. The number of such representations is $n(n+1)/2$ and they determine the form completely. These systems of numbers are set down in tabular form. Details are given for the construction of one such table. Next Korkin and Zolotarev find extremal forms whose characteristic determinants are ≤ 1 in absolute value. It turns out that all extremal forms in n variables, whose characteristic determinants are equal to 0, $+1$, or -1, are equivalent to the same form U_n. In particular, for binary forms there exists the unique extremal form

$$\sqrt{\frac{4}{3}D}(x_1^2 + x_2^2 + x_1 x_2),$$

and for the ternary forms there exists the unique extremal form

$$\sqrt[3]{2D}(x_1^2 + x_2^2 + x_3^2 + x_1 x_2 + x_1 x_3 + x_2 x_3).$$

In the second chapter of their paper Korkin and Zolotarev consider forms in four variables. For these there are two extremal forms, namely

$$2\sqrt[4]{\frac{D}{5}}(x_1^2 + x_2^2 + x_3^2 + x_4^2 + x_1 x_2 + x_1 x_3 + x_1 x_4 + x_2 x_3 + x_2 x_4 + x_3 x_4)$$

and

$$\sqrt[4]{4D}(x_1^2 + x_2^2 + x_3^2 + x_4^2 \pm x_1 x_4 \pm x_2 x_4 \pm x_3 x_4).$$

This implies that $\sqrt[4]{4D}$ is the least upper bound for the minima of quaternary forms (they give here a new proof of the theorem which was proved previously in their first joint article). For five variables they obtain three extremal forms for a given D and among them they choose a form with the greatest minimum. The quantity $\sqrt[5]{8D}$ is the least upper bound of the minima of forms in five variables.

In "On the equivalence of algebraic forms" (*Sur l'équivalence des formes algébriques.* C. r. Acad. sci. Paris, 1879, 88, pt. 1), C. Jordan applied Korkin's and Zolotarev's method of decomposition of forms by minima to proving the theorem on the finiteness of the class number of reduced forms with integral algebraic coefficients and determinant $D \neq 0$. He noted that for quadratic forms in n variables a similar theorem was proved earlier by Hermite and also by Korkin and Zolotarev "in a remarkably simple way". Jordan also wrote about the Korkin-Zolotarev reduction procedure in the more detailed *Mémoire sur l'équivalence des formes* (Acta math., 1882). H. Poincaré applied the Korkin-Zolotarev method of reduction of forms while investigating cubic ternary forms. In 1933 N. Hofreiter determined all classes of extremal forms for $n = 6$. Among them he found the class with the minimum $\sqrt[6]{64/3}$, singled out by Korkin and Zolotarev. In 1934 H. Blichfeldt proved that

$$\mu(6) = \sqrt[6]{64/3}, \quad \mu(7) = \sqrt[7]{64}, \quad \mu(8) = 2,$$

using the Korkin-Zolotarev method as well as some other relationships established by them. It turned out that the minima of the limit forms introduced by Korkin and Zolotarev were the largest values of the minima. Hermann Minkowski also found the limit forms for $n = 6$. He often used Korkin's and Zolotarev's results and referred to their work. G.F. Voronoĭ solved the problem of finding all limit forms for any number of variables.

Archive materials provide additional information on Korkin's and Zolotarev's investigations in the theory of quadratic forms. In one of his manuscripts, Korkin determines the least upper bounds of the minima for ternary quadratic indefinite forms. Korkin suggested a similar problem to A.A. Markov, who solved it.

The Investigations of A.A. Markov

Markov's papers in number theory concerned mainly the theory of indefinite quadratic forms in three and four variables. Almost all of them deal with the determination of limit forms of a given discriminant D. Korkin and Zolotarev were the first to study this question for indefinite quadratic forms. They obtained the first two extremal forms and discovered the principal distinction between the least upper bounds for definite and indefinite forms.

In his Master's dissertation "On binary quadratic forms with a positive determinant" (St. Petersburg 1880), Markov considered the problem of finding minima for indefinite binary quadratic forms and in subsequent papers he examined this problem for forms in three and four variables. Korkin informed him that $\sqrt{D/2}$ is the minimum for forms equivalent to the form

$$f_1 = \sqrt{D/2}(x^2 - 2xy - y^2).$$

Markov showed that $\sqrt{100D/221}$ is the least upper bound for the minima of all binary forms not equivalent to $f_0 = \sqrt{\frac{4}{5}D}(x^2 + xy - y^2)$ or to f_1, and is the minimum for forms equivalent to the form

$$f_2 = \sqrt{4D/221}(5x^2 - 11xy - 5y^2).$$

He did not stop here. Using continued fractions, he proved that if l is a given number larger than $2/3$, then there is only a finite number of classes of forms of a given discriminant D whose values are smaller in absolute value than $l\sqrt{D}$. If l approaches $2/3$, then the number of classes of forms of given discriminant D increases beyond all bounds.

Markov applied his method for finding minima to solving the indeterminate equation

$$x^2 + y^2 + z^2 = 3xyz.$$

Markov's investigations were continued by G. Frobenius in "On the reduction of indefinite binary forms" (*Über die Reduction der indefiniten binären quadratischen Formen.* Sitzungsber. Preuss. Akad. Wiss., 1913) and "On Markov's numbers" (*Über die Markoffschen Zahlen.* Sitzungsber. Preuss. Akad. Wiss., 1913) and by I. Schur in "On the theory of indefinite binary quadratic forms" (*Zur Theorie der indefiniten binären quadratischen Formen.* Sitzungsber. Preuss. Akad. Wiss., 1913).

In the early 1900s Markov began the investigation of similar problems for indefinite forms in three and four variables. In the paper "On indefinite ternary quadratic forms" (Izv. Peterburg. Akad. Nauk (5), 1901, 14)[12] he determined the first two extremal forms (one indicated to him by Korkin) and gave, without proof, the third form. He supplied a proof in the paper "On indefinite ternary quadratic forms" (*Sur les formes quadratiques ternaires indéfinies.* Math. Ann., 1903, 56). Markov returned to this problem in "A table of indefinite ternary quadratic forms" (Zap. Akad. Nauk, (8), 1909, 23), where he gave without proof a fourth extremal form and noted two more forms shown later by B.A. Venkov[13] to be the sixth and seventh extremal forms. For forms in four variables Markov found the first two extremal forms ("On Indefinite Quadratic Forms in Four Variables". Izv. Peterburg. Akad. Nauk (5), 1902, 16, No. 3).

In the paper "On Three Indefinite Ternary Quadratic Forms" (Izv. Peterburg. Akad. Nauk (5), 1902, 17, No. 2) Markov examines three quadratic forms and establishes for each of them the possibility of choosing from different representations of a number, representations in which the variables x, y, z are bounded by several inequalities and therefore can have only a finite number of different values. He employs Chebyshev's findings in "On Quadratic forms", the properties of the Legendre-Jacobi symbol, and Dirichlet's theorem on arithmetic progressions, and proves that the form $x^2+xy+y^2-2z^2$ can represent any odd number $\pm C$ not divisible by 3. Specifically, the equation

$$x^2 + xy + y^2 - 2z^2 = C$$

admits at least one solution satisfying the inequalities

$$0 \leq z \leq \frac{1}{2}(x + y/2), \quad x \geq y \geq 0, \quad x + y/2 \leq \sqrt{2C},$$

---

12 Markov, A.A., *Collected works.* Moscow 1951, pp. 143–163 (in Russian).

13 Venkov, B.A., *On Markov's extremal problem for the indefinite ternary forms.* Izv. AN SSSR, ser. matem., vol. 9, 1945, pp. 429–494 (in Russian).

and the equation

$$x^2 + xy + y^2 - 2z^2 = -C$$

at least one solution satisfying the inequalities

$$x \geq y \geq 0, \quad x + y/2 \leq z \leq \sqrt{3/2\,C}.$$

Markov proves similar assertions concerning other forms considered in this paper.

B.N. Delone made a deep analysis of Markov's work and gave a systematic geometric interpretation of the results obtained by Markov and his followers.[14]

The subject of Markov's investigations attracted in the past (and continues to attract) the attention of many mathematicians, among them L. Mordell, H. Davenport, K. Mahler, C.L. Siegel, B.N. Delone and others.

Vladimir Andreevich Markov (1871–1897), A.A. Markov's brother, proved Eisenstein's unproved formulas for the number of classes of ternary quadratic forms with given determinant in "On the number of classes of ternary quadratic forms with given determinant" (Soobshch. Khar'k. matem. ob-va (2), 1893, 4). V.A. Markov's second work, "On positive ternary quadratic forms", published posthumously (St. Petersburg, 1897), is also related to Eisenstein's work and is devoted to the problems of the theory of the representations of numbers by forms and the theory of equivalence of the forms dealt with.

Even before V.A. Markov, E.V. Borisov published "On the reduction of positive ternary quadratic forms by Selling's method" (St. Petersburg, 1890), where he gave a critical review of various reduction theories, including the theory of E. Selling (1834–1920) which he simplified. Hermite's student L. Scharve devoted his investigations to the reduction of positive quaternary quadratic forms.

In 1918 L. Mordell reproved Eisenstein's formulas in the article "On the Class Number for Definite Ternary Quadratics" (Messenger of Math., 1918, 47, 65–78). Apparently, he was not familar with V.A. Markov's article.

Later, Ya.V. Uspenskiĭ (1883–1947) also studied the theory of quadratic forms. His Master's dissertation was devoted to the application of Hermite's method of continuous parameters. Uspenskiĭ's papers contain proofs of formulas establishing the class number of positive binary quadratic forms. From the time of Kronecker's publication of the first formulas of this type in 1857–1860, researchers in this field used two methods, namely those of Kronecker and

---

14 Delone, B.N., *The Petersburg school of number theory*. Moscow-Leningrad 1947, pp. 141–193 (in Russian).

Hermite. Both methods were analytic, although Kronecker was able to obtain part of his results arithmetically. Liouville suggested another arithmetic method. Ya.V. Uspenskiĭ improved Liouville's method, derived Kronecker's and Liouville's formulas arithmetically, and proved many other results, including Gauss' formulas for the number of representations of an integer as the sum of three squares. A historical survey of investigations on the derivation of Liouville's formulas for the number of representations of integers by quadratic forms is given in the book of L.A. Kogan.[15]

2 Geometry of Numbers

Origin of the Theory

Delone justly considers as the first investigation on the geometric theory of numbers Lagrange's paper "Analytic solution of some problems of triangular pyramids" (*Solution analytique de quelques problèmes sur les pyramides triangulaires*. Nouv. mém. Acad. sci. Berlin, 1773), in which Lagrange examined various properties of tetrahedrons given by the coordinates of three vertices, the fourth vertex of each of these tetrahedrons lying at the origin.[16] Mention should also be made of the work of Ludwig August Seeber (1793–1855), professor of physics at the University of Freiburg: the article "An attempt to clarify the inner structure of solids" (*Versuch einer Erklärung des inneren Baues der festen Körper*. Ann. Phys. und Chem., 1824, 76) and the book "A Study of Properties of Positive Ternary Quadratic Forms" (*Untersuchungen über die Eigenschaften der positiven ternären quadratischen Formen*, Freiburg, 1831).

In the former work, Seeber, following the famous crystallographer R.J. Haüy, studied the partition of space into identical parallelepipeds. The examination of the square of the Euclidean distance between vertices of such a parallelepiped led Seeber to a geometric interpretation of positive ternary quadratic forms. Seeber noted that the theory of positive ternary quadratic forms was a useful tool in crystallography. Indeed his article was written in connection with a course on crystallography which he taught at the university.

In his book, Seeber considered the theory of the reduction of positive ternary quadratic forms. In a special note on Seeber's book, Gauss in 1831 cited the main problems solved by Seeber: (1) for each given positive quadratic form one can find an equivalent reduced form; (2) two nonidentical reduced forms cannot be equivalent or, what is the same, in each class there is only one reduced form; (3) to decide if a given form contains another form

---

15 Kogan, L.A., On the Representation of whole numbers by quadratic forms with positive determinat. Tashkent, "Fan", 1971.

16 Delone, B.N., "Gauss' work on number theory". In: *Carl Friedrich Gauss*. Moscow 1956, p. 63 (in Russian).

not equivalent to it; (4) to determine all possible transformations of a given form into another form, equivalent to the first or contained in it; (5) to find all possible classes of positive ternary forms with a given discriminant.

To solve the last problem Seeber used the theorem which asserts that the product abc of the coefficients of the squares of the variables of a binary quadratic form

$$ax^2 + by^2 + cz^2 + 2a'yz + 2b'xz + 2c'xy$$

does not exceed $2D$, that is, twice the discriminant of the form (we take the absolute value of the discriminant).

In his note Gauss gave a geometric interpretation of a positive binary quadratic form

$$f(x, y) = ax^2 + 2bxy + cy^2,$$

where a, b, c are real and x, y are integers. This is the squared distance between some point P and the origin, provided that the coordinate axes form an angle whose cosine is $b/\sqrt{ac}$. Since x, y are integers, we can associate to the form $f(x, y)$ the system of points of intersection of two systems of lines, parallel to the axes Ox and Oy, respectively. The distance between the lines of the first system is $\sqrt{a}$, and the distance between the lines of the second system is $\sqrt{c}$. Thus, the entire plane is partitioned into equal parellelograms, and furthermore, the vertices of these parallelograms are the points whose coordinates are integral multiples of $\sqrt{a}$ and $\sqrt{c}$ and make up the system of points referred to as a "lattice" of points. The absolute value of the discriminant $D = |ac - b^2|$ is equal to the square of the area of an elementary parallelogram.

If we consider ternary rather than binary forms, then the parellelograms must be replaced by parallelepipeds obtained by intersecting three systems of parallel equidistant planes. Their vertices form a lattice in space. Gauss notes that using these geometric representations it is possible to explain the meaning of many concepts and results of the theory of quadratic forms in two and three variables; in this way it is possible to explain the geometric significance of the remaining principal aspects of the theory of ternary forms: "the property of a form being contained in another, of representation of a definite number or an indefinite binary form by means of a ternary form..."[17].

Concluding his note, Gauss outlined his own (arithmetic) proof of Seeber's theorem.

Lejeune-Dirichlet continued Gauss' geometric researches. In his article "Study of various applications of infinitesimal analysis to the theory of numbers" (*Recherches sur diverses applications de l'analyse infinitésimale à la*

---

17 Gauss, C.F., *Werke*, Bd. 2. Göttingen 1866, p. 195.

théorie des nombres. J. für Math., 1839, 18), discussing the theory of binary forms, Dirichlet mentions Gauss' comment on Seeber's book and gives a geometric interpretation of some results obtained analytically. Using geometric representations, he solves, for example, the following problem: represent geometrically

$$ax^2 + 2bxy + cy^2 = m \tag{2}$$

(for $a > 0$, $b > 0$, $c < 0$ this is the equation of a hyperbola), and choose among the infinite number of solutions of Eq. (2) one satisfying certain specified conditions. Another problem: determine the number of points whose coordinates x and y have the form

$$x = 2D_1 v + \alpha, \quad y = 2D_1 w + \beta, \quad (D < 0, D_1 = -D), \tag{3}$$

where α, β, v and w are integers, and are inside and on the contour of the ellipse:

$$ax^2 + 2bxy + cy^2 = \sigma.$$

Here Dirichlet employs the lemma on asymptotic representation of the number of integral points lying in a region bounded by a closed curve when the dimensions of the figure increase beyond all bounds and the boundary curve remains self-similar. Dirichlet solves a similar problem concerning the number of integral points with coordinates of the form (3), lying inside and on the boundary of a hyperbolic sector.

Apparently, similar geometric considerations enabled Dirichlet to establish an asymptotic formula for the sum of the number of divisors (see "On the determination of asymptotic laws in number theory" (*Über die Bestimmung asymptotischer Gesetze in der Zahlentheorie*. Bericht. Verhandl. Akad. Wiss., 1838)[18] and "On the determination of mean values in number theory" (*Über die Bestimmung der mittleren Werthe in der Zahlentheorie*. Abhandl. Preuss. Akad. Wiss., 1849).[19] This formula was later improved by G.F. Voronoï.

Finally, in the article "On the reduction of positive quadratic forms with three undetermined integers" (*Über die Reduction der positiven quadratischen Formen mit drei unbestimmten ganzen Zahlen*. J. für Math., 1850, 40)[20] Dirichlet turns to the geometric interpretation of the theory of binary and ternary forms proceeding from Gauss' remarks on Seeber's work. The complexity of Seeber's method motivated Dirichlet to seek other methods for reducing ternary forms. In his article Dirichlet provides a geometric interpretation of Seeber's results:

---

18 Lejeune-Dirichlet, P.G., *Werke*, Bd. 1. Berlin 1889, pp. 351–356.

19 Ibid., Bd. 2. Berlin 1897, pp. 49–66.

20 Ibid., Bd. 2. Berlin 1897, pp. 21–48.

1. each parallelepiped lattice of points can be rearranged so that the corresponding elementary parallelepiped has edges not larger than the diagonals of the sides and the diagonals of the parallelepiped itself;
2. this can be done for a given system in one and only one way. In other words, each class of forms can have only one reduced form.

Dirichlet considers the ternary quadratic positive form

$$ax^2 + by^2 + cz^2 + 2a'yz + 2b'xz + 2c'xy = F(x, y, z). \qquad (4)$$

The coefficients a, b, c are real and positive, and the combinations of the coefficients

$$a'^2 - bc, \ b'^2 - ac, \ c'^2 - ab, \ -D = aa'^2 + bb'^2 + cc'^2 - abc - 2a'b'c'$$

are negative. The discriminant $-D$ is negative. One thus obtains a trihedral angle with plane angles whose cosines are given by

$$\cos \lambda = \frac{a'}{\sqrt{bc}}, \quad \cos \mu = \frac{b'}{\sqrt{ac}}, \quad \cos \nu = \frac{c'}{\sqrt{ab}},$$

and satisfy the inequality

$$\cos^2 \lambda + \cos^2 \mu + \cos^2 \nu - 2 \cos \lambda \cos \mu \cos \nu < 1 \text{ for } D > 0.$$

Out of two angles one chooses each time the angle traversed from right to left (if we look in the direction to the origin). The sides of the trihedral angle are regarded as coordinate axes. A segment $\sqrt{a}$ is laid off on the first axis, a segment $\sqrt{b}$ on the second, and a segment $\sqrt{c}$ on the third. Then the form (4) is the square of the distance between the point $P(x\sqrt{a}, y\sqrt{b}, z\sqrt{c})$ and the origin. If x, y, z assume only integral values not all zero, then one obtains a system of points formed by the intersection of three systems of parallel equidistant planes. Dirichlet examines in detail the transformation of the form (4) under the linear substitution

$$x = \alpha x' + \beta y' + \gamma z', \quad y = \alpha' x' + \beta' y' + \gamma' z',$$
$$z = \alpha'' x' + \beta'' y' + \gamma'' z' \qquad (5)$$

with integer coefficients and nonzero determinant

$$E = \alpha\beta'\gamma'' + \beta\gamma'\alpha'' + \gamma\alpha'\beta'' - \gamma\beta'\alpha'' - \alpha\gamma'\beta'' - \beta\alpha'\gamma''.$$

The geometric representation of the new form $F_1(x', y', z')$ is the system of points of intersection of three systems of parallel planes such that the distance

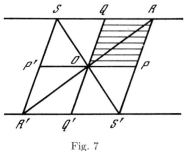

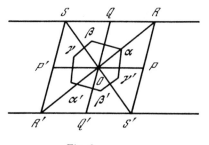

Fig. 7   Fig. 8

between them is $\sqrt{a'}$ (for the first system), $\sqrt{b'}$ (for the second system), and $\sqrt{c'}$ (for the third system). One can take as origin an arbitrary point of the system of points. We superpose the origin of the new system on the origin of the original system. Depending on the sign of E, the linear substitution allows us to obtain a system of points coinciding with the original system or symmetric to it. If the coefficients in (5) are integral and the determinant is $E = \pm 1$, then the new system of points will be integral as well. If $E = +1$, the new system can be superposed on the initial system, that is, to properly equivalent forms there correspond systems of parallelepipedly located points, which can be made to coincide.

Next Dirichlet considers the case of binary forms and shows that properly equivalent forms have systems of points which can be superposed via a motion without a rotation in a plane. For improperly equivalent forms one of the systems must be rotated. Dirichlet explains that for equivalent forms in two and three variables the elementary figures (parallelograms and parallelepipeds) have equal area (volume). This is due to the fact that the discriminants of forms connected by a linear substitution with integer coefficients and determinant $E = \pm 1$ are equal in absolute value.

Dirichlet considers the system of integral points (a "point lattice") corresponding to the binary quadratic form $ax^2 + 2bxy + cy^2$ and proves that such a system can always be divided into elementary parallelograms whose sides are not larger than the diagonals. Dirichlet calls this elementary parallelogram "reduced". Points of the system always lie pairwise at equal distances and in the opposite directions from the center O. Let P be a point of one of these pairs for which the distance from O is smaller than for any other pair. If there are several closest points, we choose any one of them. Next we take a point Q of the system, $Q \neq P$, closest to O after P (if there are several points, we choose any one of them). We thus obtain the parallelogram $POQR$ (Figure 7), in which $OP \leq OQ$, $OQ \leq OS = PQ$, $OQ \leq OR$ by construction. This is the reduced parallelogram. Let $OP = \sqrt{a}$, $OQ = \sqrt{c}$, so that $a \leq c$.

The minimum distance between the points of the system and the initial point O is $\sqrt{a}$, and the second minimum is $\sqrt{c}$. If we put $\cos POQ = b/\sqrt{ac}$, where $b \geq 0$, then $(PQ)^2 = a - 2b + c \geq c$, and therefore $2b \leq a$, $2b \leq c$,

158

$4b^2 \leq ac$. Denote by k the square of the height of the parallelogram. Then for the area of the parallelogram we have

$$\Delta = ak = ac - b^2 \geq \frac{3}{4}ac,$$

so that

$$h = \sqrt{k} \geq \frac{1}{2}\sqrt{3c}.$$

The distance between the parallel lines OP and QR is $\geq \sqrt{3c}$. In other words, the second parallel line QR is at the minimum distance $\sqrt{3c}$ from the line OP. For different arrangements the sides of the reduced parallelogram always have the values $\sqrt{a}$ and $\sqrt{c}$.

Dirichlet also solved a problem of a different kind: given a point O of a plane lattice determine the part of the plane whose interior points are closer to the point O than to any other point of the lattice. This figure is a hexagon $\alpha\beta\gamma\alpha'\beta'\gamma'$ (Figure 8). Each point inside this hexagon is closer to the point O than to any other point of the lattice.

Dirichlet proceeds next to solve the analogous problem for a third-order system (a three-dimensional lattice), that is, to construct a reduced elementary parallelepiped all of whose edges are smaller than its diagonals and the diagonals of the sides. Dirichlet constructs successive minimum distances from the initial point and chooses the coordinate axes on three consecutive minima. He then applies the geometric results thus obtained to quadratic forms: the reduced quadratic form corresponds to the reduced parallelopiped, etc. Dirichlet also used the geometric method to prove "Seeber's beautiful theorem", the analytical proof of which was given earlier by Gauss.[21]

The Gauss-Dirichlet geometric method was later presented in detail by H.J.S. Smith in his *Reports on the theory of numbers*. F. Klein studied and used this method and H. Minkowski and G.F. Voronoĭ developed it further.

The Work of H.J.S. Smith

Henry John Stephen Smith (1826–1883), son of an Irish barrister, was a student at Oxford and upon graduation in 1850 was appointed lecturer and, in 1860, professor. Smith was a man of both linguistic and mathematical abilities which made it difficult for him to decide between a career in classics versus one in mathematics. He finally chose the latter.

Smith was one of the most active Fellows of the Royal Society. He was President of the Mathematical Section of the British Association and Fellow

---

21 Seeber's theorem was also proved by other mathematicians, including Hermite, Korkin and Zolotarev. Zolotarev proved it in his master's thesis "On a certain indeterminate cubic equation" (1869), and Korkin and Zolotarev in the paper "On quadratic forms" (1873).

H.J.S. Smith

of Corpus Christi College, Oxford. It was before the British Association that Smith presented his *Reports on the theory of numbers* between 1859 and 1865, in which he gave a complete and systematic description of the history of number theory beginning with Legendre (1798) and Gauss (1801) and ending with the 1860s. Moreover, in the article *On the present state and prospects of some branches of pure mathematics* (Proceed. Lond. Mathem. Soc., 1876, vol. 8, pp. 6–29; Collect. Mathem. Papers, 1894, vol. 2, pp. 166–190) he extended the historical survey of number theory up to 1876 (see bibliography). The works of Smith also contained a comprehensive discussion of the theory of quadratic forms, including geometrical methods, as well as some crucial results obtained by Chebyshev, Korkin, and Zolotarev.

Smith worked in the theory of quadratic forms, the theory of elliptic functions, geometry, and other fields.

In 1883 the Paris Academy of Sciences awarded its Grand Prix des Sciences Mathématiques to two candidates: Smith and H. Minkowski, a student at the University of Königsberg, for an essay on the theory of representation of integers as sums of five squares. Smith died only a few days before the award of the prize: it was awarded officially on February 26 and Smith died

160

on February 9. For the 18-year-old Minkowski this work was the beginning of his scientific career. In fact, the problem of representation of an integer as a sum of five squares had already been solved by Smith in 1867. In the article *On the order and genera of ternary quadratic forms* (Philos. Trans., 1867) he showed that the various theorems of Jacobi, Eisenstein, and Liouville on the representation of an integer as a sum of four squares and other simple quadratic forms can be derived in a unique way from the principles formulated in this article. The basic theorems on the representation of an integer by a sum of five squares were given by Eisenstein in "New Theorems of Higher Arithmetic" (*Neue Theoreme der höheren Arithmetik*. J. für Math., 1847, 35), but he considered only squarefree numbers. In this connection, in the article *On the order and genera of quadratic forms containing more than three indeterminates* (Proc. Roy. Soc., 1867, 16) Smith elaborated Eisenstein's research and gave a complete proof for the case of five squares. Moreover, he added corresponding theorems for the case of seven squares.

Smith's works did not receive wide recognition on the continent, and the French academicians apparently did not know about them until the 1882 competition was announced. When they did, they invited Smith to present his proof and thus participate in the competition.

Geometry of Numbers: Hermann Minkowski

Hermann Minkowski (1864–1909) was born in the town of Aleksoty in Russia (now the Lithuanian SSR), but his familiy moved to Königsberg when he was still a child. At the age of fifteen he graduated from the local gymnasium and studied at the Universities of Königsberg and Berlin, attending the lectures of Helmholz, Hurwitz, Lindemann, Kronecker, Kummer, H. Weber, Weierstrass, Kirchhoff and others. Beginning in 1881 he got interested in the theory of quadratic forms and entered the competition announced by the Paris Academy of Sciences. His article made a profound impression on the jury which included Hermite, Bertrand, Bonnet, Bouquet, and Jordan who made the formal presentation of the article. The judges were so impressed that the author was forgiven for writing his paper in German, which was against the rules of the Academy.

By that time Minkowski had mastered the various methods of number theory, such as Dirichlet's series and Gauss' trigonometric sums. He had studied deeply the works of Dirichlet, Gauss, and Eisenstein. The starting point of his article was the idea that the representation of an integer as a sum of five squares had to depend on quadratic forms in four variables just as the representation of an integer as a sum of three squares depends on quadratic forms in two variables, as shown in Gauss' proof.

Minkowski considered n-ary quadratic forms and formulated the main concepts and definitions of the theory of these forms. He generalized Gauss' method by investigating the representation of forms in fewer variables by

Hermann Minkowski

forms in a larger number of variables. The solution of the problem for the Competition easily followed from the general theory. Having studied Hermite's articles in Crelle's Journal and the *Note sur la réduction des formes quadratiques positives quaternaires* (Ann. Ecole Norm., 1882, 11) of Hermite's student L. Scharve, Minkowski presented his idea in the article *Sur la réduction des formes quadratiques positives quaternaires* (C. r. Acad. sci. Paris, 1883, 96).

Using the investigations of Hermite, Smith, and Poincaré, Minkowski published articles on the theory of positive quadratic forms in n variables. These papers served as the subject of his doctoral dissertation that he defended at the University of Königsberg in 1885. The dissertation included the idea of justifying mathematical investigations by using the space representation of a quadratic form.

Minkowski began teaching at Königsberg in 1885 and in 1887 he moved to Bonn. In 1892 he became an Adjunct and later in 1894 a full professor. In 1895 he replaced his university colleague and friend D. Hilbert (who moved to Göttingen) at the University of Königsberg. He stayed there two years and in 1896 he accepted the invitation of the Ecole Polytechnique at Zurich.

162

From the fall of 1902 until the end of his life he was professor at Göttingen University.

The article "On positive quadratic forms and on algorithms similar to that of continued fractions" (*Über die positiven quadratischen Formen und über kettenbruchähnliche Algorithmen*. J. für Math., 1891, 107) opens the series of Minkowski's researches in the geometry of numbers. Studying the problem of reduction of quadratic forms in any number of variables, he realized that this problem, as well as other problems of the theory of quadratic forms, becomes remarkably simple and clear when approached via geometric methods. Following Gauss and Dirichlet, Minkowski examines number lattices,[22] generalizing this to the case of any number of variables. To this notion he adds another. For the positive binary form

$$F(x, y) = ax^2 + 2bxy + cy^2 \tag{6}$$

with positive coefficients a, b, c, the geometric image is that of an ellipse. The expression "Form (6) assumes the value m for integral $x = p$, $y = q$" means that the ellipse passes through the point (p, q). For a positive quadratic form in three variables $F(x, y, z)$ with discriminant equal to 1, the equation

$$F(x, y, z) = ax^2 + by^2 + cz^2 + a'xz + b'yz + c'xy = m,$$

where $m = \text{const} > 0$, yields an ellipsoid centered at the origin. Minkowski proceeds from a general geometric lemma. In the two-dimensional case his lemma states that the region R in the plane xOy always contains a point (p, q) with integer coordinates, different from the origin, if this region satisfies the following conditions:

1. the region R is symmetric with respect to the origin, that is, whenever it contains (x, y) it also contains $(-x, -y)$;

2. the region R is convex, that is, if (x_1, y_1) and (x_2, y_2) are points of the region R, the entire segment

$$[\lambda x_1 + (1 - \lambda)x_2, \lambda y_1 + (1 - \lambda)y_2] \quad (0 \le \lambda \le 1)$$

connecting these points is contained in R;

3. the area of R is greater than 4.

Any ellipse $F(x, y) \le m$ satisfies conditions 1 and 2. It also satisfies condition 3 provided $m\pi > 4\sqrt{D}$, where D is the absolute value of the discriminant D_1, of the form $F(x, y)$. According to Cassels, "the whole of the

---

22 F. Klein also studied number lattices, but stated: "I restricted myself at the time to explaining certain known principles geometrically, whereas Minkowski undertook to discover new results." (See Klein, F., *Vorlesungen ...*, T. 1. Berlin 1926, p. 328.)

geometry of numbers may be said to have sprung from Minkowski's convex body theorem".[23] Here is the general formulation of this theorem: each convex body centered at the origin of a parallelopipedal system of points the volume of which is more than 2^n times (if the space is n-dimensional) larger than the volume of the principal parallelopiped of the system contains, in addition to the central point, at least one more point of the system.

Using this lemma, Minkowski proved, for example, that there exists an upper bound for the minima of a positive quadratic form in n variables with given discriminant D:

$$M < A(n) \sqrt[n]{D}, \tag{7}$$

where

$$A(n) = \frac{4}{\pi} \left(\Gamma \left(\frac{n}{2} + 1 \right) \right)^{2/n},$$

and M is the minimum of the form. Minkowski obtained this estimate using geometric considerations which can, roughly speaking, be expressed as follows: if solid, not interpenetrating, balls are placed in a closed box, then the total volume occupied by the balls is smaller than the volume of the box, because there is empty space between the balls, and, in any case, the volume occupied by the balls cannot be larger than the volume of the box.[24]

Minkowski said that his investigations on the geometry of numbers began when he read Hermite's letters to Jacobi, containing Hermite's theorem on upper bounds for the minima of quadratic forms, and Dirichlet's article "On the Reduction of Positive Quadratic Forms in Three Indeterminates" in the fortieth volume of Crelle's Journal. Comparing these papers, Minkowski decided to represent Hermite's statement geometrically. Later he discovered that this property of ellipsoids is based on the fact that the ellipsoid is a convex body with a central point and went on to derive his fruitful principle which he then applied to various convex bodies. The result in inequality (7) is more precise than that of Hermite (see p. 142).

In his paper *Über positive quadratische Formen* (J. für Math., 1886, 99) Minkowski considered the minima of such forms in 2, 3, 4, 5 variables, and introduced the notion of a "limiting form". Having read the papers of Korkin and Zolotarev, he concluded that

> the classes of these limiting forms are identical to the "extremal forms", which Korkin and Zolotarev introduced in their interesting articles in the sixth and eleventh issues of *Mathematische Annalen*. By this they meant positive forms having the property that their minimum decreases when the coefficients are subject to infinitesimal changes that do not increase the determinant of the form. The close connection of these

---

23 Cassels, J.W.S., *An introduction to the geometry of numbers.* Berlin etc. 1959, p. 64.

24 Delone, B.N., "Hermann Minkowski". UMN, 1936, #2, 34–35 (in Russian).

forms with the greatest minimum is characteristic of Hermite's reduction method.[25]

Minkowski cites Korkin and Zolotarev on many other occasions as well, for example, in the article "On positive quadratic forms and on algorithms similar to that of continued fractions". We quote:

Korkin and Zolotarev discovered new points of view. They made use of special forms, which they completely determined for up to five variables, for which the ratio mentioned in Hermite's fundamental theorem (of the minimum attained for integers to the n-th root of the determinant) is a maximum.[26]

It was here that Minkowski proved geometrically that the minimum M of a positive form in n variables satisfies the inequality (7), and, using the asymptotic expression for the gamma-function, found the estimate

$$M < \frac{2n}{\pi e} \sqrt{n\pi e^{1/3n}} \sqrt[n]{D},$$

where $2/\pi e \approx 0.234\ldots$. From his lemma Minkowski obtained many results relevant to the theory of algebraic units, the theorem on the finiteness of the number of classes of integral quadratic forms with given discriminant, the approximation of several real values by rational fractions with identical denominators, the theory of continued fractions and its generalizations, etc.

In 1896 Minkowski published the book *Geometrie der Zahlen* (Leipzig 1896) in which he systematically explored these results. The summary (in French) of this work is in his letter to Hermite to whom he dedicates his book.[27]

In several subsequent articles Minkowski applies his results to various areas of number theory. In particular, he generalizes and makes more precise Chebyshev's and Hermite's inequalities. In his article "On a Certain Arithmetic Question" (1866) Chebyshev proved that there exist infinitely many pairs of integers x, y satisfying the inequality

$$|x - ay - b| \le \frac{1}{2|y|}.$$

Hermite (J. für Math., 1880, 88) showed that

$$|x - ay - bz| \le \sqrt{\frac{2}{27}} \frac{1}{|yz|},$$

25 Minkowski, H., *Gesammelte Abhandlungen*, Bd. 1. Leipzig-Berlin 1911, p. 156.

26 Ibid., p. 245.

27 Minkowski, H., *Extrait d'une Lettre adressée à M. Hermite*. In: *Gesammelte Abhandlungen*, Bd. 1. Leipzig-Berlin 1911, pp. 266–270.

whence

$$|x - ay - b| \leq \sqrt{\frac{2}{27}} \frac{1}{|y|}.$$

In *Diophantische Approximationen* (Leipzig, 1907) Minkowski showed that there exist infinitely many integers x, y for which

$$|(\alpha x + \beta y - \xi_0)(\gamma x + \delta y - \eta_0)| < \frac{1}{4},$$

where ξ_0, η_0 are any given numbers, α, β, γ, δ, ξ_0, η_0 are assumed real, and in the case where the determinant $\varepsilon = |\alpha\delta - \beta\gamma| \neq 1$ the right-hand side is to be multiplied by $|\varepsilon|$, $|\varepsilon| \neq 0$. The book is based on the lectures that Minkowski delivered in Göttingen in 1903 and 1904, taken down and edited by A. Axer. Here the fundamentals of the geometry of numbers and its applications are given more clearly and in more detail, with many drawings.

From the geometry of numbers Minkowski went on to the geometry of convex bodies and to the theory of polyhedra. He was also deeply interested in problems of mechanics and physics. Minkowski gave an important interpretation of the kinematics of the special theory of relativity, based on the synthesis of previously separate notions of space and time in a single four-dimensional spacetime manifold with a hyperbolic metric (1909).

Another mathematician who worked successfully in the field of the geometry of numbers was G.F. Voronoĭ.

The Works of G.F. Voronoĭ

Georgiĭ Feodos'evich Voronoĭ (1868–1908) was born into the family of a professor of Russian literature at the Lyceum in Nezhinsk. After graduating from the gymnasium in Priluki in 1885, Voronoĭ enrolled in the mathematics section of the Faculty of Physics and Mathematics of the University of St. Petersburg. When he was still a student at the gymnasium he wrote his first article in algebra, published in the *Journal of Elementary Mathematics* (1885). In the same period he made an attempt to find the integral solutions of the diophantine equation $x^2 + y^2 + z^2 = 2mxyz$, where $m > 0$ is a given integer, but he soon discovered that the problem was too complicated.

During his university years, Voronoĭ investigated the properties of Bernoulli numbers and reported his discovery in this field at a meeting of the mathematical circle headed by A.A. Markov. In 1889 he presented his work on Bernoulli numbers, which was published in 1890. On the recommendation of Markov, Korkin, and other professors, Voronoĭ was retained at the university to prepare for the position of a professor.

Voronoĭ's first works were devoted to the theory of algebraic numbers, which many mathematicians in St. Petersburg were working on at that time.

G.F. Voronoĭ

Voronoĭ looked for a generalization of the algorithm of continued fractions to be applied to cubic irrationalities. Lagrange had found that quadratic irrationalities can be expanded into an infinite periodic continued fraction. The problem consisted in finding a periodic algorithm for cubic irrationalities. This problems had been studied by Jacobi, Lejeune-Dirichlet, and other prominent mathematicians.

In his Master's dissertation "On integral algebraic numbers depending on a root of an equation of the third degree" (St. Petersburg, 1894) Voronoĭ established the form of algebraic integers depending on a root of an irreducible equation of the third degree $\rho^3 = r\rho + s$. A.A. Markov had earlier examined numbers depending on a root of the equation $\rho^3 = s$. Voronoĭ considered the more general case. The resolution of all the problems in his thesis is based on the investigation of solutions of congruences of the third degree with respect to simple and composite moduli. He introduced "complex numbers modulo p", that is, numbers of the form $X + Yi$, where i is the imaginary solution of the congruence $i^2 \equiv N \pmod{p}$, p is a prime, and N is a quadratic nonresidue modulo p.

After defending his Master's dissertation, Voronoĭ became professor at the University of Warsaw, where he worked, except for an interval of one year, until the end of his life.

In his doctoral dissertation "On a generalization of the algorithm of continued fractions" (St. Petersburg, 1896), Voronoĭ gives a generalization of continued fractions, which, in application to numbers depending on a root of a third-degree equation, has the property of periodicity and can serve for finding base units and solving other problems in this theory.

Jacobi developed Euler's ideas and established an algorithm for the simultaneous approximation of two real numbers by fractions M_k/N_k, M'_k/N'_k, ... where M_k, N_k, ... are ordinary integers. He applied his algorithm to algebraic integers depending on roots of a third-degree equation and, using particular examples, discovered their periodicity. Other generalizations of continued fractions had been known at that time (Dirichlet's, Hermite's, Kronecker's) but all this was a long way from practical applications. Applying Jacobi's algorithm to forms whose coefficients are algebraic numbers depending on roots of an irreducible cubic equation, Voronoĭ noticed that in several cases the forms thus obtained recur periodically. There arose the question: does one always obtain periodically recurring forms as a result of applications of Jacobi's algorithm?

Voronoĭ suggested a new generalization of continued fractions based on the notion of relative minima of a lattice. Suppose that the points of a space of dimension $n = k + 2m$ (over the field of real numbers) are characterized by k real and m complex coordinates. An n-dimensional lattice with no points in coordinate subspaces is given in this space. (Apparently, Voronoĭ thought geometrically, but used algebraic terminology, referring to the projections of points onto coordinate axes and planes as covariant forms.) The relative minimum is a point of the lattice with coordinates $(x_1^0, \ldots, x_k^0, \xi_{k+1}^0, \ldots, \xi_m^0)$ such that the lattice has no points with coordinates of smaller absolute value. For each relative minimum there are natural definitions of adjacent minima in the direction of real axes and complex planes. Voronoĭ considered three cases:

1) $k = 2$, $m = 0$ (real quadratic fields);

2) $k = 1$, $m = 1$ (cubic fields with negative discriminant);

3) $k = 3$, $m = 0$ (cubic fields with positive discriminant).

For the first an adjacent relative minimum is obtained with the aid of the algorithm for continued fractions. In the second and third cases Voronoĭ suggests algorithms for calculating adjacent relative minima. When applied to lattices in cubic extensions of the field of rational numbers, the algorithms turn out to be periodic and one can use them to obtain algebraic solutions of the problem of equivalence of two ideals and for the determination of units — problems which can be solved for real quadratic fields by using the algorithm for continued fractions. In the more difficult of the two cases of a cubic extension, the completely real one, the fundamental units can be constructed on the basis of a deep investigation of the configuration of relative minima, which, in contrast to the two other cases, is fairly complicated.

In 1896 the St. Petersburg Academy of Sciences awarded Voronoǐ the Bunyakovskiǐ prize for both his dissertations, and in 1907 he was elected corresponding member of the Academy. During the next five years of his short life Voronoǐ was involved in studies of mainly two topics: the theory of quadratic forms and analytic number theory.

In 1907, 1908, and 1909 an extensive study by Voronoǐ, *Nouvelles applications des paramètres continus à la théorie des formes quadratiques* (J. für Math., 1907, 133; 1908, 134; 1909, 136) was published, in which he exploited his geometrical apparatus.[28]

In the first part of this study "On some properties of positive perfect quadratic forms" (*Sur quelques propriétés des formes quadratiques positives parfaites*. J. für Math., 1907, 133) Voronoǐ continued Korkin's and Zolotarev's researches from the new standpoint, using as the basis the property of limiting forms of being determined by the representations of their minima. Voronoǐ called forms having this property perfect. Thus each limiting form is perfect, but not each perfect form is limiting. Voronoǐ associated perfect forms with the faces of a certain infinite polyhedron in a cone of positive quadratic forms. Next he proved that there are only finitely many nonequivalent faces as well as nonequivalent perfect forms, and gave an algorithm for the transition from face to adjacent face. He thereby constructed an (admittedly extremely cumbersome) algorithm for the construction of all inequivalent perfect forms and along with them all limiting forms in the Korkin-Zolotarev sense.

In the second part of his study entitled "Researches on primitive parallelehedrons" (*Recherches sur les parallelöedres primitifs*. J. für Math., 1908, 134; 1909, 139), Voronoǐ associates to a lattice defining a class of positive quadratic forms its "Dirichlet-Voronoǐ domain", that is, the set of points of the space closer to a given lattice point than to all others. These domains turned out to be polyhedrons. Obviously, the images of one of them under all translations by vectors of the lattice are space-filling. Voronoǐ refers to polyhedra having this property as parallelehedrons. It turned out that for a given n there are finitely many types of parallelehedrons. He gave an algorithm for constructing all types of primitive ("general position") parallelehedrons. These investigations of Voronoǐ are to a certain extent connected with the work of the Russian crystallographer E.S. Fedorov, who investigated parallelehedrons in Euclidean space in his "Elements of the theory of figures" (St. Petersburg 1885).

Voronoǐ also obtained important results in the theory of indefinite quadratic forms. They were published posthumously in his "Collected works" (Kiev

---

28 Delone, B.N., Faddeev, D.K., *Theory of cubic irrationalities*. Moscow-Leningrad 1940 (in Russian). The book includes a geometric account of the dissertation of Voronoǐ.

1952–1953). The editors of this collection think that these results are only a small portion of Voronoï's heritage in this area of mathematics.

B.A. Venkov compared the nature of Minkowski's investigations with those of Voronoï in the theory of quadratic forms. He noted that the geometric formulation of Hermite's theorem on an upper bound for the minima of a positive quadratic form with real coefficients and integer values of the variables (not all zero) is due to Minkowski who regarded this problem as a problem of discrete geometry pertaining to the densest disposition of n-dimensional balls: "However, such an interpretation, while best in terms of simplicity and intuitive appeal, provides no means to carry the solution of the problem forward further than was done by, say, Korkin and Zolotarev, that is, beyond $n = 5$. In the paper 'New applications of continuous parameters to the theory of positive quadratic forms', discussed above, Voronoï, using the geometry of the space of coefficients of a form, gives an algorithmic solution of the problem for any n.— Owing to the boldness of his thought (in the context of the geometry of his time) and to the simplicity and naturalness of construction, Voronoï's work forged a new and important step in the solution of the problem of minima of a positive quadratic form."[29]

The Master's thesis of Ya.V. Uspenskiĭ, "Some applications of continuous parameters in the theory of numbers" (St. Petersburg, 1910), followed the researches of Voronoï and Minkowski and was devoted to applications of Hermite's principle to various problems of number theory.

The investigation of the theory of quadratic as well as other homogeneous forms continued successfully in different directions in the 19th century. More general classes of forms, definite and indefinite, were considered, new proofs of results were given, more precise estimates were calculated, etc. A new elementary proof of Dirichlet's formulas (for a number of cases) is due to B.A. Venkov. The geometry of numbers was successfully developed in the Soviet Union by B.N. Delone, B.A. Venkov, A.Z. Val'fish, and their students. Outside the USSR numerous results in the theory of homogeneous forms and in the geometry of numbers were obtained by L.J. Mordell, H. Davenport, K. Mahler, H. Weyl, N. Hofreiter, A. Oppenheim, J.W.S. Cassels, J.F. Koksma, and others.[30]

G.F. Voronoï was also involved in the study of problems of analytic number theory. He published two articles on the subject and presented a report at the International Congress of Mathematicians in Heidelberg in 1904. Additional

---

29 Venkov, B.A., On the paper "On certain properties of positive, perfect quadratic forms". In: Voronoï, G.F., *Collected works*, vol. 2. Kiev 1952, p. 379 (in Russian).

30 For a detailed survey of results in the area of the geometry of numbers see Keller, O., *Geometrie der Zahlen*. In: *Enzyclopädie der mathematischen Wissenschaften*, Bd. 1, 2, 1959, and the books: Cassels, J.W.S., *Introduction to the geometry of numbers*. Berlin etc. 1959 and Rogers, C.A., *Packing and covering*. Cambridge 1964.

information on Voronoï's research in analytic methods of number theory can be found in his unpublished manuscripts.

The main result of his article "On a problem of the theory of analytic functions" (*Sur un problème du calcul des fonctions analytiques.* J. für Math., 1903, 126) is the improvement of the remainder in the Dirichlet asymptotic formula for the sum $\sum_{k \leq n} \tau(k)$, where $\tau(k)$ is the number of divisors of k. For this purpose Voronoï used geometric considerations, special sequences of numbers that are modifications of the Farey series and N.Ya. Sonin's summation formula. The first papers of Voronoï's student at the University of Warsaw, W. Sierpiński (1882–1969), who later became an outstanding Polish mathematician, and of Academician I.M. Vinogradov are related to this line of research.

In the second article "On a transcendental function and its applications to the summation of some series" (*Sur une fonction transcendente et ses applications à la sommation de quelques séries.* Ann. Ecole Norm., III sér., 1903, 20) Voronoï applied Bessel functions to number theory for the first time.

Voronoï's manuscripts show that he was also engaged in the study of the theory of the Riemann zeta-function, the summation of series dependent on prime numbers, and other problems.

3 Analytic Methods in Number Theory

Lejeune-Dirichlet and the Theorem on Arithmetic Progressions

The works of Leonhard Euler were the foundation for the creation of analytic methods in number theory. Euler's identity for the zeta-function

$$\zeta(s) = \sum_{n=1}^{\infty} \frac{1}{n^s} = \prod_{p} \left(1 - \frac{1}{p^s}\right)^{-1}, \tag{8}$$

where n denotes a natural number, the product on the right is taken over all primes p, and s is a real number greater than 1, played a crucial role in the investigation of the distribution of primes. Euler also studied other relations between series and infinite products, which served as the source of many theorems in number theory. Euler's ideas were further developed by Legendre, Dirichlet, Jacobi, Chebyshev, Riemann, and other mathematicians of the 19th and 20th centuries.

P.G. Lejeune-Dirichlet

The most significant investigations in the area of analytic methods in number theory in the first half of the 19th century were those of Gustav Peter Lejeune-Dirichlet (1805–1859). H.J.S. Smith wrote in 1859:[31]

> His original investigations have probably contributed more to its advancement [i.e., the advancement of number theory (A.S.)] than those of any other writer since the time of Gauss; ... He has also applied himself (in several of his memoirs) to give an elementary character to arithmetical theories which, as they appear in the work of Gauss, are tedious and obscure; and he has thus done much to *popularize* the theory of numbers among mathematicians — a service which it is impossible to appreciate too highly.

A descendant of a French family which emigrated to Germany, Gustav Peter Lejeune-Dirichlet graduated from the gymnasium in Bonn at the age

31 Smith, H.J.S., *Report on the theory of numbers*. Pt I (1859). In: *The collected mathematical papers of H.J.S. Smith*, vol. 1. Oxford 1894, p. 72.

of sixteen and decided to continue his education in Paris: the young man was deeply interested in mathematics, and, at that time, the Paris school of mathematics was the best in the world. Here Dirichlet attended lectures at the Faculté des Sciences of the Sorbonne and the Collège de France, making his living as a private tutor.

Dirichlet's first interest in mathematics was number theory. This interest had been awakened in him by an earlier study of Gauss' *Disquisitiones Arithmeticae* (1801).

In 1825 Dirichlet presented to the Paris Academy of Sciences his first paper, *Mémoire sur l'impossibilité de quelques équations indéterminées du cinquième degré*,[32] in which he proved the unsolvability in integers of some equations of the form $x^5 + y^5 = Az^5$. This paper, the first of his many number-theoretic researches, was approved by the Academy but was published only three years later in Crelle's Journal (vol. 3, 1828). The paper made a strong impression on Abel who visited Paris in 1826, where the two young mathematicians had the opportunity to meet and exchange ideas.

In 1826 Dirichlet returned to Germany and in 1827, owing to Alexander von Humboldt's recommendation, was appointed Privatdozent at the University of Breslau. On his way to Breslau (now Wrocław) Dirichlet visited Gauss in Göttingen. Gauss received him very cordially. They corresponded afterwards. The scientific atmosphere in Breslau was provincial, and the following year, again with the support of Humboldt, Dirichlet moved to Berlin to become first assistant and then full professor at the University of Berlin, a position he held for the following twenty-seven years. In 1831 Dirichlet became a member of the Berlin Academy of Sciences. Gauss died in 1855 and Dirichlet was asked to take his place. He did. The last years of his life were spent in Göttingen.

The main directions of Dirichlet's investigations were formed partly under the influence of J.B. Fourier, under whose guidance Dirichlet began his work in mathematics in Paris (Dirichlet wrote fundamental papers on the theory of Fourier series and in mathematical physics), and partly under the influence of works of his predecessors in number theory, primarily Gauss' *Disquisitiones arithmeticae*, which became his everyday reading. Dirichlet exerted a considerable influence on these branches of mathematics and even on the entire development of mathematics. This was due to his pedagogical talent as well as to the topicality of his work. Minkowski succinctly described Dirichlet's style of writing in these words: "He possessed the art of blending a minimum of sightless formulas with a maximum of insightful thoughts."[33] Minkowski was referring to Dirichlet's Principle in potential theory and calling his style the true Dirichlet principle.

---

32 Lejeune-Dirichlet, P.G., *Werke*, Bd. 1. Berlin 1889, pp. 1–20.

33 Minkowski, H., *Gesammelte Abhandlungen*, Bd. 2. Leipzig-Berlin 1911, pp. 460–461.

To this we should add that by the style of his thinking, by the level of rigor and exactitude, Dirichlet was probably the foremost representative of those new trends which were gradually establishing themselves in 19th century mathematics. Jacobi wrote in one of his letters to A. von Humboldt:

> Dirichlet alone — rather than I, Cauchy or Gauss — knows what is a completely rigorous mathematical proof; we have learned it first from him. When Gauss claims to have proved something then I know that it is very likely so. When Cauchy says it then one can bet either way. But when Dirichlet says it then it is beyond doubt.[34]

Following the example of Euler, Dirichlet successfully applied the methods of mathematical analysis to number theory. The most important investigations of Dirichlet in number theory are: the proof of the theorem on arithmetic progressions, which he found analytically and then carried over to progressions and quadratic forms with complex-valued coefficients; the determination of the class number of binary quadratic forms with given determinant; and the development of the theory of algebraic integers of higher degree discussed above.

Dirichlet introduced into number theory the new concepts of "character" and "Dirichlet's series", formulated for the first time in general form the concept of an asymptotic law, and gave a number of examples of proving asymptotic formulas.

Dirichlet's "Proof of a theorem on the arithmetic progression" (*Beweis eines Satzes über die arithmetische Progression*. Bericht. Verhandl. Preuss. Akad. Wiss., 1837)[35] contained only a brief sketch of the first exact proof of the fundamental theorem that bears his name: any arithmetic progression, whose first term and difference are relatively prime must include an infinite number of primes — the first rigorously proved theorem on the distribution of primes after the theorem on the infinitude of primes established by the ancients.[36] The proof is carried out for the case where the difference of the progression is an odd prime, and, as Dirichlet points out, is similar to Euler's reasoning in Section 229 of Chapter 15 in Volume I of "An introduction to the analysis of infinites" (1748). Soon after, he wrote a more detailed "Proof of a theorem that each infinite arithmetic progression, whose first term and difference are integers having no common divisor, must include an infinite number of primes", (*Beweis des Satzes, dass jede unbegrenzte arithmetische Progression, deren erstes Glied und Differenz ganze Zahlen ohne*

---

34 Biermann, K.R., Die Mathematik und ihre Dozenten an der Berliner Universität, 1810–1920. Berlin 1973, p. 31.

35 Lejeune-Dirichlet, P.G., *Werke*, Bd. 1. Berlin 1889, pp. 307–312.

36 We recall that Euler stated this proposition as a hypothesis in 1783, and Legendre tried to prove it in 1798 (HM, vol. 3, pp. 109–120).

gemeinschaftlichen Factor sind, unendlich viele Primzahlen enthält. Abhandl. Preuss. Akad. Wiss., 1837).[37]

As Euler before him, Dirichlet establishes a connection between a series and an infinite product. But instead of the series

$$\sum_{n=1}^{\infty} \frac{1}{n^s} = \zeta(s)$$

Dirichlet considers the series

$$\sum_{n=1}^{\infty} \frac{\chi(n)}{n^s} = L(s, \chi),$$

where $\chi(n)$ is a numerical function (the Dirichlet "character") which splits the set of prime numbers into disjoint residue classes. This series $L(s, \chi)$ is now known as the Dirichlet series. To illustrate Dirichlet's method, we prove his theorem for the special arithmetic progression with general term $4n + 1$. Our exposition is based on Ayoub.[38]

We define the character $\chi(n)$ as follows:

$$\chi(n) = \begin{cases} (-1)^{(n-1)/2} & \text{for odd } n, \\ 0 & \text{for even } n. \end{cases}$$

Then $\chi(n) = 1$, if $n \equiv 1 \pmod 4$, and $\chi(n) = -1$, if $n \equiv 3 \pmod 4$. It is easy to verify that $\chi(mn) = \chi(m)\chi(n)$, that is, $\chi(n)$ is completely multiplicative.

Consider the series $L(s, \chi)$ for real s:

$$L(s, \chi) = \sum_{n=1}^{\infty} \frac{\chi(n)}{n^s} = 1 - \frac{1}{3^s} + \frac{1}{5^s} - \frac{1}{7^s} + \cdots,$$

which is convergent for $s > 0$, absolutely convergent for $s > 1$ and divergent for $s < 0$. By the property of multiplicative functions[39] we have, for $s > 1$,

$$L(s, \chi) = \prod_{\substack{p \text{ are the} \\ \text{odd prime} \\ \text{numbers}}} \left(1 - \frac{\chi(p)}{p^s}\right)^{-1}$$

---

37 Lejeune-Dirichlet, P.G., *Werke*, Bd. 1, pp. 313–342.

38 Ayoub, R., *An introduction to the analytic theory of numbers.* Providence 1963, pp. 6–8.

39 If a set-theoretic function $f(n)$ is multiplicative, then $\sum_{n=1}^{\infty} f(n) = \prod_p (1 + f(p) + f(p^2) + \ldots + f(p^k) + \ldots) = prod_p \sum_{k=0}^{\infty} f(p^k)$. If $f(n)$ is completely multiplicative, then $\sum_{n=1}^{\infty} f(n) = \prod_p (1 - f(p))^{-1}$. In both cases the product is taken over all primes p. It is assumed that both sides of these equalities converge absolutely.

Taking logarithms on both sides of this equality, we obtain

$$\ln L(s, \chi) = \sum_{\substack{p \text{ are the} \\ \text{odd prime} \\ \text{numbers}}} \sum_{k=1}^{\infty} \frac{(\chi(p))^k}{kp^{ks}}$$

$$= \sum_{\substack{p \text{ are the} \\ \text{odd prime} \\ \text{numbers}}} \frac{\chi(p)}{p^s} + \sum_{\substack{p \text{ are the} \\ \text{odd prime} \\ \text{numbers}}} \sum_{k=2}^{\infty} \frac{(\chi(p))^k}{kp^{ks}}$$

$$= \sum_{p \equiv 1 (\mathrm{mod}\, 4)} \frac{1}{p^s} - \sum_{p \equiv 3 (\mathrm{mod}\, 4)} \frac{1}{p^s} + R_1(s),$$

where the second sum is denoted by $R_1(s)$. As in Euler's proof, we have for $\zeta(s)$:

$$|R_1(s)| \leq \frac{1}{2} \sum_{\substack{p \text{ are the} \\ \text{odd prime} \\ \text{numbers}}} \sum_{k=2}^{\infty} \frac{1}{p^{ks}} < \frac{1}{2} \frac{1}{1 - 2^{-s}} \zeta(2s).$$

Thus, $R_1(s)$ remains bounded as $s \to 1 + 0$. To separate the primes in the arithmetic progression $p \equiv 1(\mathrm{mod}\, 4)$, that is, $p = 4m + 1$, we add the sum over all primes p. Since

$$\ln \zeta(s) = \sum_p \frac{1}{p^s} + R(s)$$

$$= \frac{1}{2^s} + \sum_{\substack{p \text{ are the} \\ \text{odd prime} \\ \text{numbers}}} \frac{1}{p^s} + R(s)$$

$$= \frac{1}{2^s} + \sum_{p \equiv 1 (\mathrm{mod}\, 4)} \frac{1}{p^s} + \sum_{p \equiv 3 (\mathrm{mod}\, 4)} \frac{1}{p^s} + R(s),$$

where $R(s)$ remains bounded as $s \to 1 + 0$, we obtain

$$\ln L(s, \chi) + \ln \zeta(s) = \frac{1}{2^s} + 2 \sum_{p \equiv 1 (\mathrm{mod}\, 4)} \frac{1}{p^s} + R_1(s) + R(s),$$

where $\ln \zeta(s) \to \infty$ as $s \to 1 + 0$, and the terms $R_1(s)$ and $R(s)$ remain bounded as $s \to 1+0$. Since $L(s, \chi)$ is also convergent for $s \to 1+0$, $L(s, \chi) \to L(1, \chi)$, the theorem will have been proved if we can prove that $L(1, \chi) \neq 0$. In our case, this can be done in many ways. For example,

$$L(1, \chi) = 1 - \frac{1}{3} + \frac{1}{5} - \frac{1}{7} + \cdots = \pi/4 \neq 0.$$

There is no simple proof in the general case of an arithmetic progression $kn + a$, where $(k, a) = 1$. The proof of the assertion that $L(1, \chi) \neq 0$ was the main difficulty in Dirichlet's proof of this theorem. Dirichlet relied on the fact that for real characters $L(1, \chi)$ is a factor in the expression for the class number of binary forms belonging to a certain determinant, and the number of such classes cannot be equal to zero.

Also, a connection was established between the class number and other problems of number theory: for negative determinants — with quadratic residues and nonresidues, and for positive determinants — with solutions of Fermat's (Pell's) equation and with the theory of cyclotomy.

In the article "On a property of quadratic forms" (*Über eine Eigenschaft der quadratischen Formen*. Bericht. Preuss. Akad. Wiss., 1840)[40] the theorem on arithmetic progressions was extended to quadratic forms: each quadratic form whose three coefficients have no (nontrivial) common divisor, represents infinitely many prime numbers. Finally, in the paper "Investigations on the theory of complex numbers" (*Untersuchungen über die Theorie der complexen Zahlen*. Abhandl. Preuss. Akad. Wiss., 1841)[41] Dirichlet carried the theorem over to complex integers: if k, a are complex integers and $(k, a) = 1$, then the expression $kn + a$ contains infinitely many complex primes.

Asymptotic Laws of Number Theory

Another group of Dirichlet's researches is connected with establishing asymptotic laws for number-theoretic functions. The first paper in this group was "On the determination of asymptotic laws in number theory" (*Über die Bestimmung asymptotischer Gesetze in der Zahlentheorie*. Bericht. Verhandl. Preuss. Akad. Wiss., 1838).[42] Dirichlet pointed out that, for values of the independent variable increasing beyond all bounds, it is frequently possible to approximate a complicated function by a simpler one, much as it is possible to approximately represent a curve by its asymptote for points far from the origin. By analogy with geometry, Dirichlet proposed to call the simpler function an asymptotic law for the complicated function if the ratio between the two functions tends to 1 as the argument increases beyond all bounds.

As a very old example of an asymptotic law Dirichlet cites a formula for binomial coefficients of very large even degree, obtained by J. Stirling from the infinite-product formula given by J. Wallis for the number π. Later studies produced a number of results of this kind, especially important for probability theory. Dirichlet notes that similar asymptotic laws exist for number-theoretic

40 Lejeune-Dirichlet, P.G., *Werke*, Bd. 1. Berlin 1889, pp. 497–502.

41 Ibid., pp. 503–532.

42 Ibid., pp. 351–356.

functions. By way of an illustration he gives Legendre's asymptotic formula for the number of primes not exceeding a given bound x:

$$\pi(x) \approx \frac{x}{\ln x - 1.08366}, \tag{9}$$

and Gauss' formula in the theory of quadratic forms for the asymptotic representation of the mean value of the class number and the number of orders of such forms. Both Gauss and Legendre obtained their formulas by induction, without rigorous proof. Dirichlet set for himself the goal of developing methods which, when specialized, would provide the proofs for the results obtained by Gauss and Legendre. He applied such a method to the problem of the number of divisors of a given natural number n. Today this function is denoted by $\tau(n)$ but Dirichlet used the symbol b_n. He noticed that the function $\tau(n)$ varies irregularly, hence it is better to consider instead its mean value $\frac{1}{n}\sum_{k \leq n} \tau(k)$. Using the properties of the definite integral[43]

$$\Gamma(k) = \int_0^\infty e^{-x} x^{k-1} dx,$$

Dirichlet found an asymptotic law for $\frac{1}{n}D(n)$, where

$$D(k) = \sum_{k \leq n} \tau(k),$$

and from it the result

$$D(n) = \sum_{k=1}^{n} \tau(k) = \left(n + \frac{1}{2}\right) \ln n - n + 2Cn, \tag{10}$$

where C is the Euler constant. In a similar way Dirichlet established several other asymptotic laws for number-theoretic functions. Later, in "On the determination of mean values in the theory of numbers" (*Über die Bestimmung der mittleren Werthe in der Zahlentheorie*, 1849),[44] Dirichlet once again returned to the problem of the number of divisors, but this time he argued differently. Let $D(n) = \sum_{k=1}^{n} \tau(k)$. The number of multiples of an integer s, where $s \leq n$, is $[n/s]$ ($[x]$ denotes the integral part of the number x) and therefore

$$D(n) = \sum_{s=1}^{n} [n/s].$$

---

43 We note that, following Dirichlet, P.L. Chebyshev, B. Riemann and other mathematicians used the gamma function in problems of analytic number theory.

44 Lejeune-Dirichlet, P.G., *Werke*, Bd. 2. Berlin 1897, pp. 49-66.

178

Hence, since $n/s - [n/s] < 1$, we have, with error $O(n)$,

$$D(n) = n \sum_{s=1}^{n} \frac{1}{s}.$$

But

$$\sum_{s=1}^{n} \frac{1}{s} = \ln n + C + \frac{1}{2n} + \cdots,$$

where C is the Euler constant. Therefore $D(n) = n \cdot \ln n + O(n)$. For a more exact expression for $D(n)$ Dirichlet uses the following identity: if μ is an integer,

$$\mu^2 \geq n, \quad \mu(\mu+1) > n, \quad [n/\mu] = \nu, \quad \psi(s) = \sum_{k=1}^{s} \phi(k),$$

$$[n/p] = q,$$

then

$$\sum_{s=1}^{p} [n/s]\phi(s) = q\psi(p) - \nu\psi(\mu) + \sum_{s=1}^{\mu} [n/s]\phi(s) + \sum_{s=q+1}^{\nu} \psi([n/s]).$$

Hence for $p = n$, $q = 1$ we have

$$\sum_{s=1}^{n} [n/s]\phi(s) = -\nu\psi(\mu) + \sum_{s=1}^{\mu} [n/s]\phi(s) + \sum_{s=1}^{\nu} \psi([n/s]),$$

and for $\phi(s) = 1$, $\psi(s) = s$

$$D(n) = \sum_{s=1}^{n} [n/s] = -\mu\nu + \sum_{1}^{\mu} [n/s] + \sum_{1}^{\nu} [n/s].$$

The new formula for the sum of divisors $D(n)$ is thus:

$$D(n) = n \cdot \ln n + (2C - 1)n + O(\sqrt{n}), \tag{11}$$

where C is the Euler constant.

In later years the problem of making the remainder in the formula (11) more exact attracted the attention of several mathematicians. Voronoĭ a-

179

chieved the greatest success. In his paper "On a certain problem in the determination of asymptotic functions" (1903)[45] he proved that the formula (11) can be made more exact:

$$D(n) = n \ln n + (2C - 1)n + O(\sqrt[3]{n} \ln n).$$

Continuing the investigation of properties of infinite series and products similar to those of Euler in Volume I of his "An introduction to infinitesimal analysis", Dirichlet concludes that they are applicable to many problems in number theory. He writes:

> My method seems to me to be especially noteworthy because it establishes a connection between infinitesimal analysis and transcendental arithmetic.[46]

In the article "On the application of infinite series in the theory of numbers" (*Sur l'usage des séries infinies dans la théorie des nombres.* J. für Math., 1838, 18)[47] he generalized the Euler identity (8), considering instead of $\zeta(s)$ the series $\sum_{n=1}^{\infty} \left(\frac{n}{q}\right) \frac{1}{n^s}$, where $\left(\frac{n}{q}\right)$ is the Legendre symbol, and more general series of the form $\sum_{n-1}^{\infty} \frac{F(n)}{n^s}$, where $F(n)$ are various numerical functions.

Using these series ("Dirichlet series") he finds asymptotic laws for the mean values of some number-theoretic functions. Dirichlet wrote that he applied similar principles to the proof of the Legendre formula for the number of primes not exceeding a given bound. However, this proof has not been found in Dirichlet's published papers or in his *Nachlass*.

Dirichlet used his series in "Research on various applications of infinitesimal analysis to number theory" (*Recherches sur diverses applications de l'analyse infinitésimale à la théorie des nombres* (J. für Math., 1839, 19; 1840, 21).[48] Here he solves the problem of determining the number of classes of quadratic forms with given determinant D.

Moreover, Dirichlet discovered that his methods enable one to prove in a simple way many theorems stated by Gauss in the second half of Part V of *Disquisitiones arithmeticae*. Dirichlet said that it was because of the difficulty of this particular part of Gauss' book that Legendre refused to include all of Gauss' results in the second and third editions of *The theory of numbers* (1808, 1830).

---

45 Voronoĭ, G.F., *Collected Works*, vol. 2. Kiev 1952, pp. 5–50 (in Russian).

46 Lejeune-Dirichlet, P.G., *Werke*, Bd. 1. Berlin 1889, p. 360.

47 Ibid., pp. 357–374.

48 Ibid., pp. 411–496.

In "Investigations of quadratic forms with complex coefficients and variables" (*Recherches sur les formes quadratiques à coefficients et à indéterminées complexes*. J. für Math., 1842, 24)[49] Dirichlet extends the methods and results of the just-quoted 1839–1840 paper to quadratic forms with complex integral coefficients. Here he formulates the basic theorems of the theory of complex integers and the most essential theorems on quadratic forms, and gives a classification of forms. Using Dirichlet series and the unique factorization of complex integers, he obtains a new generalization of the Euler identity. He introduces an analogue of the Legendre symbol. Then he defines the class number of quadratic forms with complex coefficients for a given real integral determinant and for the same determinant multiplied by i.

These investigations led Dirichlet to his famous theorem on complex units. In a letter to Liouville, Dirichlet wrote that while searching for the number of forms with integral complex coefficients with given determinant, he arrived at the conclusion that this number depends on the division of a lemniscate, whereas in the case of real coefficients and positive determinants it is connected with the theory of cyclotomy (see *Sur la théorie des nombres*. C. r. Acad. sci. Paris, 1840, 10)).[50]

Dirichlet's famous paper "On the theory of complex units" (1846)[51] was based on these investigations. In addition to all these, Dirichlet's "Lectures on the theory of numbers", edited, supplemented and published by Richard Dedekind (*Vorlesungen über Zahlentheorie*. Braunschweig 1863; 2. Aufl. 1871; 3. Aufl. 1879; 4. Aufl. 1894), exerted a tremendous influence on the development of number theory and on the education of several generations of mathematicians. The analytic and algebraic methods developed by Dirichlet were further elaborated by Dedekind, Kummer, Kronecker, and Riemann, and mathematicians of the late 19th and early 20th centuries developed the theory of characters and the theory of Dirichlet series by considering Dirichlet series with complex exponents and complex coefficients. Following Dirichlet and Liouville, many mathematicians employed infinite series to obtain various number-theoretic identities (Bunyakovskiĭ, Cesàro, Bugaev and his students). Following the trail blazed by Euler, Legendre, and after him Jacobi and other mathematicians, used infinite series and products in the theory of elliptic functions to prove various theorems in additive number theory.

Chebyshev and Riemann continued Dirichlet's investigations in the field of analytic number theory.

---

49 Ibid., pp. 533–618.

50 Ibid., p. 619.

51 Ibid., pp. 639–644.

Chebyshev and the Theory of Distribution of Primes

The biography of Pafnuty Lvovich Chebyshev will be given in Chapter Four of this book. Here we discuss a few of his excellent and famous works in number theory. The main directions of Chebyshev's investigations in this area were the theory of distribution of primes, essentially advanced by him, the theory of series whose general term element is a function of prime numbers, problems in the theory of quadratic forms, Diophantine approximations and the generalization of the algorithm for continued fractions.

Chebyshev began these investigations when he was asked by V.Ya. Bunyakovskiĭ to help prepare for publication Euler's works on number theory. In collaboration with Bunyakovskiĭ, Chebyshev prepared a "Systematic index" of Euler's works on number theory with annotations on many articles. Chebyshev reconstructed several of Euler's manuscripts and notes and corrected errors. These errors in Euler's works can be explained by the fact that Euler's students wrote down what he dictated or carried out his instructions. But Euler was blind at the time and so could not personally check their work.

This edition (L. Euleri *Commentationes arithmeticae collectae* (vol. 1–2. Petropoli 1849)) strongly influenced the further development of number theory in the 19th century. Jacobi helped the publication by his counsel, and by gathering information on Euler's work at the Berlin Academy of Sciences. Immediately after the publication of these two volumes of Euler's works, Jacobi and Dirichlet discovered many interesting facts in the reissued materials. In the same year, 1849, Chebyshev defended his doctoral dissertation "Theory of congruences" at Petersburg University. It was published as a separate book in 1849 (St. Petersburg 1849).[52]

The knowledge of Euler's works and the study of the writings of Lagrange, Legendre, Gauss, Dirichlet, and other mathematicians served as a basis for Chebyshev's further investigations in the theory of numbers. Bunyakovskiĭ and P.N. Fuss, reviewers of "The theory of congruences", noted that Chebyshev's work was the first monograph in Russian on this important subject and that it was distinguished by its strict consistency, its simplicity of presentation, and the elegance of methods devised by the author. Later, this book was reprinted several times and translated into German (1888) and Italian (1895). The St. Petersburg Academy of Sciences awarded Chebyshev the Demidov Prize for this work.

Chebyshev's article "On the determination of the number of primes not exceeding a given number" was an appendix to "The Theory of Congruences". The French translation of this article was published in *Mémoires des savants étrangers* (vol. 6, 1851) of the St. Petersburg Academy of Sciences and in Liouville's Journal in 1852.[53]

---

52 Chebyshev, P.L., *The complete collected works*, vol. 1. Moscow-Leningrad 1944, pp. 10–172 (in Russian).

53 Ibid., pp. 173–190.

Problems in the application of mathematical analysis to number theory attracted Chebyshev's interest when he studied Euler's work in arithmetic and, in this connection, Dirichlet's articles "On the Application of Infinite Series in the Theory of Numbers", "Investigations of Various Applications of Analysis in the Theory of Numbers", and "On the Determination of Asymptotic Laws in the Theory of Numbers". In the latter article, Dirichlet, as noted earlier, gave Legendre's formula (9) for the number $\phi(x)$ of primes not exceeding x. Chebyshev checked this formula and found that it did not approximate the function $\phi(x)$ sufficiently closely (following Landau's suggestion of 1909, this function is now denoted by $\pi(x)$). Chebyshev studied extensively various properties of the function $\phi(x)$. The first theorem of the article stated: "if $\phi(x)$ is the number of primes less than x, n is an integer, and ρ is greater than 0, then the sum

$$\sum_{x=2}^{\infty} \left[\phi(x+1) - \phi(x) - \frac{1}{\log x} \right] \frac{\log^n x}{x^{1+\rho}}$$

yields a function which tends to a finite limit as ρ tends to 0."[54] The proof of this theorem is based on the use of the Euler zeta-function $\zeta(s)$ for real values of the argument (HM, vol. 3, p. 109), namely on the behavior of $\zeta(s)$ near its pole $s = 1$. From this Chebyshev deduced a second theorem, dealing with a property of the function $\phi(x)$ of fundamental importance in the theory of distribution of prime numbers, namely that it oscillates about the function $\operatorname{Li} x = \int_2^x \frac{dx}{\ln x}$ (HM, vol. 3, pp. 360–361), or, as Chebyshev put it, that in the range $x = 2$ to $x = \infty$ the function $\phi(x)$ satisfies infinitely many times both the inequality

$$\phi(x) > \int_2^x \frac{dx}{\ln x} - \frac{\alpha x}{\ln^n x},$$

and the inequality

$$\phi(x) < \int_2^x \frac{dx}{\ln x} + \frac{\alpha x}{\ln^n x},$$

for arbitrarily small positive α, and arbitrarily large n. In turn, this implied a third theorem, to the effect that as $x \to \infty$ the difference $x/\phi(x) - \ln x$ cannot tend to a limit other than -1. According to the Legendre formula, as $x \to \infty$ the difference $x/\phi(x) - \ln x$ tends to the limit -1.08366 rather than -1. In his proof Chebyshev used estimates of definite integrals, series, and differentiation, that is, the tools of mathematical analysis.

---

54 Ibid., p. 173. In Chebyshev's writings $\log x$ denotes the natural logarithm.

Chebyshev's results implied that if the limit as $x \to \infty$ of the ratio of $\phi(x)$ to $x/\ln x$ — or in modern notation: of $\pi(x)$ to $x/\ln x$ — exists, then it is 1.

Perfecting the notion of an asymptotic law, Chebyshev introduced the notion of a function representing another function "correctly up to an amount of order $x/\ln^n x$ inclusively". In the fifth theorem he showed that if the function $\phi(x)$ can be expressed algebraically, correctly up to the order $x/\ln^n x$, in terms of x, e^x, and $\ln x$, then this expression is $\operatorname{Li} x = \int_2^x \frac{dx}{\ln x}$. Chebyshev took a major step towards proving the asymptotic law of distribution of prime numbers, conjectured by other mathematicians before him.[55] Having derived a new asymptotic formula for $\phi(x)$, Chebyshev replaced all the formulas obtained by Legendre by his own. Instead of Legendre's formula

$$\sum_{2 \le p \le x} \frac{1}{p} \approx \ln(\ln x - 0.08366) \quad (p \text{ are prime numbers})$$

Chebyshev obtains

$$\sum_{2 \le p \le x} \frac{1}{p} \approx C + \ln \ln x; \tag{12}$$

instead of Legendre's formula

$$\prod_{2 \le p \le x} \left(1 - \frac{1}{p}\right) \approx \frac{C_0}{\ln x - 0.08366}$$

he has

$$\prod_{2 \le p \le x} \left(1 - \frac{1}{p}\right) \approx \frac{C_0}{\ln x}. \tag{13}$$

Chebyshev's paper brought him wide recognition which increased even more upon the publication of the French version *Sur les nombres premiers* (1852), in Liouville's Journal.[56]

This paper is remarkable for the simplicity of methods developed by Chebyshev and also for the significance of his results. He considers the functions

$$\theta(x) = \sum_{p \le x} \ln p, \qquad \psi(x) = \sum_{p^\alpha \le x} \ln p,$$

---

55 The asymptotic law $\pi(x) \approx \int_2^x \frac{dx}{\ln x}$ was already familiar to Gauss; see his letter to Enke of December 24, 1849, published in *Werke*, Bd. 2. Göttingen 1863, p. 444. Gauss didn't leave any published results, nor did he have a proof of this law.

56 Chebyshev, P.L., *The complete collected works*, T. 1. Moscow-Leningrad 1944, pp. 191–207 (in Russian).

where p are the prime numbers from 2 to x, $\alpha \geq 1$. Proceeding from the identity for $\theta(x)$, derived in an original way, he arrives at "Chebyshev's fundamental identity":

$$\sum_{n \leq x} \psi\left(\frac{x}{n}\right) = T(x), \tag{14}$$

where

$$T(x) = \ln 1 \cdot 2 \cdot 3 \cdot \ \cdots \ \cdot [x].$$

From this he obtains inequalities for $\psi(x)$:

$$\psi(x) \geq T(x) + T\left(\frac{x}{30}\right) - T\left(\frac{x}{2}\right) - T\left(\frac{x}{3}\right) - T\left(\frac{x}{5}\right),$$
$$\psi(x) - \psi\left(\frac{x}{6}\right) \leq T(x) + T\left(\frac{x}{30}\right) - T\left(\frac{x}{2}\right) - T\left(\frac{x}{3}\right) - T\left(\frac{x}{5}\right).$$

Replacing $T(x)$ by Stirling's approximate formula, Chebyshev finds that

$$\psi(x) > Ax - \frac{5}{2}\ln x - 1, \quad \psi(x) - \psi\left(\frac{x}{6}\right) < Ax + \frac{5}{2}\ln x,$$

where $A = \ln \frac{2^{1/2} \cdot 3^{1/3} \cdot 5^{1/5}}{30^{1/30}} = 0.92129202\ldots$ Then he establishes bounds for the function $\theta(x)$:

$$Ax - \frac{12}{5}A\sqrt{x} - \frac{5}{8\ln 6}\ln^2 x - \frac{15}{4}\ln x - 3$$
$$< \theta(x) \tag{15}$$
$$< \frac{6}{5}Ax - A\sqrt{x} + \frac{5}{4\ln 6}\ln^2 x + \frac{5}{2}\ln x + 2 \quad \text{for } x > 160.$$

In particular, these estimates enabled Chebyshev to prove "Bertrand's postulate", which he stated as follows: for any number $n > 3$ there always exists a prime number larger than n and smaller than $2n - 2$.[57]

Other results in the same article deal with the convergence of series whose terms depend on primes. The first discoveries in this area belong to Euler, whom Chebyshev mentions at the beginning of his article. Euler proved that the series $\sum_{p \geq 2} \frac{1}{p^k}$ and $\sum_{n=2}^{\infty} \frac{1}{n^k}$ converge for $k > 1$ and diverge for $k \leq 1$. In general, however, the convergence of the series $\sum_{n=2}^{\infty} u_n$ is not necessary

---

57 Bertrand, J., *Mémoire sur le nombre de valeurs que peut prendre une fonction quand on y permute les lettres qu'elle renferme.* J. Ec. Polyt. Paris, 1845, 18, 123–140. Bertrand used this postulate to prove a certain proposition in the theory of groups of substitutions; he checked it for all numbers up to six million.

for convergence of the series $\sum_{p\geq 2} u_p$. For example, as Chebyshev shows, the series $\sum_{p\geq 2} \frac{1}{p\ln p}$ is convergent (its sum is equal to 1.63 to within two significant figures) whereas the series $\sum_{n=2}^{\infty} \frac{1}{n\ln n}$ is divergent.

Chebyshev established the following criterion: if for sufficiently large x the function $F(x) > 0$ and $F(x)/\ln x$ decreases, then a necessary and sufficient condition for the convergence of the series $\sum_{p\geq 2} F(p)$ is the convergence of the series $\sum_{n=2}^{\infty} \frac{F(n)}{\ln n}$. The proof is based on estimating the sum $\sum_p F(p)$ from above and from below; here the prime numbers p are between bounds ℓ and L.

In the special case $F(x) = 1$, $\ell = 2$, $L = x$, Chebyshev found bounds for the function $\pi(x)$ which expresses the number of primes not exceeding x. For this purpose he wrote $\sum F(p)$ as

$$\sum_{\ell \leq p \leq L} F(p) = \sum_{k=\ell}^{L} \frac{\theta(k) - \theta(k-1)}{\ln k} F(k),$$

used the estimates for $\theta(k)$, and put $F(x) = 1$, $\ell = 2$, $L = x$. His estimates for $\pi(x)$ are

$$A\frac{x}{\ln x} < \pi(x) < \frac{6}{5}A\frac{x}{\ln x}$$

or

$$0.92129 < \frac{\pi(x)}{x/\ln x} < 1.10555.$$

Chebyshev's discoveries attracted the attention of the great Cauchy. In this connection Hermite wrote to Chebyshev:

> I would like you to know that Cauchy, whom I saw last Thursday, was extremely interested in your discoveries in the theory [of prime] numbers... and was particularly impressed by the proof of Bertrand's postulate and the results pertaining to the convergence of the series $u_1, u_2, u_3, \ldots, u_p, \ldots$ (p integral).[58]

Another French mathematician, J.A. Serret, included the results of Chebyshev's second paper in the second volume of his well-known *Cours d'algèbre supérieure* (T. 2, Paris 1866, pp. 203–216).

In addition to those cited above, there are several other papers of Chebyshev dealing with problems of analytic number theory.[59] In these papers Chebyshev considers formulas for the inversion of series whose terms depend

---

58 Chebyshev, P.L., *The complete collected works*, vol. 5. Moscow-Leningrad 1951, p. 425 (in Russian).

59 They were all included in the first volume of his *Complete collected works*.

on prime numbers contained in the arithmetic progressions $4n+1$ and $4n+3$ and not exceeding a certain large number x. He tries to establish the connection between the identity

$$\sum_{n\leq x}\psi\left(\frac{x}{n}\right)=\sum_{n\leq x}\ln n$$

and Euler's identity for the zeta-function

$$\sum_{n=1}^{\infty}\frac{1}{n^s}=\prod_{p}\left(1-\frac{1}{p^s}\right)^{-1}.$$

At the same time as Chebyshev, A. de Polygnac (1826–1863), a French officer and mathematician, studied similar problems involving prime numbers. After he had acquainted himself with Chebyshev's methods he admitted their superiority and later used them in his research.

Chebyshev's articles led to many investigations by other mathematicians. They improved estimates for the Chebyshev functions, estimated other similar functions, generalized theorems, and provided new proofs of his results. One of the first mathematicians who turned to Chebyshev's results was F.K.J. Mertens (1840–1927), a professor of mathematics at Berlin University, and from 1869, at Cracow University. In the article "A note on analytic number theory" (*Ein Beitrag zur analytischen Zahlentheorie*. J. für Math., 1874, 78, 46–62), he employed Chebyshev's results and replaced Chebyshev's asymptotic expressions for $\sum_{p\leq x}\frac{1}{p}$ and $\prod_{p\leq x}\left(1-\frac{1}{p}\right)$ (p are prime numbers) by more exact expressions:

$$\sum_{p\leq x}\frac{1}{p}=\ln\ln x+B+O\left(\frac{1}{\ln n}\right);\quad \prod_{p\leq x}\left(1-\frac{1}{p}\right)=\frac{e^{-C}}{\ln x}\left\{1+O\left(\frac{1}{\ln x}\right)\right\};$$

here C is the Euler constant, and B is a constant. He calculated the values of the constants with high accuracy. In the course of his proof, Mertens also derived an asymptotic formula for $\sum_{p\leq x}\frac{\ln p}{p}$:

$$\sum_{p\leq x}\frac{\ln p}{p}=\ln x+O(1).$$

In this article and in the article "On some asymptotic laws of the theory of numbers" (*Über einige asymptotische Gesetze der Zahlentheorie*. J. für Math., 1874, 77) Mertens obtained additional results, of which we mention the asymptotic formulas for the sum of the reciprocals of the primes in arithmetic progressions of the form $4n+1$, $4n+3$, and $k\ell+m$; for $\sum_{m\leq x}\phi(m)$, where $\phi(m)$ is the Euler function, etc.

J. Sylvester improved somewhat the bounds for the ratio of $\pi(x)$ to $x/\ln x$ and proved that this ratio lies between 0.95695 and 1.04423. He also made the Bertrand postulate more precise by proving that for all sufficiently large n there exists at least one prime number between n and $1.092n$ (Amer. J. Math., 1881, 4; Messenger of Math., ser. 2, 1891, 21).

The attempt to extend Chebyshev's theorems to complex prime numbers is due to Henri Poincaré (see *Sur la distribution des nombres premiers*. C. r. Acad. sci., Paris, 1891, 113; *Extension aux nombres premiers complexes des théorèmes de M. Tchébicheff*. J. math. pures et appl., sér. 4, 1892, 8, 25–69). Instead of Chebyshev's functions $\theta(x)$ and $T(x)$ he examined the function $T^*(x)$, equal to the sum of the logarithms of the norms of all ideals with norm $\leq x$, and the function $\theta^*(x)$ equal to the sum of logarithms of the norms of all prime ideals with norm $\leq x$, but failed to obtain analogues of Chebyshev's inequalities (15) for these functions. In his articles Poincaré established some other results concerning the sum of logarithms of prime numbers of the form $4n+1$ and the number of primes of that form for $4n+1\leq x$.

In an article "On prime numbers of the forms $4n+1$ and $4n-1$" (Sb. In-ta inzh. puteĭ soobshcheniya, vyp. 50, St. Petersburg, 1899) V.I. Stanevich applied the method devised by P.L. Chebyshev in his "On prime numbers" in order to determine the bounds for the number of primes of the forms $4n+1$ and $4n-1$ ($\leq x$) for sufficiently large x and obtained results similar to those in Chebyshev's article. In particular, he obtained analogues of the Bertrand postulate: between a and $2a$ there is at least one prime number of the form $4n+1$ (if $a>15/2$) and at least one prime number of the form $4n-1$ (if $a>9/2$).

A survey of works relevant to the determination of the number of primes not exceeding a given number is given in an extensive article by G. Torelli, "On the totality of prime numbers smaller than a given bound" (*Sulla totalità dei numeri primi fino a un limite assegnato*. Atti Accad. sci. fis. e mat. Napoli, ser. 2, 1902, 11) and in the doctoral dissertation of I.I. Ivanov, "On some problems connected with the calculation of prime numbers" (St. Petersburg, 1901). A very detailed presentation of results in this field with proofs and extensive bibliography is found in E. Landau's "A manual on the theory of distribution of prime numbers" (*Handbuch der Lehre von der Verteilung der Primzahlen*. Bd. 1, 2. Berlin 1909). For a long time after the publication of Chebyshev's papers in 1849 and 1852 there was no progress along the lines suggested by Chebyshev, and the outlook for further development in this direction appeared dim. This view was refuted only in 1949 by A. Selberg and P. Erdös who revived "elementary" methods in analytic number theory — elementary in the sense that they did not involve the use of functions of a complex variable.

Long before this happened, an analytic proof of the law of distribution of prime numbers was found. This was made possible by new methods initiated by Riemann.

The Ideas of Bernhard Riemann

We will mention Riemann's name many times in the sequel, and his biography will be given in the section dealing the history of the theory of analytic functions. Here we need only emphasize the fact that it was Riemann's work in this area that opened new horizons in analytic number theory in general and in the theory of the distribution of prime numbers in particular.

Riemann was a pupil and follower primarily of Gauss while in Göttingen, and of Dirichlet while in Berlin (he was a student of the Universities at Göttingen and Berlin); for a few years Riemann worked in Göttingen at the same time as Dirichlet, for whom, as Klein wrote, he felt a deep inner sympathy that derived from the similarity of their modes of thought.[60]

Both Gauss and Dirichlet had a great influence on Riemann; we do not, however, know what specific circumstances made him write the article "On the number of primes not exceeding a given bound" (*Über die Anzahl der Primzahlen unter einer gegebenen Grösse*).[61] Riemann submitted the article in 1859 to the Berlin Academy of Sciences as a token of gratitude for his election as a Corresponding Member, and it was published in the same year in the "Monatsberichte".

The starting point of Riemann's work, which belongs equally to the theory of functions[62] — one of the main areas of Riemann's work — and to number theory — the study of which was only a short episode in Riemann's life — was Euler's identity

$$\sum_{n=1}^{\infty} \frac{1}{n^s} = \prod_{p} \left(1 - \frac{1}{p^s}\right)^{-1}, \tag{16}$$

where p ranges over the primes and n over the positive integers. Riemann regarded the variable s as a complex variable, $s = \sigma + it$. The function defined by (16), which Riemann denoted by $\zeta(s)$, exists in the case of convergence of both sides of the equality, that is, for $\sigma > 1$. But Riemann points out that it is possible to give a representation of the function $\zeta(s)$ which is meaningful for all values of s. Riemann establishes several deep properties of the zeta-function, although he does not always give comprehensive proofs. He represents $\zeta(s)$ as a contour integral over an infinite path enveloping the sin-

---

60 Klein, F., *Vorlesungen* ... , Bd. 1. Berlin 1926, p. 250.

61 Riemann, G.F.B., *Gesammelte mathematische Werke*. Dover, New York, 1953.

62 Together with Cauchy and Weierstrass, Riemann is the most eminent representative of the theory of analytic functions throughout its periodical development.

gular points of the integrand. He derives in two ways a functional equation for the zeta-function[63]

$$\zeta(1-s) = 2(2\pi)^{-s} \cos \frac{1}{2}\pi s \Gamma(s)\zeta(s), \tag{17}$$

establishes that $\zeta(s)$ has a simple pole at $s = 1$, "trivial" zeros at $s = -2, -4, -6, \ldots$, and, furthermore, zeros in the strip $0 \leq \sigma \leq 1$, located symmetrically with respect to the straight line $\sigma = 1/2$. Riemann conjectured that all nontrivial zeros of $\zeta(s)$ lie on the line $\sigma = 1/2$. This is the famous "Riemann hypothesis", which still has not been proved or refuted.

The starting point for the analytic proof of the law of distribution of primes used by Riemann is the formula (16) for complex $s = \sigma + it$, $\sigma > 1$. Following Euler, he takes the logarithm of this identity,

$$\ln \zeta(s) = -\sum_p \ln \left(1 - \frac{1}{p^s}\right) = \sum_p \sum_{k=1}^{\infty} \frac{1}{k p^{ks}},$$

and chooses the branch of the logarithm which is real for real s. To write down the right-hand side as a Dirichlet series one must introduce the function $\Lambda_1(m)$:[64]

$$\Lambda_1(m) = \begin{cases} 1/k, & \text{if } m = p^k, \ p \text{ prime}, \\ 0 & \text{otherwise}. \end{cases}$$

Then

$$\ln \zeta(s) = \sum_{m=1}^{\infty} \frac{\Lambda_1(m)}{m^s}.$$

Put

$$\sum_{m \leq x} \Lambda_1(m) = f(x).$$

Using summation in the Abel sense, we obtain

$$\sum_{m=1}^{k} \frac{\Lambda_1(m)}{m^s} = \frac{f(k)}{k^s} + s \int_1^k f(x)x^{-s-1}dx.$$

For $k \to \infty$ this implies

$$\frac{\ln \zeta(s)}{s} = \int_1^{\infty} f(x)x^{-s-1}dx.$$

---

63 Except for notation, eq. (17) was already known to Euler (HM, Vol. 3, p. 338).

64 Ayoub, R., *An introduction to the analytic theory of numbers*. Providence 1963.

Riemann's next step is to invert this relation (which for $s = \sigma + it$ can be regarded as a Fourier transform). As a result we obtain for $a > 1$

$$f(x) = \frac{1}{2\pi i} \int_{a-i\infty}^{a+i\infty} \ln \zeta(s) \frac{x^s}{s} ds. \tag{18}$$

Riemann points out that

$$f(x) = \sum_n \frac{1}{n} \pi(x^{1/n}),$$

where $\pi(x)$ denotes the number of primes $\leq x$, $x > 1$. Using the Möbius inversion formula, we can, in turn, express the function $\pi(x)$ in terms of $f(x)$:

$$\pi(x) = \sum_{n=1}^{\infty} \frac{\mu(n)}{n} f(x^{1/n}), \tag{19}$$

where $\mu(n)$ is the Möbius function.

Riemann replaces $\ln \zeta(s)$ in (18) by

$$\ln \zeta(s) = \ln \xi(0) + \sum_{\rho} \ln \left(1 - \frac{s}{\rho} \right) - \ln \prod \left(\frac{s}{2} \right) + \frac{s}{2} \ln \pi - \ln(s-1), \tag{20}$$

where ρ ranges over the nontrivial zeros of $\zeta(s)$ and the function $\xi(t)$ is related to $\zeta(s)$ by

$$\xi(t) = \prod (s/2) \pi^{-s/2} (s-1)\zeta(s), \qquad \prod(s) = \Gamma(s+1), \qquad s = \frac{1}{2} + it.$$

Also,

$$\xi(t) = \xi(0) \prod_{\alpha} \left(1 - \frac{t^2}{\alpha^2} \right),$$

where α ranges over the positive zeros of $\xi(t)$, ρ and α are connected by the relation

$$\alpha = (2\rho - 1)/2i,$$

and $|\operatorname{Im} t| < 1/2$.

Instead of (18) Riemann obtains an expression for $f(x)$ consisting of several terms. In order to find the final expression for $f(x)$, he needs to estimate each of these terms. But the main term of this representation of $f(x)$ is obtained from the term $-\ln(s-1)$ in (20). This is

$$\frac{1}{2\pi i}\frac{1}{\ln x}\int_{a-i\infty}^{a+i\infty}\frac{d}{ds}\left[\frac{\ln(s-1)}{s}\right]x^s ds. \tag{21}$$

Riemann shows that for $x > 1$ this is a logarithmic integral and using (19) he obtains $\pi(x) \approx \mathrm{Li}\,x$. Moreover, he shows that the order of the difference $\mathrm{Li}\,x - \pi(x)$ depends on the location of the nontrivial zeros of $\zeta(s)$. If Riemann's hypothesis that all nontrivial zeros of $\zeta(s)$ lie on the line $\sigma = 1/2$ holds, then $|\pi(x) - \mathrm{Li}\,x| < c\sqrt{x}\ln x$, where c is a constant.

Riemann also indicated an approximation for the number of roots $N(T)$ of the equation $\zeta(s) = 0$ in the rectangle $0 \le t \le T$, $0 < \sigma < 1$:

$$N(T) = \frac{T}{2\pi}\ln\frac{T}{2\pi} - \frac{T}{2\pi} + O(\ln T).$$

Detailed calculations made by Riemann concerning the behavior of $\zeta(s)$ were discovered in the University Library at Göttingen a few years after his death.[65]

Proof of the Asymptotic Law of Distribution of Prime Numbers

On the initiative of C. Hermite mathematicians began studying Riemann's methods,[66] hoping to fill the gaps in his proofs.

In his letter of March 28, 1889, Hermite asked the Dutch mathematician T.J. Stieltjes (1856–1894), who was professor at the University of Toulouse since 1886, to communicate to him the content of Riemann's article on number theory (it was in German, and Hermite could not read German). In 1890 the Paris Academy of Sciences announced the program of awards for the next several years. Among them was "A State Grand Prize in the Mathematical Sciences" for 1892 on the topic suggested by Hermite:[67] "The determination of the number of primes not exceeding a given bound". The essence of the topic and the solution methods were explained:

---

65 Siegel, C., *Über Riemanns Nachlass zur analytischen Zahlentheorie.* In: *Quellen und Studien zur Geschichte der Mathematik, Astronomie und Physik,* Band 2, Ht 1. Berlin 1932, pp. 45–80.

66 For details on Riemann's work and the evolution of his ideas see Edwards, H.M., *Riemann's zeta function.* New York 1974; Titchmarsh, E.C., *The theory of the Riemann zeta function.* Oxford 1951.

67 See Hermite's letter to Stieltjes dated 17. Jan. 1891. In: *Correspondance d'Hermite et de Stieltjes,* T. 1. Paris 1905.

A new approach to the solution of this important problem was discovered by Riemann in a famous memoir which attracted a great deal of attention. But the work of the great geometer contains in some essential points results which he only sketched and whose proofs would be of great interest. The Academy sets the problem of filling these gaps through a deeper study of the function denoted in Riemann's memoir by $\zeta(s)$.[68]

Hermite invited Stieltjes to take part in this contest but Stieltjes' illness and other personal circumstances prevented him from doing so. On January 17, 1893, Hermite informed Stieltjes that the Grand Prize of the Academy was awarded to J. Hadamard and, also, wrote him about E. Cahen's paper on the subject (C. r. Acad. sci. Paris, 1893, 116). The results of Hadamard's paper — "The study of the properties of entire functions, and, in particular, of a function considered by Riemann" (*Etude sur les propriétés des fonctions entières, en particulier d'une fonction considérée par Riemann.* J. math. pures et appl. (4), 1893, 9, 171–215) — were based on a deep study of the properties of integers. These results and the results in Cahen's papers were soon used by Hadamard and also by Vallée-Poussin to prove the asymptotic law of the distribution of prime numbers.[69]

The main researches of the French mathematician Jacques Hadamard (1865–1963) and the Belgian mathematician Charles Jean de la Vallée-Poussin (1866–1962) were not concerned with number theory, and we shall encounter their names many times in the sequel in other contexts. Here we will dwell only on their contribution to the theory of the distribution of primes. They proved independently and simultaneously the asymptotic law that Riemann came close to proving by means of the theory of analytic functions. Vallée-Poussin did this in his *Recherches analytiques sur la théorie des nombres* ((I pt). Ann. Soc. sci. Bruxelles, 1896, 20, N 2, 183–256) and Hadamard in *Sur la distribution des zéros de la fonction $\zeta(s)$ et ses conséquences arithmétiques* (Bull. Soc. math. France, 1896, 24, 199–220). The proof given by Hadamard was a little simpler, but in a second article "On the Riemann function and the number of primes not exceeding a given limit" (*Sur la fonction de Riemann et le nombre des nombres premiers inférieurs à une limite donnée.* Mém. couronnées Acad. Belgique, 1900, 59) Vallée-Poussin investigated in detail the problem of the accuracy of the approximation to the function $\pi(x)$.

---

68 See: C. R. Acad. Sci. Paris, 1890, 111, pt 2, 1090–1092.

69 A detailed survey of results on analytic number theory, and, in particular, of results pertaining to an analytic proof of the law of distribution of prime numbers, is found in the fundamental work: Landau, E., *Handbuch der Lehre von der Verteilung der Primzahlen*, Bd. 1, 2. Leipzig 1909.

Vallée-Poussin's formula is:

$$\pi(x) = \int\limits_{2}^{x} \frac{dx}{\ln x} + O(xe^{-\alpha\sqrt{\ln x}}), \quad \alpha > 0.$$

In the 1930s, the estimates of the remainder in this formula were considerably sharpened by the use of methods developed by I.M. Vinogradov.

Later the analytical proof of the law was considerably simplified and no longer involved the theory of entire functions. The general scheme of the proof is the following: the function $\zeta(s)$ is extended by analytic continuation to the left of the line $\sigma = 1/2$. Then, $\zeta(s)$ is estimated to the left of $\sigma = 1$. The latter operation requires the knowledge of the behavior of the zeros of $\zeta(s)$, since the zeros of $\zeta(s)$ are singular points of the integrand $\ln \zeta(s)$.

It is possible to show that the magnitude of the remainder in the asymptotic law for $\pi(x)$ depends directly on the greatest lower bound of the values of σ $(0 \le \sigma \le 1)$ for which $\zeta(\sigma+it) \ne 0$. The Riemann hypothesis states that all complex zeros of $\zeta(s)$ have the real part $\sigma = 1/2$. The bounds for $|\zeta(s)|$ to the left of $\sigma = 1$ were the subject of many articles up to very recently. Making the bounds for $|\zeta(s)|$ more exact enables us to make the remainders more exact in other asymptotic formulas related to the law of distribution of primes.

The proof of the fact that $\zeta(1 + it) \ne 0$ is fundamental for all analytic proofs of the law of prime numbers. Much later, in 1927–1932, Norbert Wiener (1894–1964), famous for his work in cybernetics, showed that this is the only essential necessary assertion concerning $\zeta(s)$ required to prove the law of distribution of prime numbers.

In the 20th century other analytic proofs of the law of distribution of prime numbers and similar assertions were obtained. As mentioned above, in the middle of the 20th century it became possible to prove the asymptotic law of distribution "in an elementary way", that is without invoking the theory of functions of a complex variable. This was done by A. Selberg and P. Erdös in 1949.[70]

Some Applications of Analytic Number Theory

Methods of analysis are employed in other problems of the theory of numbers as well, one important application being to problems of partitioning numbers (*partitio numerorum*).

---

70 Selberg, A., *An elementary proof of the prime number theorem.* Ann. Math. (2), 1949, 50, 305–313; Erdös, P., *On a new method in elementary number theory.* Proc. Nat. Acad. USA, 1949, 35, 374–384.

Euler obtained interesting theorems of this kind by expanding infinite products in power series. His theory of pentagonal numbers (see HM, vol. 3, p. 108) is especially well known.

Euler's investigations were continued by Legendre who obtained in his works (*Essai sur la théorie des nombres*, Paris, 1798, 2d. edition, 1808; *Théorie des nombres*, Paris, t. 1, 2, 1830) additional results on partition of numbers. The technique used by Legendre was that of expanding infinite products in power series and equating the coefficients of terms of the same degree.

Jacobi used series and products of elliptic functions in order to establish number-theoretic identities, including identities pertaining to the partition of numbers into components of a special kind. In Russia similar problems were tackled by V.Ya. Bunyakovskiĭ, N.V. Bugaev and his students P.S. Nazimov and others. In the 20th century analytic methods in the theory of partitions were applied by Ya.V. Uspenskiĭ. Asymptotic formulas for the number of partitions were found by G.H. Hardy and S. Ramanujan (1918), and by Ya.V. Uspenskiĭ (1920).

We note that problems of this kind were investigated also by mathematicians who worked in the field of combinatorial analysis. In particular, many results belong to J. Sylvester, A. Cayley, and P.A. MacMahon.

In the 20th century Ramanujan, Hardy, Littlewood, Uspenskiĭ, and I.M. Vinogradov successfully applied analytic methods in additive number theory.

The modern method of trigonometric sums originated in the 19th century and was developed by Gauss, Dirichlet, Landau, and later by J.G. van der Corput, I.M. Vinogradov and his students, A.Z. Val'fish and others.

Analytic methods penetrated also the geometry of numbers. The classical problems of number theory are problems of finding the number of integral points in various regions in the plane as well as in 3-space. The problem of the number of integral points in a circle $x^2 + y^2 \leq r^2 = R$ consists in estimating, as closely as possible, the difference

$$\Delta(R) = A(R) - S(R) \quad \text{or} \quad \Delta(R) = A(R) - \pi R, \tag{22}$$

where $A(R)$ is the number of integral points inside the circle, and πR is the area of the circle. The determination of the best remainder in (22) is referred to as the problem of the number of integral points in a circle. This problem is closely connected with the problem of divisors. After G.F. Voronoĭ established a new remainder in the formula for the sum of the number of divisors, his student at the University of Warsaw, W. Sierpiński, improved upon it. (*O pewnem zagadnieniu z rachunku funkcyj asymptotycznych.* Prace mat.-fiz, Warszawa 1906, 17, 77–118). Later, this problem had many generalizations.

Elementary methods play a major role in the theory of numbers. Having proved a result by a method outside the province of number theory, mathematicians usually try to prove it in an "elementary" way by using only methods of number theory. Such are the sieve methods, the method of numerical identities, and the methods of congruence theory.

We will not dwell on the sieve methods and their improvement in applications, which were particularly successful at the beginning of the 20th century. We will only mention that already in the 19th century various sieve methods were used by Legendre, V.Ya. Bunyakovskiĭ, and P.S. Poretskiĭ, who made a comparative study of various sieve methods,[71] and in the 20th century by J. Merlin, V. Brun, A. Selberg, Yu.V. Linnik, A.A. Bukhshtab, and V.A. Tartakovskiĭ. Sieve methods were usually employed by compilers of tables of primes.

We will consider in some detail several aspects of the method of numerical identities. In addition to Gauss, Dirichlet, Jacobi, Liouville, Kronecker and other foreign scientists, Russian mathematicians actively contributed to this area of mathematics.

Arithmetic Functions and Identities. The Works of N.V. Bugaev

Arithmetic functions play an important role in number theory. An arithmetic function that occurs especially frequently is the Möbius function, defined as follows:

$\mu(n) = 0$, if n is divisible by the square of a number other than 1,

$\mu(n) = 1$, if $n = 1$,

$\mu(n) = (-1)^k$ for n equal to the product of k different prime numbers.

This function appears in inversion formulas which play an important role in many number-theoretic identities:

$$f(n) = \sum_{d|n} f_1(d), \quad f_1(n) = \sum_{d|n} \mu\left(\frac{n}{d}\right) f(d);$$

each of these formulas is a consequence of the other. Euler had already known the function $\mu(n)$.[72] The systematic investigation of this function and the associated inversion formulas began with the work of A.F. Möbius (1832),[73] better known for his research in geometry. In 1851 P.L. Chebyshev published three pairs of inversion formulas in his "Note on a series".[74]

---

71 Poretskiĭ, P.S., *On the study of primes*. Soobshch. i protokoly sektsii fiz.-matem. nauk. Kazan', 1888, 6, vyp. 1–2, 1–142 (in Russian).

72 Euler, L., Goldbach, Ch., *Briefwechsel 1729–1764*. Herausgegeben und eingeleitet von Juškevič, A.P. und Winter, E. Berlin 1965 pp. 71–73.

73 Möbius, A.F., *Gesammelte Werke*, Bd. 4, pp. 589–612.

74 Chebyshev, P.L., *Complete collected works*, vol. 1. Moscow-Leningrad 1944, pp. 229–236 (in Russian).

N.V. Bugaev

Dirichlet considered a series of number-theoretic identities in the 1849 paper "On the determination of the mean values in the theory of numbers" (see p. 156). Later many mathematicians used transformations of the identities due to Dirichlet. Liouville established a large number of identities in his papers, *Sur quelques fonctions numériques* and *Sur quelques formules générales qui peuvent être utiles dans la théorie des nombres*, published in Liouville's Journal from 1857 to 1860.

The most systematic and elementary construction of arithmetic identities on the basis of the concept of arithmetic derivative and inversion formulas were given by N.V. Bugaev in his theory of arithmetic derivatives. The subsequent theory of multiplicative arithmetic identities is, in the main, an elaboration of Bugaev's theory.

Nikolai Vasil'evich Bugaev (1837–1903) graduated from Moscow University in 1859, where he studied under the guidance of N.D. Brashman. After he defended his Master's thesis on the theory of series (1863), Bugaev was sent abroad to prepare himself for a professorship. In Berlin he attended lectures given by Kummer, Kronecker, and Weierstrass, and in Paris lectures of Liouville, Lamé, and Duhamel. After defending his doctoral dissertation

197

(1866), Bugaev became professor at Moscow University. He was one of the founders of the Moscow Mathematical Society and from 1891 was its president. He enrolled in the society accomplished mathematicians as well as students. Many well-known mathematicians were Bugaev's students, among them N.Ya. Sonin, D.F. Egorov, L.K. Lakhtin, K.A. Andreev, P.S. Nazimov, and others. Bugaev worked in number theory, in differential equations, in approximate computations, etc.

Bugaev's central concern was to create in number theory methods as general as those of analysis. This objective was served by his work in the theory of arithmetic derivatives and integrals over divisors, on identities involving the function $E(x)$, his investigations on the application of elliptic functions in number theory, and on identities related to problems of number partitions.

The strongest influence on Bugaev's investigations was exerted by Dirichlet's investigations in analytic number theory (see p. 154) and by the lectures of Liouville which expounded the idea that not only can mathematical analysis be used for investigating problems in number theory but, conversely, number-theoretic functions can be useful for investigating problems of mathematical analysis. Bugaev devoted his doctoral dissertation "Arithmetic identities connected with the properties of the symbol E" (Moscow, 1866) to this group of problems. It was followed by articles "A general theorem of number theory involving an arbitrary function" (Matem. Sb., 1867, 2, otd. 1, 10–16) and "Some particular theorems for arithmetic functions" (Matem. Sb., 1868, 3, otd. 1, 69–78). In these articles Bugaev considered general arithmetic identities obtained by comparing different representations of the same function in the form of a series. As examples he obtained the unproved Liouville formulas in the articles published in Liouville's Journal (J. math. pures et appl., 1857, 2). Among the general identities derived by Bugaev in his papers, there is, for example, the following:

$$\sum_{d\delta=n} \theta(d) \sum_{d'|\delta} \psi(d') = \sum_{d\delta=n} \psi(d) \sum_{d'|\delta} \theta(d'),$$

where $\psi(n)$, $\theta(n)$ are multiplicative arithmetic functions.

These were the first in a cycle of Bugaev's works on the theory of arithmetic derivatives and integrals. He presented the basic methods and results in his work "The theory of arithmetic derivatives" (Matem. Sb., 1870, 5, otd. 1, 1–63; 1872–1873, 6, otd. 1, 133–180, 201–254, 309–360); a summary of the latter appeared in Darboux's journal (Bull. sci. math. et astr. (1), 1876, 10).

Bugaev calls the sum $\sum_{d|n} \theta(d) = \psi(n)$ the arithmetic integral of the function $\theta(n)$ over the divisors and calls $\theta(n) = D\psi(n)$ the arithmetic derivative of the function $\psi(n)$ over the divisors. Next he introduces the notions of an arithmetic integral and an arithmetic derivative over the natural numbers: if $f(n)$ is an arbitrary numerical function, the sum $\sum_{k\leq n} f(k) = F(n)$ is called the arithmetic integral of the function $f(n)$ over the natural numbers, and the expression $f(n) = F(n) - F(n-1)$ is called the arithmetic derivative of the function $F(n)$ over the natural numbers. Bugaev calls arithmetic sums of the form

$$F(n) = \sum_{k \leq n} Q(k)E(n/k),$$

where $E(x)$ (or $[x]$) is the integral part of x, an arithmetic series over $E(n/k)$. He defines the coefficients of this series as follows: $Q(n) = D(F(n) - F(n-1))$.

Using his general identities Bugaev deduced many corollaries, including known as well as new arithmetic identities. The coefficients of an arithmetic series are arithmetic functions, such as the Möbius function, that play an important role in number theory.

Among the arithmetic functions considered by Bugaev is the convolution of two arbitrary functions

$$\sum_{d\delta=n} \theta(\delta)\chi(d) = \psi(n).$$

If the functions $\theta(n)$ and $\psi(n)$ are known, then $\chi(n)$ can be found by arithmetic differentiation (it is assumed that the function $\theta(n)$ is multiplicative). Bugaev considered the law of arithmetic differentiation and integration of a convolution $\sum_{d\delta=n} \theta(\delta)\chi(d)$. Putting $\sum_{d|n} \theta(d) = D^{-1}\theta(n)$, he obtains for any nonzero integer k the general identity:

$$D^k \sum_{d\delta=n} \chi(\delta)\theta(d) = \sum_{d\delta=n} \chi(\delta)D^k\theta(d) + \sum_{d\delta=n} \theta(d)D^k\chi(\delta).$$

Next he gives some applications such as the use of properties of arithmetic derivatives for the summation of certain series, for inversion of series and infinite products, and for the expansion of functions in arithmetic series.

While studying the function $H_1(n)$ that gives the number of "primary numbers" $\leq n$, that is, the number of squarefree natural numbers > 1 and $\leq n$, Bugaev found several interesting relations, later derived independently by L. Gegenbauer[75] using the Dirichlet series:

$$H_1(n) = \sum_{k=1}^{n} \sum_{u=1}^{E(\sqrt{n/k})} \mu(u), \quad H_1(n) = \sum_{k \leq \sqrt{n}} \mu(k)E\left(\frac{n}{k^2}\right),$$

$$\sum_{k \leq \sqrt{n}} H_1\left(\frac{n}{k^2}\right) = n.$$

75 Gegenbauer, L., *Über die Divisoren der ganzen Zahlen*. Sitzungsber. Akad. Wiss. Wien, Math.-Naturwiss. Kl., 1885, 91, Abt. 2, 600–621.

Bugaev obtained similar relations for the functions $H_2(n)$, $H_3(n)$, $\ldots$ $H_m(n)$, where $H_2(n)$ is the number of natural numbers not divisible by cubes, $H_3(n)$ the number of natural numbers not divisible by fourth powers, etc. These results of Bugaev attracted the attention of Gegenbauer, who proved and generalized many of Bugaev's results. P.L. Chebyshev became interested in Bugaev's report at the Third Congress of Russian Natural Scientists and Physicians (1873) and applied to the function $H_1(n)$ his own method presented in "On prime numbers" (see p. 184) and found bounds for $H_1(n)$. Among the numerical functions considered by Bugaev there is the function "the order of the number n", designated by $\Lambda(n)$ and usually called the function of H. von Mangoldt. This function is the arithmetic derivative of the logarithm: $\sum_{d|n} \Lambda(d) = \ln n$. Using the sum $\sum_{d|n} \Lambda(d) = \ln n$, Bugaev derives, among others, the Chebyshev identity: Since $\sum_{d|n} \Lambda(d) = \ln n$, the logarithm of the product $\prod(n)$ of consecutive natural numbers from 2 to n is equal to

$$\ln \prod(n) = \sum_{k=1}^{n} \sum_{m=1}^{E(n/k)} \Lambda(m),$$

whence, passing from the logarithms to numbers, Bugaev obtains the Chebyshev identity.

Continuing the analogy with mathematical analysis, Bugaev considers also "definite arithmetic integrals over divisors" and establishes the connection between the integral over divisors and the definite integral over divisors, etc.

Bugaev's research was continued by his pupils, S.I. Baskakov, N.V. Bervi, A.P. Minin, P.S. Nazimov, (partly) N.Ya. Sonin, and others. The influence of Bugaev's ideas was obvious in the work of foreign mathematicians such as L. Gegenbauer, J. Gram, M. Cipolla, F. Pellegrino, and others.

At the same time as Bugaev, similar problems occupied several other mathematicians, such as E. Cesàro, who sometimes rediscovered results obtained earlier by Bugaev from general considerations.

Later, mathematicians continued efforts aimed at developing an operational calculus for number-theoretic functions, including those in several variables (E.T. Bell, R. Vaidyanathaswamy), and at applying the operational calculus of Bugaev-Cipolla (with or without mentioning their names) to obtain new results or to prove known assertions, such as the famous Selberg identity (I. Popken, C. Yamamoto, and others). New types of operational calculi were developed, for example, by A.L. Amizur. These works, however, belong to the 20th century.

Bugaev's student, N.Ya. Sonin (1849–1915) published fundamental investigations on special functions and other problems of analysis as well as the paper "On Arithmetic Identities and Their Application to the Theory of Infinite Series" (Varshavsk. univ. izv., 1885, #5, 1–28) and several works related to the use of the arithmetic function $[x]$ (Bugaev called it $E(x)$). The

200

application of this function to the integral calculus led Sonin to a generalization of the well-known Euler-MacLaurin formula (see "On a Definite Integral Containing the Arithmetic Function [x]" (Varshavsk. univ. izv., 1885, #3, 1–28)).

Sonin's work had a marked influence on the work of G.F. Voronoĭ, who worked together with Sonin at Warsaw University. Voronoĭ applied Sonin's formula as well as cylindrical functions, which constituted the main subject of Sonin's investigations, to problems in number theory. Later the formulas of Sonin and Voronoĭ were used by I.M. Vinogradov.

4 Transcendental Numbers

The Works of Joseph Liouville

We conclude the discussion of analytic number theory with a survey of one of its areas that evolved in the 19th century and soon became an essential subdiscipline of number theory. This is the theory of transcendental numbers. We recall that an algebraic number is a root of an algebraic equation with rational coefficients and a transcendental number is a nonalgebraic irrational number.

Conjectures on the transcendence of some mathematical constants were made as early as the 17th century. In 1656 J. Wallis advanced the idea of the special nature of the number π, different from ordinary irrationalities. A hundred years later, in 1758, Euler expressed the same idea in a somewhat different form, and later noted that the impossibility of expressing π in terms of "radical quantities" had not been established (1775, published in 1785). J.H. Lambert, who proved (not completely rigorously) in 1776 the irrationality of e and π (published in 1768 and 1770, respectively) was deeply convinced that neither number belongs to "radical irrational quantities". This opinion was shared by A.M. Legendre, who in 1800, filled the gap in Lambert's proof. Legendre was the first to formulate the notion of a transcendental number in modern terms, adding that it was apparently very difficult to prove the transcendence of π. In the 18th century mathematicians studied the transcendence of entire classes of numbers of a certain type. Thus, Euler wrote in Volume I of his "An introduction to the analysis of infinites" (1748) that all the numbers $\log_a b$, where a and b are rational and b is not a rational power of a, are transcendental (see, HM, vol. 3, pp. 110–114).

The first proof of the existence of transcendental numbers belongs to Liouville (1844, 1851).

Joseph Liouville (1809–1882), was a student and later professor at the Ecole Polytechnique. He taught at the Collège de France and, after 1857, at the Paris Faculty of Sciences. He was a very active, versatile and influential scientist; this was facilitated by his election to the Paris Academy of Sciences in 1839. In 1836 Liouville began to publish one of the major mathematical

J. Liouville

journals of the 19th century, *Journal de mathématiques pures et appliquées.* He published 39 volumes of this journal. Another editor succeeded him in 1874 when Liouville resigned the chair of rational mechanics at the Sorbonne. He continued lecturing at the Collège de France, where professors could lecture on their own research, until 1879.

Liouville published about four hundred articles on many topics, including the theory of integration of algebraic functions, doubly periodic functions, ordinary differential equations (his name is connected with the important class of Sturm-Liouville equations and the Liouville-Ostrogradskiĭ formula, well known in the theory of linear equations), mathematical physics, differential geometry, etc. As already noted, to Liouville goes the credit for the correct evaluation of the significance of Galois' works, then unknown to the mathematical public, and for their first publication in 1846.

We have also discussed earlier Liouville's contribution to the theory of algebraic numbers.

One of the characteristics of Liouville's work was his permanent interest in proofs of unsolvability (in one sense or another) of various problems, and, conversely, in proofs of the existence of various objects. While it is true

202

that existence theorems and proofs of unsolvability were of great concern to mathematicians of that time, Liouville conducted especially extensive investigations of problems of this kind. Such were, for example, his investigations of the problem of integrability in closed form of algebraic functions, the conditions of solvability in quadratures of the Riccati equation, and a number of works in number theory. Such were also his fundamental researches in a then unknown field — the theory of transcendental numbers.

Liouville presented his basic result containing the proof of the existence of an infinite number of transcendental numbers first in his brief notes in the 1844 *Comptes rendues* of the Académie des Sciences and in more detail in the 1851 article "On very large classes of quantities the value of which is neither algebraic nor reducible to algebraic irrationalities" (*Sur des classes très étendues de quantités dont la valeur n'est ni algébrique, ni même reductible à des irrationelles algébriques*. J. math. pures et appl., 1851, 16, 133–142).

We will present the main content of Liouville's works following A.O. Gel'fond, the greatest expert on the history of this question, who wrote in one of his articles in 1930: "The trait of transcendence of a number established in these works is based on the fact that the deviation of an algebraic number α from a rational fraction p/q which approximately represents α cannot be smaller than a certain number depending only on the denominator of the fraction and the degree of the equation which α satisfies. The exact formulation of this property of algebraic numbers is the following:

Let α satisfy an n-th degree equation with integral coefficients. Let p and q be coprime integers. Then the absolute value of the difference $|\alpha - p/q|$ must satisfy the inequality

$$\left| \alpha - \frac{p}{q} \right| > \frac{1}{Aq^n},$$

where A is a constant not depending on q.

The proof is extraordinarily simple. Let

$$f(x) = a_0 x^n + a_1 x^{n-1} + \cdots + a_n = 0$$

be an irreducible equation of nth degree with integral coefficients satisfied by α.

By the Lagrange formula of the differential calculus

$$f\left(\frac{p}{q}\right) - f(\alpha) = \left(\frac{p}{q} - \alpha\right) f'\left[\alpha + \theta\left(\frac{p}{q} - \alpha\right)\right], \quad |\theta| < 1,$$

or, since $f(\alpha) = 0$,

$$\frac{p}{q} - \alpha = \frac{f(p/q)}{f'[\alpha + \theta((p/q) - \alpha)]} = \frac{a_0 p^n + a_1 p^{n-1} q + \cdots + a_n q^n}{q^n f'[\alpha + \theta((p/q) - \alpha)]}.$$

Since the numerator is an integer and the denominator is not zero it follows that

$$\left|\alpha - \frac{p}{q}\right| > \frac{1}{Aq^n}, \quad A > \max_{|\theta| \leq 1}\left|f'\left[\alpha + \theta\left(\frac{p}{q} - \alpha\right)\right]\right|.$$

This Liouville inequality immediately implies that if for any real number ω there exists a sequence of integers p and q such that

$$\lim_{q \to \infty} \frac{\ln|\omega - p/q|}{\ln q} = -\infty, \tag{23}$$

the number ω is necessarily transcendental. The equality (23) is therefore a sufficient criterion of transcendence. Starting from this equation it is possible to construct a simple example of a transcendental number. Let $\ell > 1$ be an integer. Consider the number ω given by the series

$$\omega = 1 + \frac{1}{\ell} + \frac{1}{\ell^{1 \cdot 2}} + \cdots + \frac{1}{\ell^{1 \cdot 2 \cdot 3 \cdots n}} + \frac{1}{\ell^{1 \cdot 2 \cdot 3 \cdots n(n+1)}} + \cdots. \tag{24}$$

Multiply both sides of (24) by $\ell^{n!}$. Then

$$\ell^{n!}\omega = \sum_{k=0}^{n} \ell^{n! - k!} + \frac{1}{\ell^{n!n}}\left[1 + \ell^{-(n+1)!(n+1)} + \cdots\right].$$

Setting $\ell^{n!} = q$, $\sum_{k=0}^{n} \ell^{n! - k!} = p$ and noting that $1 + \ell^{-(n+1)!(n+1)} + \cdots < 2$, we obtain

$$|\omega q - p| < 2/q^n, \tag{25}$$

where $\lim_{q \to \infty} n = \infty$.

Comparing the inequality (25) with the criterion (23) we conclude that the number ω is transcendental.

The numbers ω obtained from the series (24) were found by Liouville. They were the first examples of transcendental numbers."[76]

Liouville was also interested in the arithmetic nature of the number e. Before he wrote the paper on the theory of transcendental numbers he had proved that the number e cannot be the root of any quadratic or biquadratic equation with integral coefficients (J. math. pures et appl., 1840, 5). The next brilliant result — the proof of the transcendence of e — was obtained by Hermite who applied for this purpose the tools of classical analysis.

---

76 Gel'fond, A.O., *Selected Works*. Moscow 1973, pp. 16–17 (in Russian). The quote is from the paper "Sketch of the history and present state of the theory of transcendental numbers", first published in 1930.

204

Charles Hermite and the Proof of the Transcendence of the Number *e*; A Theorem of Ferdinand Lindemann

Among Hermite's first investigations was a new proof of the irrationality of the number e, which he reported to the British Association for the Advancement of Science in 1873. Hermite[77] starts from the series for e^x:

$$e^x = 1 + \frac{x}{1} + \frac{x^2}{1 \cdot 2} + \frac{x^3}{1 \cdot 2 \cdot 3} + \cdots .$$

He denotes the partial sum of this series by

$$F(x) = 1 + \frac{x}{1} + \frac{x^2}{1 \cdot 2} + \cdots + \frac{x^n}{1 \cdot 2 \cdots n}$$

and considers the ratio of the difference of $e^x - F(x)$ to x^{n+1}:

$$\frac{e^x - F(x)}{x^{n+1}} = \sum_{k=0}^{\infty} \frac{x^k}{(n+k+1)!}. \tag{26}$$

Differentiating (26) n times, Hermite arrives at the equality

$$e^x \Phi(x) - \Phi_1(x) = \frac{x^{2n+1}}{n!} \sum_{k=0}^{\infty} \frac{(k+n)!}{(n+1)(n+2)\cdots(2n+k+1)} \cdot \frac{x^k}{k!}, \tag{27}$$

where $\Phi(x)$ and $\Phi_1(x)$ are polynomials with integral coefficients. Assuming that e^{x_0} is rational, $e^{x_0} = b/a$, for some integer $x = x_0$, Hermite sets in (27) $x = x_0$. He thus obtains on the left side of (27) a number larger than $1/a$, and on the right side a quantity which can, as $n \to \infty$, become smaller than any arbitrarily small given number. The resulting contradiction proves that e^x cannot be rational for any integer x. In particular, e is an irrational number.

Hermite used a similar method for proving the transcendence of e, which he reported in the article *Sur la fonction exponentielle* (C. r. Acad. sci. Paris, 1873, 77).[78]

We give here the proof of Hermite's theorem in its latest simplified version. A.O. Gel'fond noted that the idea behind it is due to A. Hurwitz (1859–1919), a German mathematician, who worked at the Universities of Göttingen, Königsberg and Zürich. The proof was carried out by K.A. Posse (1847–1928), professor at the Petersburg University.[79]

77 Hermite, Ch., *Œuvres*, T. 3. Paris 1912, pp. 127–134.

78 Ibid., pp. 150–181.

79 Gel'fond, A.O., *Selected Works*. Moscow 1973, p. 19 (in Russian).

Let $f(x)$ be a polynomial in x, let $f^{(k)}(x)$ be the kth derivative of $f(x)$, and let $F(x) = \sum_{k=0}^{\infty} f^{(k)}(x)$, a polynomial of the same degree as $f(x)$. Integration by parts yields the identity

$$e^x F(0) - F(x) = e^x \int_0^x e^{-t} f(t) dt. \tag{28}$$

Assume that e satisfies an algebraic equation with integral rational coefficients

$$a_0 + a_1 e + a_2 e^2 + \cdots + a_n e^n = 0 \quad (a_0 \neq 0), \tag{29}$$

that is, that e is an algebraic number. Setting in (28) $x = k$, multiplying the resulting equality by a_k, taking successively $k = 0, 1, \ldots, n$ and adding up the resulting equalities we obtain

$$F(0) \sum_{k=0}^{n} a_k e^k - \sum_{k=0}^{n} a_k F(k) = \sum_{k=0}^{n} a_k e^k \int_0^k e^{-t} f(t) dt. \tag{30}$$

Since e satisfies the equality (29), we have

$$a_0 F(0) + \sum_{k=1}^{n} a_k F(k) = - \sum_{k=0}^{n} a_k e^k \int_0^k e^{-t} f(t) dt. \tag{31}$$

Take $f(x)$ to be

$$f(x) = \frac{1}{(p-1)!} x^{p-1} \prod_{k=1}^{n} (k - x)^p,$$

where $p > n + |a_0|$, and p is a prime. Now we use the same technique as that used by Hermite to prove the irrationality of e. It is easy to show that the right side of the equality (31) tends to zero as $n \to \infty$, and, for sufficiently large n, the absolute value of the left side is greater or equal to 1. Thus the assumption that e is an algebraic number yields a contradiction. This proves that e is a transcendental number.

Hermite believed that it would be difficult to construct a similar proof for the number π. But in 1882 Ferdinand Lindemann used Hermite's idea and proved the transcendence of π.

Ferdinand Lindemann (1852–1939), a student of the geometer Clebsch, taught at several universities in Germany, for the longest period of time at the University of Königsberg (from 1883), and from 1893 at the University of Munich. It was from Clebsch that Lindemann inherited his interest in

F. Lindemann

the theory of elliptic and Abelian functions as well as in the geometry of curves. He also worked in the theory of conformal mappings, on the solution of algebraic equations using transcendental functions, on problems of geometric representation of quadratic forms, in the theory of invariants, infinite series, differential equations, and also in the history and philosophy of mathematics and some areas of physics.

Lindemann became best known for his proof of the transcendence of the number π. This settled the ancient problem of the squaring the circle, for it proved the impossibility of constructing a square whose area is that of a given circle using ruler and compass (and even using any algebraic curves).

Lindemann derived the proof of the transcendence of π as a simple corollary of a theorem generalizing Hermite's result. Lindemann communicated his proof in a letter to Hermite who immediately reported it to the Academy and published the letter under the title "On the relation between the circumference and the diameter and on logarithms of commensurable numbers or algebraic irrationalities" (*Sur le rapport de la circonférence au diamètre et sur les logarithmes néperiens des nombres commensurables ou des irrationelles algébriques.* C. R. Acad. sci. Paris, 1882, 95, 2, 72–74). Apparently, this title was due to Hermite. At the same time Lindemann published a more detailed presentation of his finding in the article "On the number π" (*Über die Zahl π. Math. Ann.*, 1882, 20).

The result obtained by Hermite can be formulated as follows:

There exists no equality

$$N_1 e^{x_1} + N_2 e^{x_2} + \cdots + N_k e^{x_k} = 0, \tag{32}$$

where all N_j are integers (not all equal to zero) and all x_j are distinct integers. Lindemann proved the impossibility of an equality

$$A_1 e^{\omega_1} + A_2 e^{\omega_2} + \cdots + A_k e^{\omega_k} = 0, \tag{33}$$

where all A_j are algebraic numbers, and all ω_j are distinct algebraic numbers. It follows that e^ω, where ω is a nonzero algebraic number, is transcendental, and therefore, the logarithm of an algebraic number not equal to 1 is also transcendental. Since, by the Cotes-Euler formula, $e^{\pi i} = -1$, πi is transcendental. Since i is algebraic, π is transcendental.

Later, new proofs of Hermite's and Lindemann's results were suggested; we have delineated above one such proof of Hermite's theorem. In 1885 Weierstrass gave a new, simpler derivation of Lindemann's theorem and pointed out that $\sin \omega$ is a transcendental number for a nonzero algebraic ω. In Russia Hermite's and Lindemann's works found a response in A.A. Markov's "The proof of transcendence of the numbers e and π" (St. Petersburg, 1883).

A somewhat special place belongs to a result, published the same year as Hermite's, by G. Cantor in the article "On a property of the totality of real algebraic numbers" (*Über eine Eigenschaft des Inbegriffs aller reellen algebraischen Zahlen. J. für Math.*, 1873, 77) where he proved the existence of transcendental numbers using set theory. Specifically, Cantor proved the uncountability of the set of all numbers on an interval $[0, 1]$ and the countability of the set of algebraic numbers in the same interval. This implied the existence of transcendental numbers and, furthermore, the uncountability of the set of these numbers.

Radically new ideas in the theory of transcendental numbers were set forth in the 1930s by A.O. Gel'fond, C.L. Siegel, and other mathematicians, largely in response to a problem posed by Hilbert at the 1900 Congress of Mathematicians in Paris. The problem concerned the arithmetic nature of numbers of the form α^β, where α is an algebraic number other than 0 and 1, and β is an algebraic irrational number. In 1934 A.O. Gel'fond proved Hilbert's proposition concerning the transcendence of all such numbers thereby proving a generalized Euler theorem: the logarithm of an algebraic number for any algebraic base is either rational or transcendental. Another proof of this proposition was independently published by T. Schneider at the end of 1934.

Conclusion

We summarize briefly our discussion of the works in the theory of numbers. During the 19th century new, essential branches were created and new methods of investigation were developed. Above all, for the first time number theory came to be viewed as equal to the other mathematical sciences. In the 18th century many mathematicians thought that preoccupation with number theory could serve as sophisticated entertainment for some intelligent minds, but was otherwise an idle game. In the 19th century it became obvious that the harmonious development of mathematics as a whole is impossible without an elaboration of number theory which not only uses algebraic, analytic and geometric means for the solution of its problems but also facilitates the creation of new and important notions, methods and theories in other areas of mathematics. The greatly increased role of number theory is attested by the long list of names of outstanding mathematicians who studied it in the 19th century, as well as by the appearance of scientific schools, of which the St. Petersburg school of number theory is a remarkable example. In the 19th century many old arithmetical problems were solved and many new difficult problems were posed. Nevertheless, many classical problems, such as the Goldbach hypothesis concerning the representability of all natural numbers by sums of no more than three primes, could not be solved by the tools which were at the disposal of the number theory of the last century. The solution of these problems required new, more efficient methods which were developed by 20th century mathematicians.

Chapter Four
The Theory of Probability

Introduction

In the 19th century, both the ideas and methods of the theory of probability, developed since the second half of the 17th century, received new incentives for further progress. These stimuli, differing one from another in their essence, were connected with the development of the natural sciences, with practical requirements of society and, also, with the formulation of purely mathematical problems.

Achievements in astronomy and in its applications to the needs of navigation, and (later on) successes in physics, raised with great acuteness the problem of creating a unified theory of errors of observations. Such a theory was also called for by the geodetic surveys undertaken by European countries, by the ever more precise determinations of the size and shape of the Earth through astronomical, geodetic and pendulum observations, and by the introduction of the metric system, based on the execution of an immense measurement of a quarter of the meridian through Paris.

Progress in artillery led to the formulation of numerous problems on the scatter of projectiles — a discipline founded on the ideas of the theory of probability.

Profound ideas, prompted by Kant's and Laplace's hypotheses on the origin of the solar system, provoked interest in general cosmological problems and, in particular, in the determination of the geometry of the surrounding world. In 1842 N.I. Lobachevskiĭ made the first attempt to test whether the geometry of physical space is Euclidean. This led him to consider an important problem in the theory of probability connected with the summing of independently distributed random variables.[1]

Population statistics, and therefore also the development of its methods essentially belonging to mathematical statistics, gained ever increasing national, civil, and economic significance.

The second class of problems leading to mathematical statistics was called forth by the requirements of biometry, the new scientific discipline which appeared in response to the work of Charles Darwin and was devoted to the mathematical treatment of biological observations and, also, to the study of various statistical regularities in biology.

Physics proved to be a new field for applying the theory of probability. The insight that the laws of the natural sciences are stochastic was the most important inference made in the 19th century.

In the second half of that century, in the theory of probability itself, it became natural to consider problems connected with the behavior of random variables rather than with estimating the probabilities of random events. The works of P.L. Chebyshev played a very important role in this fundamental change.

The very notion of a random variable, for a long time not even defined and considered self-evident, was formalized in the right way only in the 20th century (and even then only gradually) in the works of A.M. Lyapunov, P. Lévy, and A.N. Kolmogorov.

Laplace's Theory of Probability

Laplace's contribution to the theory of probability was partly described in Chapter 4 of vol. 3 of the "History of mathematics from Antiquity to the early nineteenth century", vols 1–3. Moscow, 1970–1972 (HM) devoted to the development of this science in the 18th century. Laplace (1749–1827) continued his creative work in the 19th century, and the source just mentioned carries a biography of this outstanding scholar (HM, vol. 3, pp. 146–148).

While formulating his deterministic creed, Laplace maintained that each state of the Universe is a consequence of its previous conditions and the cause of its future states. At the same time, he realized that the study of many very important phenomena is virtually impossible without the use of stochastic considerations. For example, he wrote:

---

1 Lobachevskiĭ, N.I., *Complete works*, vol. 5. Moscow-Leningrad 1951, pp. 333–341 (in Russian). However, this problem, mainly suggested by other astronomical requirements, was also considered in the (previous) works of T. Simpson, Lagrange, and Laplace.

This (lunar) inequality, though indicated by observations, was disregarded by most astronomers, since it did not seem to result from the theory of universal gravitation. However, upon checking its existence by means of the Calculus of Probability, it seemed to me to be indicated with such high probability that I considered it necessary to find out its cause.[2]

While continuing the work of his predecessors, Jakob and Daniel Bernoulli, A. DeMoivre, and others, Laplace, in a number of memoirs, advanced the theory of probability so much that, at the beginning of the 19th century, he set himself the task of unifying everything that had been achieved in this theory in a single work — his *Théorie analytique des probabilités* (Paris 1812). In spite of the ponderous exposition, of some vagueness in the treatment of the main concepts, and of a lack of unity of its different sections, this source became the main work in the theory of probability until the appearance of Chebyshev's writings. The title of the book reminds one of Lagrange's earlier (1788) *Mécanique analytique* and of Fourier's later (1821) *Théorie analytique de la chaleur* and points to mathematical analysis as the main tool used in the theory of probability.

During Laplace's lifetime there appeared two more editions of the *Théorie analytique*, viz., in 1814, when the author prefaced it with a popular introduction, the *Essai philosophique sur les probabilités*, and in 1820, when he added several supplements to the text. In 1886 the book appeared as tome 7 of Laplace's *Oeuvres complètes*.

A short description of Laplace's *Théorie analytique* is contained in HM (vol. 3, pp. 150–151); here, we dwell in more detail on the second of the two of its livres devoted to the theory of probability proper (livre 1 treats auxiliary materials and tools: the calculus of generating functions with its application to the solution of finite difference equations, both ordinary and partial, and to the approximate calculation of definite integrals).

Chapter 1 of livre 2 contains the classical definition of the probability of a random event, which in fact was used even by G. Cardano and expressly introduced by Jakob Bernoulli and A. DeMoivre (the probability is equal to the ratio of the number of favorable chances to the total number of equally possible chances); the addition and the multiplication theorems for the probabilities of independent events; a few theorems on conditional probabilities of events and, in addition, definitions of mathematical and moral expectations.

In Chapter 2 Laplace solved a number of problems of elementary probability theory without indicating their direct application to the natural sciences. Among them were the classical problem of the gambler's ruin, which goes back to C. Huygens, and the celebrated scheme due to Jakob Bernoulli considered later by DeMoivre.

---

2 Laplace, P.S., *Œuvres complètes*, t. 7. Paris 1886, p. 361.

Take for example the first of these problems which nowadays has important physical applications and is studied in connection with problems on random walks of particles in a one-dimensional space. Gambler A has a chips, gambler B has b chips, and the probabilities of their winning each set are p and q respectively. What is the probability that B will be ruined in not more than a given number n of sets?

Introduce the probability $y_{x,s}$ that B, having x chips, will be ruined in not more than s sets. It is easy to see that this probability obeys the partial finite difference equation

$$y_{x,s} = p y_{x+1,s-1} + q y_{x-1,s-1}.$$

It is clear that the required probability $y_{x,s}$ satisfies the following natural boundary conditions:

$$y_{x,s} = 0 \quad \text{for} \quad x > s \quad \text{and} \quad y_{x,s} = 1 \quad \text{for} \quad x = 0.$$

Laplace finds the solution of the above equation by using generating functions in two variables. He then considers several particular cases: equal initial capitals $(a = b)$; and an infinitely large capital of gambler A $(a = \infty)$. In the second instance, assuming that $p = q$, Laplace arrives at an elegant final result

$$y_{b,n} = 1 - \frac{2}{\pi} \int\limits_0^{\pi/2} \frac{\sin b\varphi (\cos \varphi)^{n+1}}{\sin \varphi} d\varphi.$$

Note that DeMoivre was the first to study the problem of B's ruin given that A's capital is infinitely large.

In Chapter 7 Laplace treated a simpler case, viz., the gambler's ruin in an infinitely long game. In this instance, the probability of B's ruin is given by the formula

$$P = \frac{p^b(p^a - q^a)}{p^{a+b} - q^{a+b}},$$

due to Jakob Bernoulli. Laplace first obtained it in §3 of his early "Memoir on probabilities" (*Mémoire sur les probabilités* (1778), 1781)[3] by reducing the problem to a simple finite difference equation.

Also in Chapter 7 of the *Théorie analytique* Laplace studied the consecutive extraction of tickets numbered 1 through n from an urn in which they had been put one after another in that order and then shuffled. He noted that it is possible for the probabilities of drawing these numbers to be unequal. However, the differences between the probabilities become smaller if

---

3 Laplace, P.S., *Œuvres complètes*, t. 9. Paris 1893, pp. 383–485.

P.S. Laplace

the tickets are put into the urn not in the assigned order, but according to their random extraction from a "preliminary" urn. These differences become still smaller if two, three, or more "preliminary" urns are used.

This example may be viewed as a special case of the reshuffling of a pack of cards, i.e., as an example of trials connected by a Markov chain. Without providing a rigorous proof, Laplace points out that there exists a limit state in which all the tickets have the same probability of being drawn.

In Chapter 3 Laplace proves the DeMoivre-Laplace limit theorems on the convergence of the binomial distribution to the normal distribution. Unlike DeMoivre, he used the Maclaurin-Euler summation formula.

Laplace obtained the integral limit theorem in the form

$$P(-l \leq \mu - np - z \leq l) = \frac{2}{\sqrt{\pi}} \int_{0}^{\frac{l\sqrt{n}}{\sqrt{2xx'}}} e^{-t^2} dt + \frac{\sqrt{n}}{\sqrt{2\pi xx'}} e^{\frac{-l^2 n}{2xx'}}.$$

Here μ is the number of occurrences of the event in n Bernoulli trials; p is the probability of its happening in any single trial, n is the number of trials;

215

z is a number less than one in absolute value, $x = np + z$ and $x' = nq - z$ $(q = 1 - p)$.

This formula gives a good estimate of the probability on the left-hand side. It may be considered the origin of later studies that found a certain completion in S.N. Bernshteĭn's paper *A return to the problem of the accuracy of Laplace's limiting formula* (Izv. Akad. Nauk SSSR, ser. mat., vol. 7, 1943, 3–16 [in Russian]).

Laplace applied the limit theorems to the solution of a number of urn problems in which he introduced partial differential equations into probability theory.

Here is one such problem, dating back to Daniel Bernoulli (1770). There are n white and n black balls in two urns, and each urn contains n balls. The balls are moved cyclically, one by one, from one urn to another. What is the probability $z_{x,r}$ that after r moves (cycles) urn A will contain x white balls?

Laplace obtained an equation in partial finite differences for the required probability, and used nonrigorous transformations to pass from it to the differential equation

$$u'_{r'} = 2u + 2\mu u'_\mu + u''_{\mu\mu} \left(u = z_{x,r}, \ r = nr', \ x = \frac{n + \mu\sqrt{n}}{2} \right). \qquad (1)$$

In solving this equation he used expressions which are now called the Chebyshev-Hermite polynomials.

Later (in 1915) this problem was considered by A.A. Markov and V.A. Steklov. In the same year, in connection with his studies of Brownian motion, M. Smoluchowski derived equation (1) in a somewhat more general form.

Both Daniel Bernoulli and Laplace noted that the above process of moving the balls leads to a stage when the numbers of white balls in the different urns almost coincide and are approximately equal to the quotient of the total number of white balls by the number of urns. Laplace proved this for arbitrary initial distributions of the balls in the urns. Since the celebrated model of P. and T. Ehrenfest (1907, see p. 271) which is accepted as the origin of the theory of random processes essentially coincides with the Daniel Bernoulli-Laplace model (problem), this theory should be dated at least from Daniel Bernoulli (1770). Also, the result of these two scholars anticipated the celebrated Markov ergodic theorem on Markov chains.

Although he failed to indicate any direct applications of his problem, Laplace foresaw its essential importance. He states:

> ... the original irregularity will disappear in time and give way to very simple order ... These results can be extended to all combinations in nature in which constant forces ... establish regular modes of action ca-

pable of calling forth, even from the depths of chaos, systems governed by admirable laws."[4]

Laplace expressed a similar opinion on limit theorems in general, justly regarding them as a mathematical tool for studying the statistical significance of observations and for revealing demographic regularities. He even believed that exactly the use of these theorems would eventually ensure the triumph of the "eternal principles of reason, justice, and humanity".[5]

This naive humanism, as well as Laplace's sounder opinion on the advantages of applying the theory of probability to human affairs, attracted general attention. In particular, these views found supporters in Russia. Influenced by these views, P.L. Chebyshev turned his attention to the theory (see p. 251).

Laplace explained the stability of the receipts from the French lottery, of the number of marriages contracted yearly in a given country, and of the number of dead letters, by the action of the same limit theorems. The first two examples, however, were known even before him.

Laplace also applied the DeMoivre-Laplace limit theorem in Chapter 9, while solving problems connected with the calculation of the value of life annuities. Here, as in the theory of errors (see p. 223), he used characteristic functions and the inversion formula, i.e. the formula for passing from characteristic functions to density functions. In one of his methodological problems, Laplace, without restricting himself to treating Bernoulli trials, studied their generalization, which is now named after Poisson. Laplace's reasoning in problems such as the stability of the receipts from the lottery shows that he understood Poisson's "law of large numbers" (see p. 232).

The study of finite random sums occupied a special place in Laplace's theory of probability. Rudiments of such studies are already found in Galileo's note *Sopra le scoperte dei dadi* ("Thoughts about games of dice"), in which he calculated the probabilities of the occurrence of various numbers of points in a throw of three dice. DeMoivre obtained more general results by using generating functions. T. Simpson, R. Boscovich, and J.L. Lagrange transferred the study of finite random sums to the mathematical treatment of observations, with Lagrange using generating functions of continuous distributions and thus anticipating the introduction of characteristic functions.

Using different methods, Laplace repeatedly deduced the laws of distribution of finite sums. He did this mostly in astronomical contexts, for example, while solving the following problem in his "Memoir on the mean inclination of the orbits of comets; on the figure of the Earth; and on functions" (*Mémoire sur l'inclinaison moyenne des orbites des comètes; sur la figure de la terre,*

---

4 Laplace, P.S., *Essai philosophique sur les probabilités*. Paris 1814. *Œuvres complètes*, t. 7. Paris 1891, p. LIV.

5 Ibid., p. XLVIII.

et sur les fonctions, 1773 (1776))[6]: The inclinations of individual comets relative to the ecliptic are random (we would say, mutually independent and uniformly distributed on the segment $[0, \pi/2]$); what is the probability that the mean inclination of n comets is contained within given bounds?

Considering this problem for $n = 2$, 3, and 4, Laplace worked out an integral relation for passing from the case of $(n-1)$ to the case n. Essentially, this relation is equivalent to the familiar formula

$$p_n(x) = \int_a^b p_{n-1}(x-z)p(z)dz,$$

which Laplace did not write down explicitly but was able to use to determine the required function $p_n(x)$ with an insignificant error.

Laplace also posed the problem (*Théorie analytique*, livre 2, Chapter 2) about the distribution and the mean value of the function $\psi(t_1, t_2, \ldots, t_n)$ in non-negative variables $t_1, t_2, \ldots, t_n$ whose sum had a given value and whose density functions $\varphi_i(x_i)$ were, in general, different. While calculating a multiple integral of such a function he used discontinuity factors and, essentially, derived the so-called Dirichlet formula.

Laplace considered the case he was directly interested in, $\psi = t_1 + t_2 + \ldots + t_n$, for distributions such as $\varphi_i(x) = a + bx + cx^2$ and

$$\varphi_i(x) = \begin{cases} \beta x, & 0 \le x \le h, \quad (\beta > 0), \\ \beta(2h - x), & h \le x \le 2h, \end{cases}$$

by actually using the formulas

(a) $\qquad p_n(x) = \dfrac{d}{dx}\left[\underset{x_1 + x_2 + \ldots + x_n \le x}{\iint \ldots \int} p(x_1)p(x_2)\ldots p(x_n)dx_1 dx_2 \ldots dx_n \right],$

(b) $\qquad p_n(x) = \dfrac{1}{dx}(I_1 - I_2).$

Here, the integral I_2 is the one written out in line (a), and the integral I_1 differs from I_2 in that its region of integration is $x_1 + x_2 + \ldots + x_n \le x + dx$.

In connection with this investigation we mention another problem solved by Laplace: a segment is divided into i equal or unequal intervals at whose end points perpendiculars are erected. The lengths of these perpendiculars make up a non-increasing sequence and their sum is equal to s. If such sequences are

---

6 Laplace, P.S., *Œuvres complètes*, t. 8. Paris 1891, pp. 279–321.

constructed many times over, what will be the mean broken line connecting the top points of the perpendiculars (or, in the continuous case, the mean curve)? This problem may be interpreted in the language of random functions: each construction is a realization of a random function, with the required mean curve being its expected value.

Suppose now that a certain event can be brought about only by i mutually exclusive causes arranged in decreasing order of their (subjective) probabilities. The mean value of the probability that any given cause occasioned the event may be determined if the relevant procedure is carried out by several people. Laplace suggested that a similar procedure should be used in courts and during elections. This recommendation was hardly ever put into practice, but Laplace's reasoning may be seen to belong to the prehistory of rank correlation and of the statistics of random processes.

Problems that are now included in the domain of mathematical statistics occupy a very important place in Laplace's *Théorie analytique*. Laplace believed that the theory of probability is a discipline belonging to the natural sciences rather than to mathematics and for this reason alone was unable to isolate mathematical statistics. It is therefore all the more interesting that Laplace ("Memoir on probabilities") mentioned "a new branch of the theory of probability" in connection with problems in mathematical statistics.[7]

We will consider a typical problem in mathematical statistics that Laplace (*Théorie analytique*, Chapter 6) solves by applying the Bayesian approach, which he introduced independently of Bayes (HM, vol. 3, pp. 137–139).

The probability x of a "simple" event, such as, for example, the probability that a newly born Parisian will be a boy, is unknown. It is required to estimate it from statistical data on births. Supposing that $z(x)$ is the prior probability distribution of x, Laplace assumes that

$$P(\theta \leq x \leq \theta') = \int_{\theta}^{\theta'} yzdx \Big/ \int_{0}^{1} yzdx \quad (0 < \theta < \theta' < 1). \tag{2}$$

We will explain the sense of the function $y(x)$. Let the number of boys (girls) born during some years be $p \approx 0.393 \cdot 10^6$ and $q \approx 0.378 \cdot 10^6$ respectively. Laplace assumes that

$$y(x) = C_n^p x^p (1 - x)^q \tag{3}$$

where $n = p + q$.

Thus Laplace's problem is one of estimating the parameter of a binomial distribution. After numerous transformations, and setting $z = 1$, he deduces

---

7 Laplace, P.S., *Œuvres complètes*, t. 9. Paris 1893, p. 383.

from (2) that

$$P(-\theta \le x - a \le \theta) \approx \frac{2}{\sqrt{\pi}} \int_0^{\tau} e^{-t^2} dt$$

where θ is a small quantity of the order of $p^{-1/2}$, a is the maximum of the function (3), and the upper limit of integration is

$$\tau = \sqrt{\frac{T^2 + T'^2}{2}}, \quad T = \sqrt{\ln y(a) - \ln y(a - \theta)},$$

$$T' = \sqrt{\ln y(a) - \ln y(a + \theta)}.$$

Of course, for a unimodal curve (3) the maximum

$$a = p/(p + q) \tag{4}$$

seems to be a natural estimator of the probability x. Nevertheless, the expected value of this probability, or rather of a random variable with distribution

$$x^p(1-x)^q / \int_0^1 x^p(1-x)^q dx,$$

does not coincide with the estimator (4) which is only an asymptotically unbiased estimator of x. Laplace did not specify this fact; he thus failed to introduce directly the notions of bias and unbiasedness.

In §18 of his "Memoir on probabilities" Laplace came close to bringing in consistent estimators and in §22 he indicated that the estimator (4) of the parameter of the binomial distribution is appropriate only when the probability $P(|x - a| < \delta)$ is sufficiently high ($\delta > 0$ is small).

We will consider one more important problem solved by Laplace in Chapter 6 of livre 2, viz., the estimation of the population of France by a sample census (devised by Laplace) covering 7 per cent of the inhabitants and using data on the number of yearly births (N) in the whole country.

Suppose that the population included in the census is m and the number of yearly births in the same group is n. Then it is natural to take $M = (m/n)N$ as the required estimator of the entire population. (The ratio of the population m to the number of yearly births n is one of the most important indicators in demography.) But what is the error, ΔM, of this measure? Basing himself on the Bayesian approach and taking $N = 1.5 \cdot 10^6$ Laplace obtained a stochastic estimate for ΔM:

$$P(|\Delta M| \le 0.5 \cdot 10^6) = 1 - 1/1162.$$

220

It seems that this was the first-ever quantitative assessment of the error inherent in sampling. We note that in the process of his calculations, in connection with the need to sum the terms of a binomial, Laplace encountered the very difficult and important problem of determining the values of the incomplete B-function.

More than a century later, in the introduction to the *Tables of the incomplete B-function* (London, 1922; second ed. 1934),[8] K. Pearson called Laplace's calculations "imperfect". Nevertheless, the very fact that no further progress was made in such calculations before Pearson allows us to value Laplace's work highly.

Pearson (*Biometrika*, vol. 20A, pt. 1–2, 1928) also criticized Laplace from a theoretical point of view, noting, for example, that Laplace had considered both pairs of numbers, m, n and M, N, as independent samples from a common infinite universe. In fact, the samples are not independent, and the very existence of such a universe is questionable.

Laplace's remark in Chapter 5 of livre 2 of the *Théorie analytique* that it is possible to estimate the number π experimentally using the "Buffon needle" should also be assigned to mathematical statistics; more specifically, to the prehistory of the method of statistical testing. If a narrow cylinder — a needle of a certain length — is repeatedly dropped on a plane ruled by a set of parallel lines a given distance apart or by a network of squares with given side, then one can show that the probability p that the cylinder will intersect some line of the ruling is a function of π. By making repeated throws of the cylinder one can calculate the statistical estimator of the probability p and thus estimate π.

In this connection Laplace wisely referred to "a special type of problems in combinations of chances";[9] note, however, that in such cases the accuracy of the method of statistical testing is not high.

Laplace paid much attention to the stochastic estimation of testimonies of witnesses and verdicts of law courts. Consider for example one such problem from Chapter 11 of livre 2: A ticket is drawn at random from an urn containing n tickets numbered 1 through n. A witness testifies that the number on the ticket drawn is i. What is the probability that this is exactly what happened assuming that the witness (a) does not deceive and is not deceived; (b) does not deceive but is deceived etc. (four cases in all) and that the drawing of each number has the same prior probability?

It turns out that the less probable the event described by the witness the more probable it is that he is making a mistake or is a deliberate fraud. On these grounds Laplace, in his *Essai philosophique*, rejected Pascal's celebrated *infini-rien* wager: the trustworthiness of witnesses who made reference to

---

8 Russian translation: Moscow 1974.

9 Laplace, P.S., *Œuvres complètes*, t. 7. Paris 1893, p. 365.

God and declared that infinite bliss was awaiting believers was infinitely small. Accordingly, the believer's expected benefit is doubtful, since it is the product of an infinitely large good multiplied by the infinitely small (not finite, as Pascal had thought) trustworthiness.[10]

While discussing the work of tribunals, Laplace proceeded from the assumption that the probability p of a just verdict, that is, of either acquitting the innocent or convicting the guilty, is the same for each judge and, in addition, exceeds one-half. Under this supposition the probability that r judges unanimously arrive at a correct decision turns out to be

$$\frac{p^r}{p^r + (1-p)^r}.$$

Comparing this result with statistical data Laplace determined an estimate of p.

These and similar considerations also due to Laplace are based on the assumption that the opinions of different judges are independent. Actually, however, this supposition is never realized, either in the weighing of material evidence, say, or in the possibly unintentional gauging of the personality of the accused. In his "Science and method" Poincaré expressed this idea in exaggerated form.[11]

Already A.A. Cournot, the French mathematician and philosopher (1801–1877), took notice of the social nature of prejudices in legal proceedings in §213 of his "Exposition of the theory of chances and probabilities" (*Exposition de la théorie des chances et des probabilités*. Paris 1843).[12]

It is possible that these applications of the theory of probability attracted public attention to the need for improving criminal statistics. Again, the theory itself advanced somewhat under the influence of such applications (p. 236).

Laplace's Theory of Errors

General notions about errors of observations have had a long history. Even Ptolemy (second century) was aware of observational errors and made recommendations on how to combine observations. In the eleventh century, Al-Biruni, in his treatises "Geodesy" and "Quanun al-Masudi", noted the ex-

---

10 Laplace, P.S., *Essai philosophique*, p. LXXXVIII.

11 Poincaré, H., *Science et méthode*. Paris 1906; see p. 92 of the edition of 1914. An English translation of this source and of several other of the author's popular contributions is in Poincaré, H., *Foundations of science*. Washington 1982.

12 Latest French edition: Paris 1984.

istence of random errors in astronomical observations and calculations. In treating observations mathematically he used a qualitative approach that reflected the stochastic properties of usual random errors. Galileo (HM, vol. 2, Chapter 5) was the first to formulate these properties explicitly. However, the theory of errors was founded only in the middle of the 18th century, largely by Simpson and Lambert. At the end of that century, Daniel Bernoulli did important work; in particular (HM, vol. 3, pp. 133–137), he separated errors of observation into random (normally distributed) and systematic (constant) ones. Nevertheless, before Laplace there was no general solution of the following key problem (which came up when, for example, one dealt with measurements of a meridian arc): Given a system of equations

$$a_i x + b_i y + c_i z + \cdots + l_i = 0 \quad (i = 1, 2, \ldots, n) \tag{5}$$

in m $(m < n)$ unknowns $x, y, z, \ldots$, choose a reasonable solution $(x, y, z, \ldots)$ leading to small residuals (v_i) and estimate the error of this solution caused by the errors in the constants l_i.

In line with the then current terminology, scientists spoke about determining the "true values" of the unknowns. This greatly impeded the establishment of connections between the theory of errors and the chapter of mathematical statistics that appeared subsequently and dealt with estimating the unknown parameters of probability distributions.

In the "Memoir on the probability of causes as determined by events" (*Mémoire sur la probabilité des causes par les événements*, 1774)[13] and in the above mentioned "Memoir on probabilities", his first works devoted to the theory of errors, Laplace considered the case of one unknown. Here he gave the impression of trying to get accustomed to the theory. He studied possible ways of using the relatively new mathematical tool of density functions and compared various more or less natural criteria for choosing an estimator for unknown parameters. Among these rules was the condition that subsequently became his main criterion — namely, the minimization of the absolute expectation.

Even for a small number of observations Laplace's practical formulas proved much too complicated. His lack of progress was also due to the fact that he assumed certain laws of distribution of observational errors without taking them from astronomical practice.

In his later writings Laplace turned to the case of a large number of observations. In the "Memoir on the approximation of formulas that are functions of very large numbers and on their application to probabilities" (*Mémoire sur les approximations des formules qui sont fonctions de très grands nombres et sur leur application aux probabilités* (1809), 1810) he begins by considering

---

13 Laplace, P.S., *Œuvres complètes*, t. 8. Paris 1891, pp. 27–65.

random variables ξ_i $(i = 1, 2, \ldots, n)$ taking values $0, \pm 1, \pm 2, \ldots, \pm m$ with the same probability $1/(2m + 1)$. Laplace forms the sum

$$
\Omega(\omega) = e^{-m\omega i} + e^{-(m-1)\omega i} + \cdots + 1 + \cdots
$$
$$
+ e^{(m-1)\omega i} + e^{m\omega i} \tag{6}
$$

and notes that in this sum, raised to the nth power, the coefficient of $\exp(l\omega i)$ is the number of combinations for which $[\xi]^{14}$ is equal to l, and that therefore the corresponding probability is

$$
\frac{1}{\pi} \int_0^{\pi} d\omega \cos(l\omega) \Omega^n(\omega). \tag{7}
$$

Modern readers will readily see that Laplace used the notion of characteristic function and applied the inversion formula, albeit in a very simple case.

Laplace is extremely careless in his reasoning and in carrying out formal transformations. He begins by considering discrete random variables ξ_i, then reasons about variables distributed uniformly on a segment $[-h, h]$ while dealing in the text itself with a variable distributed uniformly on the segment $[0, h]$. He approximates these continuous uniformly distributed random variables by discrete variables by subdividing the segment $[0, h]$ into $2m$ equal parts of "unit length". But then it turns out that by a "unit length" he means a segment of length h/m and that in fact he subdivides the segment $[-h, h]$.

By formally transforming the integrand in (7) Laplace obtains a limit theorem for the case when $n \to \infty$. This can be written as

$$
\lim_{n \to \infty} P\left(-s \leq \frac{[\xi]}{\sqrt{n}} \leq s\right) = \frac{\sqrt{3}}{h\sqrt{2\pi}} \int_0^s e^{-x^2/2\sigma^2} \cdot dx
$$

where $\sigma^2 = h^2/3$ is the variance of the variable ξ_i.

Laplace extends this result to the case of arbitrary distributions (having variances), so that the memoir under discussion may be said to be devoted to the derivation of the central limit theorem for sums of identically distributed random variables. However, this derivation is greatly lacking in rigor. Nevertheless, it is appropriate to emphasize Laplace's exceptional intuition that enabled him to arrive at correct conclusions using non-rigorous and, now and then, simply confused reasoning.

---

14 We use Gauss' notation; for example $[aa]=\sum_{i=1}^{n} a_i^2$, $[ab]=\sum_{i=1}^{n} a_i b_i$; the symbol $[a]$ stands for $a_1+a_2+\cdots+a_n$.

In his "Memoir on definite integrals and on their application to probabilities and, especially, to the determination of the mean which it is necessary to
choose from the results of observations" (*Mémoire sur les intégrales définies
et leur application aux probabilités, et spécialement à la recherche du milieu
qu'il faut choisir entre les résultats des observations* (1810), 1811) Laplace,
while applying the same mathematical tools, once more deduced the central
limit theorem, this time for a linear function $[q\varepsilon]$ of random variables ε_i distributed according to the same law, assuming that the quantities q_i were of
the same order.

Linear functions of random variables, namely, of errors of observations,
appear naturally when we multiply each equation

$$a_i x + l_i = \varepsilon_i \tag{8}$$

(cf. system (5)) by a certain factor q_i, sum the resulting equations, and choose
the estimator

$$x = -\frac{[ql]}{[qa]} + \frac{[q\varepsilon]}{[qa]}. \tag{9}$$

Assuming — in virtue of the central limit theorem — that the distribution
of $[q\varepsilon]$ is normal, and choosing as a criterion the minimum of the absolute
expectation of the estimator (9), Laplace discovered that the factors q_i and,
therefore, the estimator itself, should be determined by the method of least
squares. He then extended his account to the case of several unknowns and
obtained in the course of his transformations the bivariate normal distribution
with independent components. It should be noted that at that time, and up
to the end of the 19th century, nobody even mentioned the assumption that
the components of a random vector are independent, believing it to be self-
evident.

In the work just described Laplace used the Cauchy-Bunyakovskiĭ inequality (also known as the Cauchy-Schwarz inequality) and, while changing variables in a double integral, he effectively used the appropriate Jacobian.

In Chapter 4 of the *Théorie analytique* Laplace proved that the limit
distributions of the sums $[\xi]$, $[|\xi|]$, $[\xi\xi]$, and $[q\xi]$ are the same (namely, the
standard normal), provided that appropriate centering and norming is done
and the random variables ξ_i are independent (Laplace did not mention this
condition) and identically distributed on a segment bounded from below and
above. However, there was no need to prove these propositions since they are
a corollary of Laplace's earlier theorem on the limit distribution of sums of
identically distributed independent terms having finite variance.

Laplace added three supplements to the third edition of his *Théorie analytique*, all of them devoted to the theory of observational errors. The second
supplement is of special interest. Here Laplace studied the error of the sum

of the measured angles of a triangle. He assumed that this error is normally distributed with density function $\sqrt{h/3\pi} \cdot \exp(-hx^2/3)$. As the estimator of the measure of precision Laplace offered the expression $Eh = 3n/2\theta^2$, where θ^2 was determined by the equality

$$\theta^2 = T_1^2 + T_2^2 + \cdots + T_n^2.$$

Here T_i are the errors of the sums of the angles in the ith triangle and n is the number of independent observations (triangles).

In addition, Laplace noted that the h, considered as a random variable, had a density function proportional to

$$h^{n/2} \exp(-h\theta^2/3).$$

In the same supplement, Laplace compared the measures of precision in estimating the results of observations by their arithmetical mean and median.

In the theory of errors Laplace adhered systematically and persistently to the following idea: Suppose that it is required to estimate a certain quantity a given the results of its independent measurements ξ_i, $i = 1, 2, \ldots, n$. Then the chosen estimator should be the value $\hat{a}$ that minimizes the sum

$$\sum_{i=1}^{n} E|\xi_i - \hat{a}|.$$

This rule can be used when the errors of the relevant quantities are normally distributed. Otherwise, however, it leads to very complicated calculations. This is why Laplace's criterion was later used rather infrequently.

We see that Laplace's contribution to the theory of observational errors is very significant. He stated the exceptionally fruitful idea that the observed error is a result of summing a large number of independent elementary errors. If these elementary errors are uniformly small, then, under very general conditions, the distribution of the observed error must be close to the normal law. In our time this idea has won universal recognition. One should also note Laplace's views on the multiplicity of possible estimators of the true value associated with given results of observation and on the estimation of the measure of precision (of the variance), as well as his outline of the derivation of the central limit theorem for the case of identically distributed terms having finite variances.

Gauss' Contribution to the Theory of Probabilty

It fell to Gauss' lot to create a theory of errors that immediately found numerous followers. In the second book of his "Theory of motion of heavenly bodies rotating around the Sun in conic sections" (*Theoria motus corporum cœlestium in sectionibus conicis Solem ambientium*. Hamburg 1809)[15] Gauss proved that among unimodal, symmetric and differentiable distributions $\varphi(x - x_0)$ there exists exactly one normal distribution for which the maximum likelihood estimator of the location parameter x_0 coincides with the arithmetical mean. His derivation follows: Let M, M', M'', ... be the observations and p their arithmetical mean. Then the likelihood equation

$$\varphi'(M - \hat{x}) + \varphi'(M' - \hat{x}) + \varphi'(M'' - \hat{x}) + \cdots = 0,$$

where $\varphi'(\Delta) = d\varphi(\Delta)/\varphi(\Delta)d\Delta$ and $\hat{x}$ is the maximum likelihood estimator, must have a unique solution $\hat{x} = p$:

$$\varphi'(M - p) + \varphi'(M' - p) + \varphi'(M'' - p) + \cdots = 0.$$

Supposing that $M' = M'' = \cdots = M - \mu N$, we obtain for positive integer values of μ (μ is the number of observations, $N(\mu - 1) = M - p$),

$$\varphi'([\mu - 1]N) = (1 - \mu)\varphi'(-N), \quad \varphi'(\Delta)/\Delta = k \ (k < 0),$$
$$\varphi(\Delta) = ce^{k\Delta^2/2}.$$

As a simple corollary, Gauss found that the density function of a given set of observations attains its maximal value when the sum of the squares of the discrepancies between the observed values of the measured constant and its 'true' value attains its minimum. This was his justification of the principle of least squares.

Certain faults were inherent in this exceptionally elegant derivation and Gauss himself indicated them later on. First, only normally distributed errors were recognized as random errors of observations. In pointing out this restriction in §17 of part 1 of his "Theory of combinations of observations least prone to errors" (*Theoria combinationis observationum erroribus minimis obnoxiae*. Göttingen 1823) Gauss added that he will offer "a new discussion of the subject" so as to prove that "the method of least squares leads to the best combination of observations ... for any probability law of the errors ..."[16]

Second, as Gauss noted in a letter to Bessel in 1839, the principle of maximum likelihood is not the best possible:

I must consider it less important in every way to determine that value of an unknown parameter (Grösse) for which the probability is largest,

---

15 English translation: Boston 1857; reprint: New York 1963.

16 Gauss, C.F., *Theoria combinationis*, §17. This work is available in French (1855) and German (1887) translations.

although still infinitely small, rather than that value, by relying on which one is playing the least disadvantageous game; or if fa denotes the probability that the unknown x has the value a, then it is less important that fa should be a maximum than that $\int fxF(x-a)dx$, taken over all possible values of x, should be a minimum, where for F is chosen a function which is always positive and which always increases with increasing arguments in a suitable way.[17]

Openly acknowledging the arbitrariness of the choice of F, Gauss selected $F = x^2$, that is, the principle of minimal variance. As he proved in the "Theory of combinations of observations", among linear estimators those determined by the method of least squares have minimal variance; this was Gauss' second justification of the method. Nevertheless, until quite recently, many astronomers who considered themselves Gauss' followers linked the method of least squares with the realization of the normal distribution (and with the principle of maximum likelihood independently introduced into mathematical statistics in 1912 by R.A. Fisher). As late as 1898, in a letter to A.V. Vasil'ev, A.A. Markov had to refute this dated point of view.[18]

Gauss' idea of the preferability of an integral measure of precision ("least disadvantageous game") turned out to be exceptionally fruitful and, again independently of Gauss, such measures were later introduced into mathematical statistics.

A reappraisal of the ideas and possibilities of the classical theory of errors, that is, in essence, of Gauss' achievements, occurred only in recent decades, within the statistical theory of estimation of parameters.

The theory of probability was far from the center of Gauss' scientific interests. Nevertheless he enriched this theory by many outstanding results and influenced the subsequent development of a number of its important chapters.

The method of least squares immediately enjoyed general recognition. Gauss shares the credit for its discovery and introduction with two of his contemporaries, A.M. Legendre and Robert Adrain.

The first publication that clearly formulated the principle was Legendre's "New methods for determining the orbits of comets" (*Nouvelles méthodes pour la détermination des orbites des comètes*. Paris 1805 and 1806). Legendre maintained that the application of the principle led to a certain "equilibrium" among the residual errors and that in this case the absolute values of the maximal errors were minimal among all methods of adjusting observations.

---

17 Gauss, C.F., *Werke*, Bd. 8. Göttingen-Leipzig 1900, pp. 146–147. The English translation above is taken from Plackett, R.L., "The discovery of the method of least squares". *Biometrika*, vol. 59 pt. 2, 1972.

18 Extracts from Markov's letters to Vasil'ev are published under a general title "The law of large numbers and the method of least squares" in Markov, A.A., *Selected works. Theory of numbers. Theory of probability.* Leningrad 1951, pp. 233–251 (in Russian).

The latter statement was false, but, evidently because of the priority strife between Legendre and Gauss, it was simply forgotten.

Gauss had applied the principle of least squares — in studies unpublished at the time — long before Legendre's memoir appeared in print. This is what he wrote in this regard in his "Theory of motion of heavenly bodies" (§186): "... our principle which we have used since the year 1795 has lately been published by Legendre..."[19]

Since Legendre had offered only a qualitative justification of the principle of least squares, his merit, compared with that of Gauss, was minor and there were hardly any grounds for a priority argument. Nevertheless, Legendre, especially in the Second Supplement to the third edition of his "New methods for determining the orbits of comets", sharply protested Gauss' use of the expression "our principle" (unser Princip).[20]

In 1808 (or 1809), starting with a practical problem of land surveying, the American mathematician Robert Adrain (1775–1843) published a paper[21] which included two derivations of the normal law of distribution of observational errors and of the principle of least squares. The paper appeared in an American journal that was completely unknown in Europe and that moreover appeared for only one year. While Adrain's derivations of the normal law were flawed, and he seems to have learned of the principle of least squares from Legendre's memoirs, his results deserve to be mentioned primarily because, like Gauss, he used the principle of maximum likelihood. Also, in 1818 Adrain was the first to apply the method of least squares to the determination of the oblateness of the earth's ellipsoid of revolution from results of meridian arc measurements[22] and, in the same year, he estimated the semiaxes of this ellipsoid with a precision exceptional for his time.[23]

A number of other scholars (S.D. Poisson, A. Cauchy) also took up the theory of errors, but it is more natural to consider their work in the context of the theory of probability.

In concluding this section we point out that Gauss left isolated, unpublished notes written over the years. One of these contained the inversion formula for the Fourier transform of the density function. However, Fourier,

---

19 See note 15.

20 Legendre, A.M., *Nouvelles méthodes pour la détermination des orbites des comètes*. Second supplément. Paris 1820, pp. 79–80.

21 Adrain, R., *Research concerning the probabilities of the errors which happen in making observations*. "Analyst or math. museum", vol. 1, No. 4. Philadelphia 1808, pp. 93–109. Reprinted in *American contributions to mathematical statistics in the 19th century*. New York 1980.

22 Adrain, R., *Investigation of the figure of the earth and of the gravity in different latitudes*. Trans. Amer. Phil. Soc., vol. 1 (New ser.), 1818. Repr. in vol. 1 of *Amer. contr.*

23 Adrain, R., *Research concerning the mean diameter of the earth*. Ibid. Reprinted in vol. 1 of *Amer. contr.*

Cauchy, and Poisson introduced this formula even earlier. In a letter to Laplace dated 30 January 1812 Gauss formulated the following problem, which may be considered the first in the metric theory of numbers.[24] Let M be a certain number between 0 and 1 and let

$$M = 1/a' + 1/a'' + 1/a^{(3)} + \cdots$$

be its continued fraction expansion. What is required is the probability that the "tail" of this fraction,

$$1/a^{(n+1)} + 1/a^{(n+2)} + 1/a^{(n+3)} + \cdots,$$

is less than x. Denoting this probability by $P(n, x)$, Gauss discovered that, assuming $P(0, x) = x$,

$$\lim_{n \to \infty} P(n, x) = \frac{\ln(1 + x)}{\ln 2}.$$

Gauss himself was not satisfied with the reasoning that had led him to this result in 1799[25] and asked Laplace to find a rigorous proof. But it took more than a century before R.O. Kuzmin (1928) produced one.

The contributions of Poisson and Cauchy

In contemporary science, Poisson's name is linked with the concept of a distribution, with a process, and with the Poisson law of large numbers.

Siméon Denis Poisson (1781–1840) was the son of a petty provincial official who, during the years of the French Revolution, reached a somewhat higher station. His father wanted him to become a notary. When the boy showed little enthusiasm for the study of law, the father apprenticed him to a barber, from whom, however, he ran back home. After that the father, whose situation had meanwhile improved, agreed to send his son to school at Fontainebleu, where one of the teachers discovered the youth's mathematical gift and helped him to prepare for the entrance examination at the *Ecole Polytechnique*. There, Lagrange, Laplace, and some other very good French mathematicians guided and inspired the young man. In 1800 Poisson graduated from the *Ecole Polytechnique*, published his first scientific papers, and was appointed *répétiteur* at the Ecole. In 1806 Fourier quit teaching and Poisson replaced him as professor. In 1816, when a faculty of science was created at the Sorbonne, Poisson became professor of rational mechanics. Somewhat earlier, in 1812, he was elected to the Paris Academy of Sciences. The nature

---

24 Gauss, C.F., *Werke*, Bd. 10, Abt. 1. Göttingen – Leipzig 1917, pp. 371–374.

25 This is certified by notes made by Gauss in his manuscripts.

of his creative work marked Poisson as a typical representative of the Ecole Polytechnique. He published more than three hundred papers, a total second, it seems, only to Cauchy's. His scientific interests embraced many branches of analysis (including complex integration and multiple integrals), the theory of probability, and, above all, various fields of applied mathematics: general and celestial mechanics and mathematical physics (the theory of elasticity, heat conductivity, capillarity, electricity, and magnetism). F. Klein wrote:

> How versatile and fruitful his pursuits were may be gathered by the large number of details still connected with his name: Poisson brackets in mechanics, Poisson constants in the theory of elasticity, the Poisson integral in potential theory, and, finally, the generally known and widely used Poisson equation $\Delta V = -4\pi\rho$, which he established for the space inside an attracting body by generalizing the Laplace equation $\Delta V = 0$ valid for the space outside.[26]

This impressive list fails to mention Poisson's important "law of large numbers" (see below).

Poisson's contribution to the theory of probability is linked first and foremost with his book "Researches on the probability of sentences and verdicts in criminal and civil matters" (*Recherches sur la probabilité des jugements en matière criminelle et en matière civile*. Paris 1837) in which he continued the study of problems already considered under certain assumptions by Laplace.

Poisson's predecessors — Laplace, Gauss, and others — widely used the concept of a random variable in their works but always in connection with the results of observations and with the problems of the theory of observational errors. Poisson was probably the first who attempted to separate the concept from these problems, to regard a random variable as a general notion equally important for all natural sciences. He spoke about "anything whatever" (*une chose quelconque*) that can take values $a_1, a_2, \ldots, a_\lambda$ with corresponding probabilities. However, somewhat earlier, in the second part of his memoir "On the probability of the mean results of observations" (*Sur la probabilité des résultats moyens des observations. Connaissance des temps* (1829), 1832), he also tried to consider in the same way continuous random variables, as well as their distribution functions.[27] However, apart from the attempt to introduce a special term for a concept in actual use, Poisson's theory of random variables did not essentially differ from the store of knowledge mastered and applied by his forerunners and contemporaries.

---

26 Klein, F., *Vorlesungen über die Entwicklung der Mathematik im 19. Jahrhundert*, Tl. 1. Berlin 1926, p. 68.

27 The first part of this memoir was published in the same periodical ((1824), 1827).

S.D. Poisson

Poisson's name is linked with the introduction of the "law of large numbers" into science.[28] He saw clearly the importance of theorems such as the law of large numbers for our knowledge of the physical world, and was a passionate champion of this law. For him, the essence of this law consisted in the approximate equality of the arithmetical means of a large number of random variables and of their expectations.[29] However, he was unable to prove this proposition in all its generality. Only the generalization of the Bernoulli theorem to independent trials, where the studied event A occurs with probabilities dependent on the ordinal numbers of the trials, is still known under his name.

Poisson paid much attention to the Bernoulli scheme of trials. By tradition, he never mentioned that he studied independent trials, but he tacitly assumed this restriction. Poisson[30] wrote that if the probability that event A occurs in each trial is p, then the probability of its happening not less than m times

28 Poisson, S.D., *Recherches sur la probabilité des jugements*. Paris 1837, p. 7.

29 Ibid., pp. 138–143.

30 Ibid., p.189.

232

in $\mu = m + n$ trials is

$$P = p^m \left[1 + mq + \frac{m(m+1)}{2!} q^2 + \cdots + \frac{m(m+1)\ldots(m+n-1)}{n!} q^n \right]$$

$$= \frac{\int_\alpha^\infty X\,dx}{\int_0^\infty X\,dx},$$

$$X = \frac{x^n}{(1+x)^{\mu+1}}, \quad \alpha = \frac{q}{p}, \quad q = 1 - p.$$

(10)

On calculating these integrals he obtained asymptotic equalities for large m and n

$$P = \frac{1}{\sqrt{\pi}} \int_k^\infty e^{-t^2}\,dt + \frac{(\mu+n)\sqrt{2}}{3\sqrt{\pi\mu mn}} e^{-k^2} \quad \text{for } q/p > h, \tag{11}$$

$$P = 1 - \frac{1}{\sqrt{\pi}} \int_k^\infty e^{-t^2}\,dt + \frac{(\mu+n)\sqrt{2}}{3\sqrt{\pi\mu mn}} e^{-k^2} \quad \text{for } q/p < h, \tag{12}$$

Here, $k = \sqrt{n \ln(n/q\mu) + m \ln(m/p\mu)}$ and $h = n/(m+1)$ is the abscissa of the maximum point for the function $X(x)$.

Of course, formulas (11) and (12) are just those of the DeMoivre-Laplace integral limit theorem. All Poisson did was to write it down in a form somewhat different from Laplace's (see p. 215).

For small q he set

$$mq \approx \mu q = \omega, \quad m(m+1)q^2 \approx \omega^2, \ldots, \quad p^m \approx e^{-\omega}$$

and obtained

$$P \approx e^{-\omega} \left[1 + \omega + \frac{\omega^2}{2!} + \cdots + \frac{\omega^n}{n!} \right]. \tag{13}$$

This is the formula for the Poisson approximation of Bernoulli's distribution. Expressions such as

$$P(\xi = m) \approx e^{-\omega} \frac{\omega^m}{m!}$$

are not found in his writings.

Poisson used the limit theorem to estimate the statistical significance of the difference $n_2/\mu_2 - n_1/\mu_1$ of ratios relating to two independent samples in cases of known, and unknown, probability p. He then derived a limit theorem for an urn problem involving drawings without replacement. He also applied it to a model of an electoral system. According to his assumptions, each voter belongs to one of two existing national parties numbering a and b members respectively with $a : b = 90.5 : 100$. If party members are randomly distributed over the election districts then, as Poisson calculated, the probability of the election of a minority deputy is very small. For 459 districts and $a + b = 200,000$ this probability turns out to be 0.16. This model of an electoral system should not be taken seriously, especially since Poisson himself understood its inadequacy. For us it is more important to note his attitude towards the "representativeness" of a government so elected. In his words, a representative government is nothing but a deception (*déception*).[31]

Poisson also derived the local and integral limit theorems for a scheme of trials with variable probabilities.[32] Suppose that in μ trials an event E happens with probabilities $p_1, p_2, \ldots, p_\mu$ and the opposite event F with probabilities $q_1, q_2, \ldots, q_\mu$. Then the probability that the event E occurs m times (and the event F, $n = \mu - m$ times) is equal to the coefficient of $u^m v^n$ in the expansion

$$(up_1 + vq_1)(up_2 + vq_2)\ldots(up_\mu + vq_\mu) = X$$

and is therefore equal to

$$U = \frac{1}{2\pi}\int_{-\pi}^{\pi} Xe^{-(m-n)ix}dx = \frac{2}{\pi}\int_{0}^{\pi/2} Y\cos[y - (m-n)x]dx,$$

where

$$Y = \rho_1\rho_2\ldots\rho_\mu, \quad y = r_1 + r_2 + \cdots + r_\mu;$$

here ρ_k and r_k are, respectively, the modulus and argument of the complex function

$$(p_k + q_k)\cos x + i(p_k - q_k)\sin x = \rho_k e^{ir_k}$$

formed from the binomial $(up_k + vq_k)$ by setting $u = e^{ix}$ and $v = e^{-ix}$.

For large μ, excepting the case in which p_k (or q_k) decreases with increasing k, Poisson derived the following local limit theorem by expanding ρ_k and Y

31 Ibid., p. 244.

32 Ibid., p. 246.

in power series in x:

$$P(m = p\mu - \theta c\sqrt{\mu}, \ n = q\mu + \theta c\sqrt{\mu})$$

$$\equiv U = \frac{1}{c\sqrt{\pi\mu}}e^{-\theta^2} - \frac{h\theta}{2c^4\mu\sqrt{\pi}}(3 + 2\theta^2)e^{-\theta^2}, \tag{14}$$

$$c^2 = \frac{2[pq]}{\mu}, \quad h = \frac{4}{3\mu}\sum_{i=1}^{\mu}(p_i - q_i)p_i q_i.$$

Starting with (14), Poisson immediately obtained the integral limit theorem.

Then, making an unwarranted logical leap, he formulated a false theorem[33] and justified it by inexact reasoning. He believed that sums formed by random variables having finite variances but otherwise completely unspecified, when centered by the sums of the expectations of their terms and normed by the square root of the sums of the variances of their terms, necessarily have a distribution that is close to the standard normal. As a corollary, Poisson then "proved" the law of large numbers for arbitrary summands having finite variances.

In his book Poisson makes elementary use of generalized functions

$$f(x) = \sum_k \gamma_k \delta(x - c_k) \quad \left(\sum \gamma_k = 1, \ \gamma_k > 0\right)$$

(δ is the symbol for the Dirac function) while introducing density functions equal to zero everywhere except at a finite number of points $C_1, C_2, \ldots, C_n$. At these points

$$\int_{C_k - \varepsilon}^{C_k + \varepsilon} f(x)dx = g_k, \ k = 1, 2, \ldots, n,$$

where ε is an infinitely small positive number and

$$g_1 + g_2 + \cdots + g_n = 1.$$

Poisson did not offer any additional explanations. Note, however, that Dirac himself defined his celebrated function on the same low level of rigor.

Poisson devoted about a hundred pages of his book to applications of the theory of probability to criminal law suits. Here we find some general studies concerning the probability that the accused will be found guilty (studies of its

---

33 Ibid., pp. 271–277.

dependence on the number of jurors, especially in cases of unanimous verdicts of guilty); remarks on the relative stability of the numbers of those accused of, and condemned for, each of the two main types of offences, viz., of crimes against persons and against property; on the influence of the revolution of 1830 on the work of the law courts; and more. Poisson deliberately simplified his initial suppositions (thus, he assumed that the jurors pronounce their judgments independently), so that his research hardly yielded a real public benefit. However, his reasoning on the probabilities of convicting or acquitting an accused person,[34] as well as similar considerations due to J.A. Condorcet and Laplace, may be assigned to the prehistory of the concepts of critical region and of errors of the first and second kind.

Poisson's "Memoir on the probability of mean results of observations" is devoted to the theory of observational errors. A substantial part of it is, effectively, a commentary on the work of Laplace. Like Laplace, Poisson used characteristic functions to derive formulas for the distributions of a sum and a linear function of a large number of errors of observations. The fact that in this memoir Poisson introduced the distribution

$$f(x) = \frac{1}{\pi(1 + x^2)}, \quad |x| < +\infty, \tag{15}$$

subsequently called the Cauchy distribution, is of considerable interest for the history of the theory of probability. Poisson found that if the observations were distributed according to the "Cauchy law", then their arithmetical mean had the same distribution.

We see that Poisson discovered the Cauchy distribution and its property of being stable about twenty years before Cauchy. It is therefore natural to restore historical justice and rename the distribution (15) after its true author.

Poisson was keenly interested in expanding the field of application of stochastic methods. That is why he applied the theory of probability to problems in medicine and in population statistics. He was coauthor of a review of one of the first books on medical statistics. He wrote the "Memoir on the proportion of births of girls and boys" (*Mémoire sur la proportion des naissances des filles et des garçons*. Mém. Acad. sci. Paris, 1830, t. 9) on the sex ratio of newborns. We note that formula (13) that gave rise to the Poisson distribution first appeared in this very memoir.

In the memoir "On the advantage of the banker in the game 'thirty and forty'" (*Sur l'avantage du banquier au jeu de trente et quarante*. Ann. math. pures et appl., 1825–1826, t. 16) Poisson had to solve the following problem: An urn contains x_1 balls numbered 1, x_2 balls numbered 2, ..., x_i balls numbered i ($x_1 + x_2 + \cdots + x_i = s$). It is required to determine the probability

---

34 Ibid., pp. 388–392.

P of extracting, in n drawings without replacement, a_1 balls numbered 1, a_2 balls numbered 2, ... a_i balls numbered i $(a_1 + a_2 + \cdots + a_i = n)$, where the sum of the numbers thus drawn is

$$a_1 + 2a_2 + \cdots + ia_i = x. \tag{16}$$

While solving this problem, Poisson obtained, at first without using the condition (16), the result

$$P = \frac{n!(s-n)!}{s!} \cdot \frac{x_1!}{a_1!(x_1 - a_1)!} \frac{x_2!}{a_2!(x_2 - a_2)!} \cdots \frac{x_i!}{a_i!(x_i - a_i)!}$$

$$= (s+1) \int_0^1 (1-y)^s Y \, dy,$$

$$Y = \frac{x_1!}{a_1!(x_1 - a_1)!} \left(\frac{y}{1-y}\right)^{a_1} \frac{x_2!}{a_2!(x_2 - a_2)!} \left(\frac{y}{1-y}\right)^{a_2}$$
$$\cdots \frac{x_i!}{a_i!(x_i - a_i)!} \left(\frac{y}{1-y}\right)^{a_i}.$$

For $i = 2$ the system of probabilities thus obtained is now called the hypergeometric distribution, and the problem itself has a direct bearing on sample quality control of production.

In order to allow for the restriction (16) Poisson replaces Y by the sum of its values corresponding to sets $\{a_1, a_2, \ldots, a_i\}$ that satisfy this condition. Considering the generating function

$$\left(1 + \frac{yt}{1-y}\right)^{x_1} \left(1 + \frac{yt^2}{1-y}\right)^{x_2} \cdots \left(1 + \frac{yt^i}{1-y}\right)^{x_i}$$

he notes that the required probability is equal to the coefficient of t^x in the expression

$$(s+1) \int_0^1 (1 - y + yt)^{x_1} (1 - y + yt^2)^{x_2} \ldots (1 - y + yt^i)^{x_i} \, dy.$$

With regard to the same game of chance Poisson had to introduce a second sample of balls, $\{b_1, b_2, \ldots, b_i\}$, drawn after the first one, again without replacement, and, accordingly, to determine the joint probability that condition (16) and the equality

$$b_1 + 2b_2 + \cdots + ib_i = x'$$

are fulfilled. To this end he used a bivariate generating function.

A. Cauchy contributed greatly to the theory of probability. From 1831 through 1853 he published no less than ten memoirs on the mathematical treatment of observations and, in part, on the theory of probability. Eight of these are collected in tome 12 of the first series of his "Collected works" (*Œuvres complètes*), which appeared in 1900. Some of them were written as a result of the author's discussions with I.J. Bienaymé (1853), especially on the use of the method of least squares for solving various problems on interpolation of functions.

Cauchy studied the treatment of observations by the method of means and by the minimax principle, that is, the principle of determining residuals of maximal absolute value which are minimal for all methods of treating the observations. Nowadays, this method is used in statistical decision theory.

In his memoir "On the maximal error to be feared in a mean result, and on the system of factors that renders this maximal error a minimum" (*Sur la plus grande erreur à craindre dans un résultat moyen, et sur le système de facteurs qui rend cette plus grande erreur un minimum*, 1853; *Œuvres complètes*, t. 12. Paris, 1900, pp. 114–124) Cauchy noted that a linear function of n non-negative variables $\lambda_1, \lambda_2, \ldots, \lambda_n$ satisfying m additional linear equations $(m < n)$ attains its maximum when $(n - m)$ of these variables vanish. He thus proved one of the theorems in the theory of linear programming — whose prehistory should begin at least with Fourier (1824).[35]

The incompleteness of this memoir, its failure even to indicate the need for an effective algorithm for checking the basic solutions of the problem formulated above, is typical of the hasty exposition in many of Cauchy's writings.

In his memoir "On the mean results of observations of the same nature and on the most probable results" (*Sur les résultats moyens d'observations de même nature, et sur les résultats les plus probables*, 1853; *Œuvres complètes*, t. 12, pp. 94–104) Cauchy solves the elegant problem of determining the density function $f(\varepsilon)$ of observational errors ε_i $(i = 1, 2, \ldots, n)$ subject to the condition that the probability $P(\omega_1 < \Delta x < \omega_2)$ that the error of one of the unknowns (x) belongs to any given segment (ω_1, ω_2) is maximal. For the required even distribution $f(\varepsilon) = f(-\varepsilon)$ he obtained, for $\theta > 0$, the characteristic function

$$\varphi(\theta) = e^{-c\theta^{\mu+1}}; \tag{17}$$

here μ is real and $c > 0$.

This derivation hardly mattered in terms of direct practical use. It is interesting, however, that the function (17) is characteristic only for $-1 < \mu \leq 1$ and that the corresponding distribution is stable.

---

35 See Grattan-Guinness, I., *Fourier's anticipation of linear programming*. Operat. res. quarterly, vol. 21, 1970, pp. 361–364.

Assuming $\mu = 1$ and $\mu = 0$, Cauchy obtained, respectively, the normal distribution and the "Cauchy distribution"

$$f(\varepsilon) = \frac{k}{\pi(1 + k^2\varepsilon^2)},$$

considered earlier by Poisson (see p. 236).

The "Memoir on limiting coefficients" (*Mémoire sur les coefficients limitateurs ou restricteurs*, 1853; *Œuvres complètes*, t. 12, pp. 79–94) is devoted to discontinuity factors and especially to their use in the theory of probability. Suppose that one studies the stochastic behavior of a function $\omega(x_1, x_2, \ldots, x_n)$ of errors $x_1, x_2, \ldots, x_n$ with distribution laws $\varphi_1(x_1)$, $\varphi_2(x_2)$, ..., $\varphi_n(x_n)$ which do not vanish on the respective segments

$$[\mu_1, \nu_1], [\mu_2, \nu_2], \ldots, [\mu_n, \nu_n].$$

Then, as Cauchy notes,

$$P(\omega_1 \leq \omega \leq \omega_2) = \int\limits_{\mu_1}^{\nu_1} \int\limits_{\mu_2}^{\nu_2} \cdots \int\limits_{\mu_n}^{\nu_n} I\varphi_1(x_1)\varphi_2(x_2)\ldots\varphi_n(x_n)dx_1dx_2\ldots dx_n,$$

where

$$I = \begin{cases} 1, & \text{for } \omega_1 \leq \omega \leq \omega_2, \\ 0, & \text{for } \omega < \omega_1 \text{ or } \omega > \omega_2. \end{cases}$$

For example, Cauchy proposed to use the function

$$I = \frac{1}{2\pi} \int\limits_{\omega_1}^{\omega_2} d\theta \int\limits_{-\infty}^{\infty} e^{\theta(\tau-\omega)i}d\tau$$

as a discontinuity factor of this kind.

Laplace and Poisson used discontinuity factors in the theory of probability even before Cauchy.

Of special interest are Cauchy's papers "On the probability of errors that affect the mean results of observations of the same nature" (*Sur la probabilité des erreurs qui affectent des résultats moyens d'observations de même nature*, 1853) and the "Memoir on the mean results of a very large number of observations" (*Mémoire sur les résultats moyens d'un très grand nombre d'observations*, 1853).[36] In these writings, Cauchy proves, in roughly the same

---

36 See Cauchy, A.L., *Œuvres complètes*, t. 12, sér. 2. Paris 1958, pp. 104–114, 125–130.

way, the central limit theorem. Thus in the first work Cauchy considers the linear function

$$w = [\lambda\varepsilon] \tag{18}$$

of errors ε_i having an even density distribution $f(\varepsilon)$ which does not vanish on the segment $[-\chi, \chi]$. The characteristic function of the quantity (18) is

$$\Phi(\theta) = \varphi(\lambda_1\theta)\varphi(\lambda_2\theta)\ldots\varphi(\lambda_n\theta),$$

where $\varphi(\theta)$ is the characteristic function of the errors ε_i, so that

$$P(-\nu \leq w \leq \nu) = \int_0^\nu F(\tau)d\tau = \frac{2}{\pi}\int_0^\infty \Phi(\theta)\frac{\sin\theta\nu}{\theta}d\theta, \tag{19}$$

where $F(\nu) = \dfrac{2}{\pi}\displaystyle\int_0^\infty \Phi(\theta)\cos\theta\nu d\theta$ is the density function of (18).

Passing over to an estimation of the integral in formula (19), Cauchy notes that, for $\lambda_i = O(1/n)$, $\rho = O(\sqrt{n})$ and $\theta < \rho$

$$\Phi(\theta) = e^{-s\theta^2}, \ s = [\varphi\lambda\lambda],$$

where the φ_i are close to

$$c = \int_0^\chi \varepsilon^2 f(\varepsilon)d\varepsilon$$

(and, therefore, $2s$ is close to the variance σ^2 of the function (18)). Then it turns out that

$$F(\nu) = \frac{1}{\sqrt{\pi s}}e^{-\nu^2/4s},$$

$$P(-\nu \leq w \leq \nu) = \frac{1}{\sqrt{\pi s}}\int_0^\nu e^{-x^2/4s}dx \approx \frac{\sqrt{2}}{\sigma\sqrt{\pi}}\int_0^\nu e^{-x^2/2\sigma^2}dx.$$

Cauchy also estimated the errors due to the assumptions he made.

We may say that Cauchy used widely the "Fourier cosine transform". He devoted two memoirs, viz., "On a law of reciprocity that exists among certain functions" (*Sur une loi de réciprocité qui existe entre certaines fonctions*, 1817), and "Second note on reciprocal functions" (*Seconde note sur*

240

les fonctions réciproques, 1818),[37] exclusively to the Fourier cosine and sine transforms. In these works he referred to Poisson and to Fourier's studies on the theory of heat, unpublished at the time, but submitted by their author to the Paris Academy of Sciences in 1807 and 1811.

We have briefly described the works of the main representatives of a remarkable period in the development of the theory of probability, a period that covered almost the entire first half of the 19th century. During this time the foundations of the analytic methods used in the theory of probability were laid. Moreover, concepts of new limit theorems were provided, and the importance of these theorems for the theory of errors, for population statistics, for the study of the scatter of projectiles, and for the theory of probability itself was keenly appreciated.

Attempts were made to extend the sphere of application of the law of large numbers. They proved successful for the generalized Bernoulli scheme in which the probabilities of the occurrences of the studied events change with the number of the trial. Laplace, Gauss, and other scientists gave the theory of errors a harmonious form suitable for practical use. To a large extent this accounts for the fact that at first astronomers (and natural scientists in general), and then mathematicians began calling the normal distribution (a term introduced by K. Pearson) the Gaussian law. Historically speaking, the incorrectness of this term is obvious; for one thing the normal distribution first appeared in a work of DeMoivre long before Gauss' birth and was widely used by Laplace long before Gauss began to study mathematics and even before he went to school.

This was a heroic period in the development of the theory of probability. Even scholars of the highest caliber, such as Laplace, Poisson, and Cauchy, were anxious to express their accumulated convictions as soon as possible, and supported their ideas by analytical methods which were far from logical perfection. It is easy to detect flaws both in the formulations and in the proofs of their results. Although Laplace, Poisson, and Cauchy offered promising formulations of theorems like the central limit theorem and the law of large numbers, these results should be regarded as expressions of enthusiasm for a vast, new and promising field of mathematical studies rather than as mathematical facts. Thus it is not surprising that Poisson's general formulations of the law of large numbers and of the central limit theorem were later found to be incorrect. Considerable time and the genius of a new generation of scientists — in the first place P.L. Chebyshev, A.A. Markov, and A.M. Liapunov — were required in order to find appropriate, elegant mathematical formulations and proofs of the expectations, the conjectures, and the subjective convictions as to the wide scope and profundity of these propositions. Nevertheless, we must do justice to the genius of Laplace, Pois-

---

37 Ibid., t. 2, 1958, pp. 223–227, 228–232.

son, and Cauchy, to their profound ability to foresee the value of fundamental scientific concepts and of new directions of mathematical thought.

The seeds were sown in fertile soil and time was needed for them to sprout, to develop, and to reveal to mankind, through powerful methods of mathematical research, profound regularities in the world around us. At any rate, we see how the brilliant representatives of (mostly) French science of those times laid the basis of the mathematical science of nature and the theory of probability.

Of course, along with fundamental uses, it was inevitable that attempts be made to apply the new mathematical tool to fields of science and domains of practical activities where these applications were, to put it mildly, questionable.

The dominant philosophical idea of this period of development of probability theory was the belief in the universality of the concept of independence. Therefore, as a rule, authors did not then (or indeed up to the end of the 19th century) mention this assumption. Only in a few of Laplace's works did the idea of a Markov-type dependence begin to surface.

Social and Anthropometric Statistics

As a scientific discipline, population statistics arose in the middle of the 18th century. The first theoretical studies in this field, mainly due to Daniel Bernoulli and Laplace, are discussed in vol. 3 of the HM (pp. 130–133).

In 1800, the *Bureau de la statistique générale* was created in France (it existed until 1812 and was reestablished in 1833) and a first attempt at enumerating the population of that country was made. In 1832, after A. Quetelet had visited England, a statistical section was added to the *British Association for the Advancement of Science* and, during the next year, a permanent commission of this section was formed under the chairmanship of C. Babbage. In 1834 the *London* (now *Royal*) *Statistical Society* was established. Its aim was to collect and publish numerical data on population, industry, etc.

The *American Statistical Society* was created in 1839. The *Russian Geographical Society*, founded in 1845, became the ideological center of Russian statistical activities; but even at the beginning of the 19th century (from about 1809), the *Free Economic Society* carried out statistical and geographical studies of individual administrative provinces. In the various German states, statistical services were created relatively early. For example, the Prussian statistical bureau, which was established in 1805 and existed for only one year, was reestablished in 1810, and Austria's statistical bureau was formed in 1840.

Thus the statistical institutions or national societies which studied and developed population statistics of individual countries came into being in the main states of Europe and America in less than five decades. In addition,

242

from 1851 onward, international statistical congresses began to take place. They aimed at unifying official statistical data and, in particular, they fostered the spread of the metric system. Even in France, this system, created by the French Revolution, was finally introduced only in 1837. It acquired international status only in 1875, after seventeen countries had signed the metric convention; other states joined later. However, some countries, for example the United States, still unofficially use their own systems of measures.

It soon became apparent that the unification of statistical data pertaining to individual countries was impossible, and after 1876 the holding of official statistical congresses was discontinued. The requirements of capitalist society brought about the establishment of national statistical services, but the differences and contradictions existing in this society made it impossible to combine the separate national data into a single whole.

It is impossible to imagine the development of population statistics in the 19th century without studying the works of the Belgian geometer, astronomer, and meteorologist Lambert Adolphe Jacques Quetelet (1796–1874). These were prompted by the ideas of Laplace, Poisson, and other scientists, for example J.B. Fourier who, as editor, published a statistical description of Paris and the Département de la Seine in four volumes (1821–1829). Though he did not leave any serious legacy in the domain of science itself, Quetelet was a gifted popularizer of statistics, to which he attracted the attention of the European public. Quetelet was permanent chairman of the Belgian *Commission centrale de statistique* and the organizer of the first international statistical congress.

Quetelet's writings contain many dozens of pages devoted to the rudimentary statistical analysis of numerical data on the height of men at different ages, on the relation between the height and the weight of men, etc. He rightly considered all these data necessary both for the general study of man and for forensic medicine. Before Quetelet, only sculptors and painters (for example L.B. Alberti) had engaged in the elementary treatment of anthropometric data, and they were interested in the type of the perfect man rather than in men in general.

The main place in Quetelet's teaching was occupied by the notion of the "average man" (*L'homme moyen*) which he used extensively. (This notion turned up earlier in the works of Buffon (HM, vol. 3, p. 146).) At first (1831) he introduced it in the anthropometric sense. Then he developed and popularized this concept in many of his writings. One of the best known of these was his two-volume book "On man and on development of his faculties, or an essay on social physics" (*Sur l'homme et le développement de ses facultés, ou essai de physique sociale*, tt. 1–2. Bruxelles 1836).[38] Many mathematicians

---

38 English translation: *A treatise on man and the development of his faculties* (1842). Incorporated, with separate paging, in: *Comparative statistics in the 19th century*. Gregg Intern. Publ., 1973.

and statisticians criticized Quetelet's notion of the average man. Thus, for example, J. Bertrand, in his "Calculus of probabilities" (*Calcul des probabilités*. Paris 1888), a book about which we have occasion to speak below, wrote:

> In the body of the average man, the Belgian author placed an average soul. [The average man] has no passions or vices, he is neither insane nor wise, neither ignorant nor learned ... [he is] mediocre in every sense. After having eaten for thirty-eight years an average ration of a healthy soldier, he has to die not of old age, but of an average disease that statistics discovers in him.[39]

Of course, these cutting remarks made by the outstanding French mathematician had nothing to do with the general statistical use of mean values, whose importance nobody denied.

Quetelet's idea of studying the laws of historical development by the change of the average man from one age to another is quite interesting. However, here too his reasoning was far from lucid.

For his time, Quetelet offered an interesting interpretation of the fact that the number of crimes is constant. He wrote:

> ... there is a budget item which we pay with frightful regularity — it is that for prisons, dungeons, and scaffolds. Now, it is this item which, above all, we ought to endeavor to reduce; and every year the numbers have confirmed my previous statements to such a degree that I might have said, perhaps with more precision, 'there is a tribute which man pays with more regularity than that which he owes to nature, or to the treasury of the state, namely, that which he pays to crime'. ... We might even predict how many individuals will stain their hands with the blood of their fellowmen, how many will be forgers, how many will deal in poison, in much the same way as we can foretell the annual births and deaths.
> Society includes the germs of all the crimes committed, and at the same time the necessary facilities for their development. It is the social state, in some measure, which prepares the crimes, and the criminal is merely the instrument for their execution. Thus, every social state admits a certain number and a certain order of crimes, these being merely the necessary consequences of its organization (*Sur l'homme*, t. 1, p. 10).[40]

Given this statement, Quetelet's recommendation — evidently the only one he ventured — is rather vague: Legislators, he said, should remove the causes of crime "as far as possible".[41] To judge by his political views, Quetelet

---

39 Bertrand, J., *Calcul des probabilités*. Paris 1888, p. XLIII.

40 *A treatise on man*, p. 6. Note that Quetelet wrote this in 1836 (Editors).

41 *Sur l'homme*, t. 2, p. 341 (*autant que possible*).

A. Quetelet

was a bourgeois liberal. In 1869 K. Marx gave the following evaluation of the work of Quetelet the statistician:

He rendered important services in the past; he proved that even apparently random events in social life possess an intrinsic necessity due to their periodic recurrence and their periodic mean numbers. However, he was never able to interpret this necessity. He made no progress; he just extended his observations and calculations.[42]

From a mathematical point of view, the stability of statistical series should be scientifically substantiated (this is what Marx pointed out) whereas Quetelet restricted himself to maintaining that each man had the same potential, common to all men, to commit a crime; each man had a certain "criminality coefficient" common to all members of a given society.

A statistician could well speak about the existence of a "criminality coefficient", about the constancy of the ratio of the number of those found guilty

---

42 Marx, K., *Briefe an L. Kugelmann.* Berlin 1952, pp. 81–82.

to the number of accused, might indeed predict the number of crimes even under less restrictive conditions (Poisson).

Nevertheless — and our specifications are in keeping with the spirit of applying statistics as a scientific method — such statistical inferences can not be extended to individuals. Thus an accused cannot be held to be somewhat guilty on the strength of being charged; on the contrary, in each case, the judge must proceed from the presumption that the accused is innocent. At one time, the introduction of a single "criminality coefficient" aroused sharp opposition, especially from German statisticians. For example, G. Rümelin (1815–1889) wrote in 1867:

> If statistics, guided by its mean figures, will dare tell me that, with a probability of one over so and so much, I shall commit a crime, I shall answer it quite definitely: ne sutor ultra crepidam! [Cobbler! Stick to your last!][43]

The prominent Russian scientist, Aleksander Aleksandrovich Chuprov (1874–1926), a corresponding member of the Petersburg [National] Academy of Sciences, included this quotation in his "Essays[44] on the theory of statistics" (1909). Like the German statisticians, Chuprov criticized the metaphysical nature of Quetelet's statistical views. In the "Essays" he expounded the critique of Quetelet's propositions formulated by those who feared that references to a criminality coefficient will lead to a public excusing of crime.[45]

It should be noted that the publication of statistical data on criminality displeased officials even before Quetelet. Thus a report by the Russian economist and statistician K.F. German (1767–1838) on the statistics of murders and suicides, published as early as 1823, provoked the ire of Russia's minister for public education. For his part, German attempted to prove that these data were important for studying the moral and political state of the population.

The mathematical incompleteness of Quetelet's writings, his failure to even formulate the problem of the conditions necessary for the stability of statistical series, had negative consequences. Relying on statistical data, social scientists would often state thoughtless conclusions which compromised the statistical method itself.

One critic of this state of affairs was A.Yu. Davidov. His remarks, in which he did not mention Quetelet by name, appeared in 1855 in "Scientific and

---

43 Rümelin, G., *Über den Begriff eines socialen Gesetzes* (1867). In: *Reden und Aufsätze*. Freiburg i. Br. – Tübingen. Preface dated 1875, pp. 1–31 (p. 25). These words were presumably said by the painter Apelles to a certain mender of shoes who, after having correctly pointed out an inaccuracy in the drawing of a sandal in one of Apelles's pictures, began to criticize, with the air of connoisseur, other features of his art.

44 Chuprov, A.A., *Essays on the theory of statistics*. Moscow 1959, p. 211.

45 Ibid., pp. 210–212.

literary papers written by professors and instructors of Moscow University published on the occasion of its centenary" (we return to Davidov below). Nevertheless, it was Quetelet's works that gave rise to the so-called Continental branch of statistics, whose main object of study remained population statistics.

An eminent representative of this area of statistics in the 19th century was Wilhelm Lexis (1837–1914), professor at Strassburg, Breslau (Wrocław) and Göttingen. He introduced a quantitative measure of the stability of statistical series, isolated several types of such series according to the nature of the variability of their terms, and offered an elementary criterion for testing the mutual independence of these terms.

The works of Lexis and his followers in the 20th century, among whom we should name, above all, V.I. Bortkevich (L. von Bortkiewicz) and A.A. Chuprov, proved to be especially important for the creation of the discipline of mathematical statistics.[46]

The Russian School of the Theory of Probability. P.L. Chebyshev

The probability-theoretic studies of the Paris school of mathematicians, and especially of Laplace and Poisson, soon became known and were further developed in Russia. The need for fostering insurance, population statistics, and the mathematical treatment of observations, as well as the intrinsic logic of the development of mathematics, led to the publication, from the 1820s to the 1840s, of the first expositions in Russia of the theory of probability and of discourses on its importance and, later, of original studies in this field. Under conditions then prevailing in Russia, all this was further stimulated by the resolute anti-Kantian position held by a number of mathematicians, for example, by the Khar'kov professors T.F. Osipovskiĭ (1765–1832) and his disciple A.F. Pavlovskiĭ (1788–1856), the author of "On probability" (Khar'kov 1821), the first popular scientific booklet on this subject to be published in Russia. The idea of probability, that is, of incomplete certainty of human

---

46 Lexis' main work, "On the theory of stability of statistical series" (*Über die Theorie der Stabilität statistischer Reihen*, 1879) is available in Russian in a collection of articles "On the theory of dispersion" (Moscow, Statistika publ., 1968) with a commentary by the compiler and translator, N.S. Chetverikov. English translation: "Works Project Administration", 1941. Among Lexis' predecessors we should name I.J. Bienaymé whose role in this connection was repeatedly stressed by A.A. Chuprov (see his "Essays", 1959, p. 280). Vladislav Iosifovich Bortkevich (1868–1931), a lawyer by education and an alumnus of Petersburg University, became a professor at Berlin University in 1901 and became Ladislaus von Bortkiewicz. His works belong to statistics, mathematical statistics, and political economy. In his "Law of small numbers" (*Das Gesetz der kleinen Zahlen*. Leipzig 1898) he was one of the first to turn attention to the Poisson distribution; using Lexis' results he studied the stability of the number of occurrences of unlikely events in long series of observations.

knowledge and of scientific inferences, was quite in line with the denial of Kantian innate notions.

On June 29, 1841, during a festive meeting of Moscow University, a faculty member named N.D. Brashman (1796–1866), an expatriate from Bohemia, declared:

Who can fail to see, with extreme regret, the absolute neglect, in the academic institutions, of one of the most important branches of mathematics? Only a few universities teach the rudiments of the theory of probability and, up to now, there does not exist a single Russian work, or translation, on advanced, or even elementary probability... We hope that Russian scientists will soon try to make up this deficiency.[47]

And, indeed, just two years later, Brashman's colleague at the physical and mathematical faculty, N.E. Zernov (1804–1862) delivered a speech entitled "The theory of probability with special reference to mortality and insurance". Its published text (Moscow, 1843), 85 pages long, contained an exposition of the main concepts and elementary theorems of this science and numerous arguments in favor of various types of insurance.

Of considerable interest are the works of Viktor Yakovlevich Bunyakovskiĭ (1804–1889), a prominent organizer of scientific research and, from 1864 through 1889, vice-president of the Academy of Sciences. Between 1846 and 1860 Bunyakovskiĭ was professor at Petersburg University where, among other things, he regularly lectured on the theory of probability.

Bunyakovskiĭ's name was already mentioned in Chapter 3 and in this Chapter, and will come up in other parts of this work [in vol. 2 (1981) — once; he is not mentioned in vol. 3 (1987)]. One of his important merits is the publication of a manual on the theory of probability, "Elements of the mathematical theory of probability" (Petersburg, 1846).

Under the powerful influence of Laplace and Poisson, Bunyakovskiĭ thought that the most important goal of his manual was to simplify their conclusions and to elucidate their findings. It should be noted that A. De Morgan,[48] Cournot,[49] and, much earlier, S.F. Lacroix[50] pursued the same aim.

---

47 Brashman, N.D., *On the influence of the mathematical sciences on the development of intellectual faculties.* Moscow 1841, pp. 30–31 (in Russian).

48 De Morgan, A., *Theory of probability.* In: *Encyclopaedia metropolitana,* vol. 2. London 1845, pp. 393–490.

49 We mentioned his work (1843), intended for a broader circle of readers, on p. 222.

50 His "Elementary treatise on the calculus of probability" (*Traité élémentaire du calcul des probabilités*) ran into four editions (Paris 1816, 1822, 1833, and 1864) and was translated into German (1818).

Bunyakovskiĭ's second important goal was to work out a Russian terminology. Many of the terms he introduced became standard. On the other hand, a number of problems he considered were of no scientific interest and were even badly formulated. Many of his general statements do not stand up to criticism. According to Bunyakovskiĭ, the theory of probability studies even phenomena about which nothing at all is known. This is not so. Probability theory is no exception in the system of sciences — it does not allow scientists to draw positive inferences from complete ignorance. It is the specific nature of mass random phenomena rather than ignorance that leads to stochastic regularities. In this fundamental point Bunyakovskiĭ, compared with Laplace, took a big step backward.

Some of Bunyakovskiĭ's works belong to the prehistory of the theory of statistical testing. Following Laplace, he suggested finding the values of transcendental functions by statistical testing.[51] Also, it was Bunyakovskiĭ's works that ensured the popularity of problems on "numerical probabilities".

Here is a relevant example taken from Bunyakovskiĭ's treatise: To determine the probability that a quadratic equation with nonzero integer coefficients chosen at random has real roots. It seems that the first to propose such a problem was N. Oresme, a 14th-century scholar, who maintained that two ratios (we might say "two numbers") taken 'at random' are probably incommensurable.[52]

Finally, we note that Bunyakovskiĭ posed and solved a curious problem which would nowadays be assigned to the theory of random arrangements ("On combinations of a special type occurring in a problem concerning defects"; Supplement No. 2 to vol. 20 of the *Notes of the Academy of Sciences*, 1871). Here is the statement of the problem.

Given a number of defective copies of a booklet (for example, with equal probability that one or another page is missing in each of them). It is required to determine the probability that a definite number of complete copies can be made up from them.

Bunyakovskiĭ published a large number of works devoted to problems of Russia's population statistics, including long essays on the age distribution of the population; on the laws of mortality; and on the probable number of future conscripts. These works were of much practical importance, in particular for the development of the insurance business and of pension schemes. They also played an important part in the development of the statistical method of studying the population in Russia.

---

51 Bunyakovskiĭ, V.Ya., *On the application of the analysis of probabilities to determining the approximate values of transcendental numbers*. Mém. Acad. sci., vol. 1, No. 5, 1837 (in Russian).

52 Oresme, N., *De proportionibus proportionum* ("On ratios of ratios"). Transl. E. Grant. Madison 1966, p. 247: *Ad pauca respicientes* ("Concerning some matters"). Madison 1966.

V.Ya. Bunyakovskiĭ

Bunyakovskiĭ's contemporary, the academician M.V. Ostrogradskiĭ, published a few articles on the theory of probability and mathematical statistics. His main researches, on analysis, mathematical physics and mechanics, will be treated in the third [1987] volume of this work. One of these papers was of practical importance; it was written in connection with the establishment of a retirement fund[53] at the Navy Department ("A note on the retirement fund", 1858).[54]

In an article titled "On a problem bearing on probabilities" (1848),[55] Ostrogradskiĭ pointed out that his results can be applied to check the quality of production.

---

53 Retirement funds were set up by societies for mutual insurance. Navy officers paid yearly dues into the fund and, in return, were granted annuities on retirement. In addition, the fund gave annuities to the families of deceased officers.

54 Ostrogradskiĭ, M.V., *Complete works*, vol. 3. Kiev 1961, pp. 297–300 (in Russian).

55 *Sur une question des probabilités*. Bull. phys.-math. Acad. Imp. Sci. St-Pétersb., t. 6, No. 21–22, 1848, 321–346.

In 1858 Ostrogradskiĭ delivered a course of twenty lectures in the theory of probability at the *Ordnance Academy*. (The first three lectures were apparently published, but a search for them was unsuccessful.) Ostrogradskiĭ was also interested in the notion of "moral expectation", introduced into the theory of probability by Daniel Bernoulli and Laplace.

A report on Ostrogradskiĭ's lecture preserved in the *Recueil des actes de la séance publique de l'Acad. imp. sci. St. Pétersbourg tenue le 29 Déc. 1835* (St. Pétersbourg, 1836) attests his attempt at generalizing Laplace's assumptions by adopting a more general law of moral expectation.[56]

Of considerable methodological importance were the works of Avgust Yul'evich Davidov (1823–1885), a disciple of Brashman, a professor at Moscow University from 1853, and the author of a number of studies in mechanics and on various mathematical problems. His lithographed lectures on the theory of probability (1854), which he delivered after 1850, fully deserved to have appeared in print. In this course, as well as in his later works, Davidov paid considerable attention to the concepts of randomness and probability (definitely preferring objective probabilities) and to the law of large numbers, and regularly studied the statistical significance of observations.

The mathematical content and the main propositions of Davidov's probability-theoretic researches are close to the works of Laplace and Poisson. Davidov, however, did not engage in "moral" applications of the theory of probability. Rather, he paid much attention to its application to medicine.

In a later lithographed course on the theory of probability (1884–1885) Davidov introduced the notion of an integral law of distribution, an idea just hinted at by Poisson. Critically viewing the work of Quetelet, Davidov maintained that statistical inferences should be strictly substantiated by methods — we would now say — of mathematical statistics.

One of Davidov's remarks, made in a report at a festive meeting of Moscow University in 1857 and hardly appreciated by his contemporaries, merits special mention. Statistics, he indicated, has to do both with means obtained from observations of actually existing quantities (for example, the height of an object) and with means describing fictitious quantities (for example, a mean value of bread). However, Davidov added, this distinction is insignificant. One can only regret that, basing themselves on just this insignificant distinction, astronomers and geodesists have stubbornly refused to take notice of the achievements of mathematical statistics virtually until today.

P.L. Chebyshev, who was repeatedly mentioned in all the previous chapters, a junior contemporary of Bunyakovskiĭ and Ostrogradskiĭ, opened a new period in the development of the theory of probability.

Pafnutiĭ Lvovich Chebyshev (1821–1894), along with N.I. Lobachevskiĭ the greatest Russian mathematician of the 19th century, was born into the family of a petty landowner and received his first mathematical training from a

---

56 Russian version in: Ostrogradskiĭ, M.V., *Pedagogical legacy*. Moscow 1961, pp. 293–294.

A.Yu. Davidov

resident tutor. From 1837 through 1841 he studied at the physical and mathematical faculty of Moscow University under Brashman (mentioned above), a versatile man of great learning who directed his attention to problems of both pure mathematics and its applications and of contemporary technology. During his student days, Chebyshev wrote a paper "Calculation of the roots of equations" for a competition announced by the faculty. In view of the originality of the method used and the manner of execution of the work Chebyshev deserved the gold medal; but, because of accidental circumstances, he was awarded only the silver. This paper was published only recently.[57] In 1846 Chebyshev defended his Master's dissertation "Essay on an elementary analysis of the theory of probability" (Moscow, 1845), and published an article on the Poisson law of large numbers; see below.

In 1847 Chebyshev received an invitation to Petersburg University. There, upon defending, in the spring of the same year, a dissertation which gave him the right to deliver lectures (a thesis *pro venia legendi*) and devoted

---

57 See Chebyshev, P.L., *Complete works*, vol. 5. Moscow – Leningrad 1951, pp. 7–25 (in Russian).

to integrating algebraic irrational functions in closed form, a problem previously studied by Abel, Liouville, and Ostrogradskiĭ, he began teaching at the rank of docent (lecturer). While working on this dissertation Chebyshev encountered certain problems in the theory of numbers. His interest in this theory increased still more through his participation in preparing the publication of Euler's number-theoretic works (1849, see p. 248). At the same time Chebyshev worked on his doctoral dissertation, "The theory of congruences", which he published and defended in the same year. At that time he gained instant fame by publishing his discoveries in the theory of distribution of prime numbers (see Chapter 3). Later, he returned to the theory of numbers only infrequently.

In 1850 Chebyshev was appointed professor at Petersburg University and, in 1853, he was elected to the Academy of Sciences. For several decades he worked tirelessly both at the University and at the Academy. In addition, from the beginning of 1856 he was active for thirteen years as member of the Ordnance Department of the Military-Scientific Committee and, for seventeen years, as member of the Scientific Committee of the Ministry for public education.

At the University, Chebyshev taught various courses. First he taught the theory of numbers. Following Bunyakovskiĭ's retirement in 1860, he took over the teaching of the theory of probability. He devoted to probability two fundamental works, published in 1867 and 1887, respectively. Already while in Moscow, Chebyshev became attracted to practical (applied) mechanics, and, for a few years, he lectured on this subject in addition to lecturing on mathematics at the University and at the prestigious *Aleksandrovsky Lyceum*. Given the context of this essay, we cannot describe the many mechanisms invented by Chebyshev or his contribution to the general theory of mechanisms, but must mention that these interests helped him create his theory of best approximation of functions with its numerous ramifications. Chebyshev himself often called it the theory of functions that deviate least from zero, whereas S.N. Bernstein, in 1938, called it the constructive theory of functions.

Chebyshev devoted many works, published over forty years — from 1854 onward — to the theory of best approximation of functions. Gradually, this theory embraced a very wide group of related problems, viz., the theory of orthogonal polynomials; the method of moments; evaluation of integrals; quadrature formulas; interpolation; the theory of continued fractions, a key tool of his research, etc. The theory of best approximation of functions will be described in vol. 3 [1987] of this work.

In addition to his creative activity, Chebyshev's most important contribution was the establishment of a large mathematical school. Its members were, first, his students at Petersburg University, where he taught for thirty-five years (until 1882), the students of his students, and his numerous followers from other cities. Chebyshev was not only a good lecturer but also an outstanding scientific advisor, with a rare gift for successfully choosing and rigorously posing for young researchers new problems whose solution promised to lead to valuable discoveries.

Chebyshev's mathematical school, which came to be called the Petersburg school (also see Chapter 2), played a prominent part in the progress of mathematics not only in Russia but also worldwide. Some of his immediate students were A.N. Korkin, Yu.V. Sokhotskiĭ, Ye.I. Zolotarev, A.A. Markov, A.M. Lyapunov, I.I. Ivanov, K.A. Posse, D.A. Grave, G.F. Voronoĭ, A.V. Vasil'ev. N.Ya. Sonin, V.A. Steklov, A.N. Krylov. Other scientists also belonged to his school. (Readers either have seen the names of most of these scholars in the previous chapters or will see them below.) The subjects taken up by members of his school were not restricted to the range of topics studied by Chebyshev himself; gradually they attained extreme diversity, partly under the influence of problems posed by him. For all the diversity of their studies, the representatives of the Petersburg school were always characterized by an approach to mathematics which Lyapunov, along with Zolotarev and Markov, one of Chebyshev's most eminent students, described in these words in an obituary written soon after the death of their teacher:

> P.L. Chebyshev and his followers invariably stick to reality. They are guided by the view that only those studies are of value which arise out of applications (scientific or practical), and that only those theories are really useful which follow from considering particular instances.
>
> Detailed elaboration of problems of special importance to applications and, at the same time, offering special theoretical difficulties that require the invention of new methods and an appeal to principles of science, followed by generalization of the obtained conclusions and the creation in this manner of a more or less general theory — this is the direction of most of the works of Chebyshev and of the scientists who adopted his views.[58]

Let us now turn to Chebyshev's works on the theory of probability. The number of his main publications in this field is not large — there are only four; but it is difficult to overrate their influence on the further development of this discipline. His ideas, which stimulated the creation of the classical Russian school of the theory of probability, are still being worked out in our time, while a complete solution of problems formulated by him was only accomplished in the 1940s. A.N. Kolmogorov writes:

> From a methodological aspect, the principal upheaval due to Chebyshev consisted not only in his being the first to demand, very emphatically,

---

58 Lyapunov, A.M., *Pafnutiĭ L'vovich Chebyshev*. Quoted from a reprint of this contribution in Chebyshev, P.L., *Selected mathematical works*. Moscow – Leningrad 1946, p. 20 (in Russian).

P.L. Chebyshev

that the limit theorems be proved with absolute rigor, ... but mainly in that in each instance he strove to determine exact estimates of the deviations from limit regularities — deviations possible even in a large but finite number of trials — in the form of inequalities unconditionally true for any number of trials.

Furthermore, Chebyshev was the first to clearly appreciate and use the full power of the concepts of random variable and mathematical expectation of a random variable; "... now, we always replace the study of event A by considering its characteristic random variable ξ_A that equals unity if event A occurs and zero otherwise. Thus the probability $P(A)$ of event A is just the expectation $M\xi_A$ of the variable ξ_A. The appropriate method of characteristic functions of sets came to be systematically used in the theory of functions of a real variable considerably later."[59]

In his first work — his Master's dissertation (see above) — Chebyshev aimed at giving an account of the theory of probability that minimized the

59 Kolmogorov, A.N., *The role of Russian science in the development of the theory of probability*. Uchen. zapiski Moscow University, No. 91, 1947, p. 56 (in Russian).

use of mathematical analysis. It covered the elements of the theory of probability, binomial schemes according to Bernoulli and Poisson (with a finite number of trials), the DeMoivre-Laplace limit theorems, and the mathematical treatment of observations.

Chebyshev's methodological aim[60] caused his writing to be ponderous, especially where integration of functions was replaced by summing. However, already in this work Chebyshev consistently estimated the errors of "pre-limiting" relations — a feature which was characteristic of his creative activity and was lacking in Laplace's writings. Thus Chebyshev introduced into the theory of probability a sober "engineering" approach that became typical of his entire scientific work and of all of mathematics after him.

According to Chebyshev, the theory of probability aims at determining the probabilities of some events in terms of given probabilities of other events. This being so, he came close to rejecting the inclusion of the theory among the natural sciences. However, as long as mathematical statistics was not isolated from the theory of probability, it was impossible to directly subordinate the latter to mathematics.

In his "Elementary proof of a certain general proposition of the theory of probability" (1846)[61] Chebyshev, applying a clever algebraic method, rigorously proved the following limit theorem: If in the k-th trial a certain event E has probability p_k $(k = 1, 2, \ldots, n)$, then the total number μ of the occurrences of E obeys the equality

$$\lim_{n \to \infty} P\left(\left|\frac{\mu}{n} - \frac{[p]}{n}\right| < \varepsilon\right) = 1. \tag{20}$$

This fact was known to Poisson, but he proved it in a nonrigorous manner. In this connection Chebyshev wrote:

Regardless of the cleverness of the method used by the celebrated Geometer he does not give the limit of the error allowed by his approximate analysis, and this uncertainty in the value of the error makes the demonstration nonrigorous.[62]

In his own derivation Chebyshev determined such a "limit of error" by estimating the probability P_m that the event E occurs no less than m times

---

60 Connected with the fact that his work was intended as a manual for students of the *Demidov Lyceum* in Yaroslavl. P.G. Demidov (1738–1821) was a natural scientist and a philantropist; note that it was P.N. Demidov (1798–1841) who established a fund for the awards conferred for a long time by the Petersburg Academy of Sciences.

61 *Démonstration élémentaire d'une proposition générale de la théorie des probabilités.* J. reine und angew. Math., Bd. 33, 1846, pp. 259–267.

62 See p. 259 of the French original.

in μ trials (cf. above). In Chebyshev's own notation, this estimate, necessary for a justified use of the Poisson theorem, is expressed by the inequality

$$P_m < \frac{1}{2(m-S)} \sqrt{\frac{m(\mu-m)}{\mu}} \left(\frac{S}{m}\right)^m \left(\frac{\mu-S}{\mu-m}\right)^{\mu-m+1},$$

where $S = p_1 + p_2 + \cdots + p_\mu$.[63] Note, however, that both Poisson and Chebyshev tacitly assumed that the events were independent. This assumption seemed so natural that Chebyshev did not specify it in his subsequent works either.

As mentioned above, in 1860 Chebyshev began to lecture on the theory of probability at Petersburg University. A few years later, in 1867, he published his next contribution, "On mean quantities"[64], in the *Matematicheskiǐ zbornik* and in the *Journal des mathématiques pures et appliquées*. This paper contained two extremely important propositions:

1. Let discrete random variables $x, y, z, \ldots$, that take on a finite number of values have respective expectations $a, b, c, \ldots$, and suppose that the respective expectations of their squares are $a_1, b_1, c_1, \ldots$. Put

$$L = a + b + c + \cdots - \alpha \sqrt{a_1 + b_1 + c_1 + \cdots - a^2 - b^2 - c^2 - \cdots},$$

$$M = a + b + c + \cdots + \alpha \sqrt{a_1 + b_1 + c_1 + \cdots - a^2 - b^2 - c^2 - \cdots},$$

where $\alpha > 0$. Then

$$P(L \leq x + y + z + \cdots \leq M) > 1 - 1/\alpha^2.$$

Alternatively, assume that

$$L' = \frac{a + b + c + \cdots}{n} - \frac{\alpha}{\sqrt{n}} \sqrt{\frac{a_1 + b_1 + c_1 + \cdots - a^2 - b^2 - c^2 - \cdots}{n}},$$

$$M' = \frac{a + b + c + \cdots}{n} + \frac{\alpha}{\sqrt{n}} \sqrt{\frac{a_1 + b_1 + c_1 + \cdots - a^2 - b^2 - c^2 - \cdots}{n}},$$

where n is the number of variables $x, y, z, \ldots$. Then, obviously,

$$P\left(L' \leq \frac{x + y + z + \cdots}{n} \leq M'\right) > 1 - \frac{1}{\alpha^2}. \tag{21}$$

---

63 *Complete works*, vol. 2, p. 21.

64 French version: *Des valeurs moyennes*; t. 12, pp. 177–184 of the French periodical; Russian version: *Complete works*, vol. 2, pp. 431–437.

2. Corollaries.

a) If the quantities $a, b, c, \ldots$ and $a_1, b_1, c_1, \ldots$ are uniformly bounded, then

$$\lim_{n \to \infty} P\left(\left|\frac{x + y + z + \cdots}{n} - \frac{a + b + c + \cdots}{n}\right| < \varepsilon\right) = 1. \tag{22}$$

b) Let $b = c = \ldots = a$, $b_1 = c_1 = \ldots = a_1$. Then

$$\lim_{n \to \infty} P\left(\left|\frac{x + y + z + \cdots}{n} - a\right| < \varepsilon\right) = 1.$$

This very simple and, at the same time, extremely important corollary appeared for the first time in Chebyshev's course of lectures in 1879–1880. It was published on the basis of Lyapunov's notes in 1936.[65]

c) Now let random variables $x, y, z, \ldots$ take on the values 0 and 1 with probabilities $\bar{p}, \bar{q}, \bar{r}, \ldots$, and $p, q, r, \ldots$, respectively, so that $a = p$, $b = q$, $c = r$, ... Then we may assume that these variables and the probabilities $p, q, r, \ldots$ describe the appearance of an event in Poisson trials and that the frequency of the occurrence of this event in n trials is

$$m/n = (x + y + z + \cdots)/n.$$

According to formula (22),

$$\lim_{n \to \infty} P\left(\left|\frac{m}{n} - \frac{p + q + r + \cdots}{n}\right| < \varepsilon\right) = 1,$$

an equality that coincides with formula (20).

In fact, Item 1 is a proof of the fundamental Bienaymé-Chebyshev inequality

$$P(|\xi - E\xi| < \beta) > 1 - D\xi/\beta^2$$

($\xi = x + y + z + \cdots$, $\beta > 0$; on Bienaymé's role see below).

Corollaries 2a) and 2b) contain the Chebyshev form of the law of large numbers. The Poisson (corollary 2c)) and the Jakob Bernoulli forms of that law are special cases of the Chebyshev variant.

The modern understanding of this problem is as follows:

---

65 Chebyshev, P.L., *Theory of probability*. Lectures delivered by academician P.L. Chebyshev in 1879/1880. Edited by A.M. Krylov from notes taken by A.M. Lyapunov. Moscow – Leningrad 1936.

A sequence of random variables

$$\xi_1, \xi_2, \ldots, \xi_n, \ldots \tag{23}$$

is said to obey the law of large numbers if there exist sequences of constants $a_1, a_2, \ldots, a_n, \ldots$ and $B_1, B_2, \ldots, B_n, \ldots$ $(B_n > 0)$ such that for any $\varepsilon > 0$

$$\lim_{n \to \infty} P \left(\left| \frac{1}{B_n} \sum_{k=1}^{n} (\xi_k - a_k) \right| < \varepsilon \right) = 1.$$

The lasting scientific value of the studies devoted to the law of large numbers, accomplished by a veritable constellation of scholars from Jakob Bernoulli to Chebyshev and, later on, by Markov and subsequent mathematicians, consisted in the discovery of the general conditions which necessarily imply the statistical stability of means, that is, the regularity of randomness. In this respect, especially great praise is due to Markov who, in the first years of the 20th century, laid the foundations of a new, extensive, and extremely important branch of the theory of probability concerned with the study of dependent random variables. In his paper "Extension of the law of large numbers to quantities dependent on each other", written and published in 1907,[66] Markov wrote:

"In his derivations, Chebyshev restricted himself to the simplest and therefore the most interesting case, that of independent quantities..." He then emphasized that "Chebyshev's derivations may also be extended to some cases of a rather general nature when the quantities depend on one another".[67]

In this article Markov considerably extended Chebyshev's condition for applying the law of large numbers. He proved that the sequence (23) also obeys this law in the case of dependent variables if

$$\frac{1}{n^2} D \sum_{k=1}^{n} \xi_k \to 0 \quad (n \to \infty).\text{[68]}$$

Somewhat later, in the same edition of the just-mentioned treatise, Markov discovered that the law of large numbers holds for sums of dependent variables if for some $\delta > 0$ and all k $(k = 1, 2, 3, \ldots)$

$$E|\xi_k|^{1+\delta} \le c \quad (c > 0).\text{[69]}$$

66 Markov, A.A., *Selected works. Theory of numbers. Theory of probability.* Leningrad 1951, pp. 339–361 (in Russian).

67 Ibid., pp. 341–342.

68 Also see Markov, A.A., *Calculus of probability*, third ed. Petersburg 1913, p. 76 (in Russian).

69 Ibid., p. 84.

M.A. Tikhomandritskiĭ (1844–1921), an alumnus of Petersburg University and a professor at Khar'kov University, extended Chebyshev's law of large numbers to continuous random variables (see his "Course of lectures on the theory of probability", Khar'kov 1898). I.V. Sleshinskiĭ (1854–1931), a professor at Odessa University, achieved the same goal even earlier (see his memoir *"On the theory of the method of least squares", Zapiski mat. otdeleniya Novoross. obshchestva estestvoispitateleĭ.* Odessa, 1892, vol. 14).

One of Chebyshev's outstanding works, his "On two theorems about probabilities", appeared in Russian in 1887, in a supplement to the "Zapiski" of the Academy of Sciences, and in French, in 1890–1891, in the "Acta mathematica".[70] His propositions are true despite a certain incompleteness of this work and a few flaws in the formulation of the theorem stated in the paper. Here Chebyshev applied his results on the method of moments to prove the Central Limit Theorem, thus initiating a powerful method of proof that subsequently found many uses and was extensively developed by others. In the same paper Chebyshev notes — without offering a rigorous proof — that it is possible to make the limit theorem more precise by applying an asymptotic expansion in Chebyshev-Hermite polynomials.

The first part of the program outlined by Chebyshev in this work was largely realized by his students Markov (1898) and Lyapunov (1901). The realization of its second part was begun by H. Cramér (1928) and was continued by a number of other scientists.

Chebyshev stated the main theorem as follows: If a) the expectations $a_i^{(1)}$ of random variables $u_1, u_2, \ldots$ are equal to zero and b) the absolute values of the expectations $a_i^{(2)}, a_i^{(3)}, \ldots, a_i^{(k)}, \ldots$ of their consecutive powers do not exceed some finite bound, then, as $n \to \infty$, the probability that the expression

$$\frac{u_1 + u_2 + \cdots + u_n}{\sqrt{2(a_1^{(2)} + a_2^{(2)} + \cdots + a_n^{(2)})}} \tag{24}$$

takes on a value contained between the limits z_1 and z_2 approaches the limit

$$\frac{1}{\sqrt{\pi}} \int_{z_1}^{z_2} e^{-x^2} dx.$$

Strictly speaking, Chebyshev's conditions are not sufficient for a flawless proof. First of all, following the tradition of his time, he did not specify that the variables u_k must be independent. Also, he did not indicate that as $n \to \infty$ the expression $(1/n) \sum_{k=1}^{n} a_k^{(2)}$ may tend to zero, in which case

---

70 *Sur deux théorèmes relatifs aux probabilités.* Acta math., t. 14, pp. 305–315. Russian version: *Complete works*, vol. 3. Moscow – Leningrad 1948, pp. 229–239.

the final conclusion is wrong. Finally, condition b) of the theorem was not formulated sharply enough, since it is not necessary to have one and the same constant for all k; it can depend on the order k of the moment.

Markov made the necessary amendments in the statement of Chebyshev's theorem and introduced relevant specifications in its proof almost immediately after the work had appeared in print. We will discuss this fact at somewhat greater length below.

Chebyshev alloted so little space to his second theorem that we will now reproduce all he said:

> I wish to mention that, according to formulas that I stated in the note "On the development of functions of one variable",[71] the following expression for this probability[72] holds for any n:

$$\frac{1}{\sqrt{\pi}} \int\limits_{z_1}^{z_2} \left[1 - k_3 \left(\frac{q}{\sqrt{2}} \right)^3 \psi_3(x) + k_4 \left(\frac{q}{\sqrt{2}} \right)^4 \psi_4(x) + \cdots \right] e^{-x^2} dx.$$

> Here, $k_3, k_4, \ldots$ are the coefficients of $s^3, s^4, \ldots$ in the expansion of the function
> $$\exp \left[\frac{M^{(3)}}{\sqrt{n}} s^3 + \frac{M^{(4)}}{n} s^4 + \cdots \right]$$
> in powers of s, and $\psi_3(x), \psi_4(x), \ldots$ are polynomials obtained by using the formula
> $$\psi_l(x) = e^{x^2} \frac{d^l e^{-x^2}}{dx^l}.$$

Here, $M^{(3)}$, $M^{(4)}$ and the other coefficients of the s^k are semi-invariants for the normed sums.

A few words about the French mathematician and statistician I.J. Bienaymé (1797–1878), with whom Chebyshev maintained a scientific contact and whose work he valued. In his Memoir "On the limiting values of integrals" (1874), Chebyshev wrote:

> In a memoir, very interesting in many respects, which Bienaymé read at the [Paris] Academy of Sciences in 1833 and which was published in the *Comptes rendus* and reprinted in Liouville's journal "J. math. pures

---

71 Chebyshev, P.L., *Complete works*, vol. 3, pp. 335–341. French version: *Sur le développement des fonctions à une seule variable*. Bull. Cl. phys.-math. Acad. Imp. Sci. St.-Petersb. I, pp. 193–200.

72 That is, for the probability (24) to take a value inside the segment $[z_1, z_2]$.

et appl." (sér. 2, t. 12, 1867) under the title *Considérations à l'appui de la découverte de Laplace sur la loi des probabilités dans la méthode des moindres carrés*, the celebrated scientist presented a method that deserves special attention. This method consists in determining the limiting value of the integral $\int_0^a f(x)dx$ given the values of the integrals

$$\int\limits_0^A f(x)dx, \quad \int\limits_0^A xf(x)dx, \quad \int\limits_0^A x^2 f(x)dx, \ldots$$

where $A > a$ and $f(x)$ is an unknown function obeying only one condition, viz., of retaining the sign $+$ inside the limits of integration. A simple and rigorous proof of Jakob Bernoulli's law included in my note "On mean values" is one of the results easily obtained by Bienaymé's method. He himself used this method to arrive at the proof of a certain proposition in probabilities from which Bernoulli's law follows directly.[73]

What, then, was in this memoir of Bienaymé? First, it included an inequality which, however, was not appropriately distinguished from the general text of the work and which we will write down in the following form:

$$P(|\bar{x} - E\bar{x}| \geq \alpha) \leq D\bar{x}/\alpha^2.$$

Here $\bar{x}$ is the arithmetic mean of the results of a series of observations and $\alpha > 0$. Second, Bienaymé studied the order of the even moments of sums of random and independent errors of observations assuming that all these errors had the same order of smallness.

Chebyshev's merit consisted in his hitting on the idea of using the moments of a random variable for proving the Central Limit Theorem. Thus Chebyshev is one of the main authors of the method of moments.

Of the roughly 250 pages of the book devoted to the above-mentioned course of lectures delivered by Chebyshev in 1879–1880, only roughly the last hundred deal with probability proper. They are preceded by very large sections on definite integrals (Euler, Laplace, and Frullani integrals, discontinuous integral factors, and some integrals of complex functions) and the calculus of finite differences (for functions of one variable). Chebyshev's course in the theory of probability proper, including the justification of the method of least squares according to Gauss' memoir of 1809 and related problems in

---

73 Quoted from Chebyshev's *Complete works*, vol. 3, p. 63. French version of his contribution: *Sur les valeurs limites des intégrales*. J. math. pures et appl., t. 19, 1874, pp. 157–160.

the mathematical treatment of observations, consisted largely of known material. Typically, Chebyshev omitted all discussion of "moral" applications of probability.

One more remark. Concerning the Bayes formula for the probability of hypotheses (which, incidentally, is not contained in Bayes' memoir), Chebyshev maintained that "laws" pertaining to posterior probabilities are in fact hypotheses.[74] Chebyshev did not dwell on justifying the theory of probability or in philosophical probing of its findings, but his statement shows that he paid attention to this aspect of the subject.

Following Chebyshev's article "On two theorems about probabilities", devoted, as we saw, to the Central Limit Theorem, Markov published two papers, "The law of large numbers and the method of least squares" (Kazan', 1898; see note 18) and "On the roots of the equation $e^{x^2} \frac{d^m (e^{-x^2})}{dx^m} = 0$" (*Sur les racines de l'équation* ..., Izvestia of the Academy of Sciences, ser. 5). Petersburg, 1898) where he formulated precisely and proved a proposition similar to Chebyshev's theorem. Markov approached his problem by means of the method of moments. Here is his theorem, formulated in modern terms. If the sequence of mutually independent random variables

$$\xi_1, \xi_2, \ldots, \xi_n, \ldots \tag{25}$$

is such that for all integral values of $r \geq 3$

$$\lim_{n \to \infty} \frac{C_n(r)}{B_n^r} = 0,$$

with

$$B_n = \sqrt{D\xi_1 + D\xi_2 + \cdots + D\xi_n},$$

$$C_n(r) = E|\xi_1 - E\xi_1|^r + E|\xi_2 - E\xi_2|^r + \cdots$$

$$+ E|\xi_n - E\xi_n|^r,$$

then

$$\lim_{n \to \infty} P\left(\frac{1}{B_n} \sum_{k=1}^{n} (\xi_k - E\xi_k) < x\right) = \frac{1}{\sqrt{2\pi}} \int_{-\infty}^{x} e^{-z^2/2} dz.$$

Most of the credit for the proof of the Central Limit Theorem is due to Lyapunov, in whose total creative work the theory of probability nevertheless remained only incidental. In essence, Lyapunov published only two probability-theoretic works: "On a proposition of the theory of probability"

---

74 Chebyshev, P.L., *Theory of probability*. Lectures, p. 154.

(*Sur une proposition de la théorie des probabilités*, 1900) and "A new form of the theorem on the limit of probability" (*Nouvelle forme du théorème sur la limite de probabilité*, 1901), the first in the *Izvestia*, and the second in the *Zapiski* of the Petersburg Academy of Sciences.[75]

Believing that the method of moments is too complicated and awkward, Lyapunov proved the Central Limit Theorem using the method of characteristic functions. Both this method, and the Dirichlet discontinuity factor that Lyapunov introduced in order to simplify his transformations, were used in the theory of probability long before him, and, in principle, his approach was known. However, it was Lyapunov who spelled out precisely the conditions necessary for the validity of the Central Limit Theorem, and very accurately estimated the rapidity with which the distribution of the normed sum of random variables converges to the normal law. Thus Lyapunov was able to subordinate Laplace's method to Chebyshev's demands for rigorous proofs of the limit theorems and for estimates of the deviations from limit regularities in a finite number of trials.

Here is Lyapunov's main finding: If for the sequence of independent random variables (25) there exists at least one $\delta > 0$ such that

$$\lim_{n \to \infty} \frac{\sum_{k=1}^{n} E\xi_k^{2+\delta}}{(\sum_{k=1}^{n} D\xi_k)^{1+\delta/2}} = 0,$$

then

$$\lim_{n \to \infty} P \left(z_1 < \frac{\sum_{k=1}^{n} \xi_k - \sum_{k=1}^{n} E\xi_k}{\sqrt{2 \sum_{k=1}^{n} D\xi_k}} < z_2 \right) = \frac{1}{\sqrt{\pi}} \int_{z_1}^{z_2} e^{-z^2} dz,$$

and the convergence to the normal law is uniform for any z_1 and z_2 ($z_2 > z_1$).

Lyapunov also established the known inequalities connecting the absolute initial moments for discrete random variables

$$\nu_m^{l-n} < \nu_n^{l-m} < \nu_l^{m-n} \quad \text{if } l > m > n \geq 0$$

("A new form of the theorem..."), while in his previous memoir, "On a proposition...", at the beginning of §4, he gave the first clear, if indirect,

---

75 These memoirs overlap two short communications which he published on the same subject in the *Comptes rendus* of the Paris Academy of Sciences in 1901.

264

A.A. Markov

definition of a cumulative distribution function $F(x)$, according to which, for any u and v, $v > u$,

$$F(v) - F(u) = P(u \leq \xi < v).$$

After Lyapunov had published his work, Markov once more returned to the proof of the Central Limit Theorem, striving to reestablish the importance of the method of moments "shaken" by Lyapunov.[76] Given his goal, Markov could no longer require the existence of all the moments of random variables (because Lyapunov did not lay down this condition). It seemed that this fact was a fundamental and insurmountable obstacle to the use of the method of moments. Still, Markov coped brilliantly by curtailing the random variables — a procedure that has become standard since that time — and thus achieving the obvious existence of all moments for only these curtailed variables.

76 Markov, A.A., *Calculus of probability*, third ed. Petersburg 1913, p. 332 (in Russian).

Markov's name and his works were repeatedly mentioned in Chapters 1 and 3 and in our own exposition. Since among his studies those pertaining to the theory of probability acquired the greatest importance for the development of mathematics, we will now, in concluding this section,[77] briefly describe his life. Andrei Andreevich Markov (1856–1922), the son of a petty employee of the Forest Department who subsequently became an uncertified attorney, took a great interest in mathematics while still in school. In 1874 he entered Petersburg University, where he attended the lectures of Chebyshev, Korkin, and Zolotarev, and was an active participant in a student circle guided by the two latter scientists. On graduating from the University in 1878, Markov was retained — to use a contemporary expression — to prepare for a professorship. His Master's dissertation on the theory of minimal values of indefinite binary quadratic forms, very close to the works of Zolotarev and Korkin, was discussed in Chapter 3. Chebyshev valued this thesis highly, especially Markov's skillful use of continued fractions — which Chebyshev himself applied with such success. Markov returned to the arithmetic theory of quadratic forms in his subsequent work. In the same year (1880) in which Markov defended his Master's dissertation he began to teach at Petersburg University as docent (lecturer).

Four years later Markov defended his doctoral dissertation "On some applications of algebraic continued fractions" (Petersburg, 1884) where, among other findings, he proved and generalized the important inequalities published by Chebyshev ten years earlier in the article "On the limiting values of integrals".[78] The doctoral dissertation proved to be the first of a long series of works on the method of moments, interpolation, and the theory of best approximation of functions, which continued for almost thirty years. In 1886 Markov was elected professor at Petersburg University and, on Chebyshev's proposal, academician. He remained on the university staff until 1905, when he retired so as to give way to younger scientists. However, Markov maintained his connections with the university up to the end of his life. He regularly taught courses in the theory of probability and lectured on continued fractions as private docent (unestablished lecturer), that is, without receiving a fixed salary.

Markov's works include articles on various issues other than those mentioned above. Thus one of his articles deals with the theory of differential equations. However, from the end of the nineties onward he turned ever more frequently to the theory of probability. Earlier we described some of his papers devoted to this theory, but many of them were written in the first quarter of the 20th century. By the end of that period, in 1924, the fourth edition of his "Calculus of probability", revised by the author, appeared posthumously. The first edition of this remarkable university handbook, which, as

---

77 Lyapunov's biography will be included in vol. 3 [1987] of this work.

78 See note 73.

we have seen, included many of Markov's discoveries, was published in 1900, but its initial versions had appeared in lithographic editions from the time when Markov had begun to lecture on the theory of probability instead of the retired Chebyshev, that is, from the academic year 1882–1883.

"Undoubtedly", wrote S.N. Bernshteĭn, "the most brilliant follower of Chebyshev's ideas and trends in the theory of probability was Markov, the student closest to his teacher in terms of the nature and brilliance of his mathematical gift. Whereas Chebyshev, especially at the end of his life, but also in his lectures, sometimes deviated from the accuracy of formulations and rigor of proofs in the theory of probability that he himself insisted on, Markov's classical course in the calculus of probability and his original memoirs, being models of rigor and clarity of exposition, have contributed in the highest degree to the transformation of the theory of probability into one of the most perfect branches of mathematics and to the wide dissemination of Chebyshev's orientation and methods."[79]

Markov's most important contribution was the initiation of the study of the properties of sums and mean values of dependent random variables and the creation, in 1907 and in the following years, of the theory of so-called Markov chains, which now forms the basis of one of the main branches of the theory of probability and its applications. However, this subject goes well beyond the bounds of the period under discussion.

Markov was not only a very prominent scientist but also a model citizen. In his youth he enthusiastically read the works of Dobrolubov, Pisarev, and Chernyshevskiĭ and these evidently influenced the formation of his social views. As an adult, he courageously came out against the reactionary politics of the czarist government. Thus in 1902 he sharply protested against the revocation, at the demand of the Czar himself, of A.M. Gorkiĭ's election to honorary membership of the Academy of Sciences. In 1907 he refused, in writing, to participate in the elections to the state Duma (Russia's parliament) because of the illegal introduction of new electoral rules. In 1913 Markov effectively opposed the official celebration of the tercentenary of the House of the Romanovs by organizing a scientific jubilee on the theory of probability. To this end, according to his proposal, under his editorship and with his preface, the Academy of Sciences published a Russian translation of "Part four of Jakob Bernoulli's Ars Conjectandi" (Petersburg, 1913) with the inscription "In commemoration of the bicentenial jubilee of the law of large numbers" at the top of the cover and the title page.

---

79 Bernshteĭn, S.N., *On P.L. Chebyshev's works in the theory of probability*. In: *The scientific legacy of P.L. Chebyshev*, part 1 (mathematics). Moscow 1945, pp. 59–60 (in Russian).

New Fields of Application of the Theory of Probability.
The Rise of Mathematical Statistics

In the 18th, and the first half of the 19th century, physicists founded the elements of the kinetic theory of gases essentially without using probability-theoretic ideas or methods. The situation changed only in the second half of the 19th century beginning, perhaps, with the works of R. Clausius (1822–1888) and, especially, with J.C. Maxwell's (1831–1879) *Illustrations of the dynamical theory of gases* (1860).[80]

Maxwell assumed that, at equilibrium, the velocities of gas molecules are unequal, and he attempted to determine the law of their distribution. Supposing, in addition, that the components x, y, z, of the velocity of a molecule are independent, he obtained for the required density of distribution $\varphi(x)$ the relation

$$\varphi(x)\varphi(y)\varphi(z) = \varphi(x^2 + y^2 + z^2), \tag{26}$$

whence

$$\varphi(x) = Ce^{Ax^2} \quad (C > 0,\ A < 0).$$

The simple idea underlying this reasoning and seen in the relation (26) was not new. Already Adrain (see p. 228) used it in the theory of errors in one of his derivations of the normal law. J. Herschel (1850), W. Thomson (Lord Kelvin) and P. Tait (1867), and, later, N.Ya. Tsinger (1899), and A.N. Krylov (1932) independently used the same idea in the theory of errors.

The idea behind Maxwell's reasoning became generally known in the physical sciences as well, and the Austrian physicist L. Boltzmann (1844–1906) perfected the derivation itself in 1872. He proved that the Maxwell distribution is the only one obeying the condition of statistical equilibrium. The modern derivation of this distribution is due to Yu.V. Linnik (1952), who defined the assumptions of the derivation more precisely[81] and, in addition, repeated it under three different and weakened assumptions.

For Maxwell, the probability-theoretic method was necessary since, as he himself wrote in 1875, it is impossible to calculate the velocity of an individual molecule.

An essentially new period in the application of the theory of probabilty to physics, or, more precisely, to the kinetic theory of gases is connected with Boltzmann. Already in a paper in 1871 he noted that at least some propositions in thermodynamics should be proved with allowance for stochastic considerations. In 1872, while studying the law of distribution of the kinetic

---

80 A Russian translation is in Maxwell, J.C., *Papers and speeches*. Moscow 1968.

81 The mutual independence of the components of velocities should persist under all choices of the coordinate system.

energy of molecules (in connection with perfecting Maxwell's derivation of the law of distribution of molecular velocities) and of a certain function of this distribution, he declared that "the problems of the mechanical theory of heat are also problems of probability theory".[82]

Boltzmann's work did not win recognition. First, the theory of the atomic-molecular structure of matter was still only a working hypothesis; and second, the reasoning on the stochastic nature of the transition of substances from one state to another was at variance with the reversibility of the formulas of mechanics. A discussion of this contradiction, mainly with J. Loschmidt, forced Boltzmann[83] to abandon, in 1877, mechanical elements and to turn entirely to the probability-theoretic method.

Specifically, Boltzmann searched for a density of distribution $f(x)$ of the kinetic energy of a molecular system assuming that the system was in its most probable state (i.e., in thermal equilibrium). Suppose the kinetic energy of w_0 molecules is contained in the interval $[0, \varepsilon]$; of w_1 molecules, in the interval $[\varepsilon, 2\varepsilon]$; etc. Then, if the total number of molecules is n,

$$w_0 \approx n\varepsilon \cdot f(0), \quad w_1 \approx n\varepsilon \cdot f(\varepsilon), \quad w_2 \approx n\varepsilon \cdot f(2\varepsilon), \text{ etc.,}$$

$$w_0 + w_1 + w_2 + \cdots = n.$$

The number of possible distributions of a given state of the system is

$$p = \frac{n!}{w_0! w_1! w_2! \ldots},$$

and, as it turns out, the maximal value of p corresponds to the case of

$$f(x) = Ce^{-hx} \quad (C, h > 0).$$

Considering the distribution of the components u, v, w of the molecular velocities, Boltzmann came to calculate the integral

$$\Omega = - \iiint f(u, v, w) \ln f(u, v, w) du\, dv\, dw$$

for a given kinetic energy of the system and proved that Ω ("the measure of permutational ability" — *Permutabilitätsmaß* (p. 192) — of the system,

82 Boltzmann, L., *Weitere Studien über das Wärmegleichgewicht unter Gasmolekülen.* Wiss. Abhandl., Bd. 1. Leipzig 1909, 316–402 (pp. 316–317). English translation: "Further studies on the thermal equilibrium of gas molecules". In: Brush, S.G., *Kinetic theory*, vol. 2. Oxford a.o., 1966, 88–175 (p. 88).

83 See Boltzmann's paper *Über die Beziehung zwischen dem zweiten Hauptsatze der mechanischen Wärmetheorie und der Wahrscheinlichkeitsrechnung, resp. den Sätzen über das Wärmegleichgewicht.* Wiss. Abhandl., Bd. 2. Leipzig 1909, pp. 164–223.

L. Boltzmann

as Boltzmann called it) was the measure of the probability of its state and, moreover, unlike enthropy, it was defined for any state of the system rather than only for the state of equilibrium.

As a general conclusion, Boltzmann stochastically formulated the second law of thermodynamics: a change which can take place all by itself, Boltzmann noted, is a transfer from a less probable state to a more probable one.

As before, nobody, M. Planck being almost the only exception,[84] recognized Boltzmann's work, and it is likely that this fact played a part in his suicide.

However, during the discussions held in the last years of the 19th century Boltzmann had the opportunity for further clarification of his viewpoint. Thus, answering criticisms levelled by the German mathematician E. Zermelo (1871–1953) based on the well-known Poincaré theorem on the return of a mechanical system to its initial state, Boltzmann indicated that the period

---

84 In Russia, V.A. Mikhel'son in 1883 and N.N. Pirogov in 1886 supported the probability-theoretic point of view on the kinetic theory of gases.

of this return was unimaginably great[85] and that the fact itself of the return did not contradict probability-theoretic notions.

Later on, in the 20th century, the stochastic nature of the second law of thermodynamics was vividly illustrated by an urn model due to Boltzmann's student Paul Ehrenfest and his wife T.A. Afanasieva-Ehrenfest in their work "On two known objections to Boltzmann's H-theorem" (*Über zwei bekannte Einwände gegen das Boltzmannsche H-Theorem*, 1907).[86]

Maxwell's and Boltzmann's work laid the foundation of classical statistical physics completed by the American physicist J.W. Gibbs (1839–1903) in his *Elementary principles of statistical mechanics developed with special reference to the rational foundations of thermodynamics* (New Haven, 1902).[87] We will not dwell on Gibbs' work or on the ergodic hypothesis stated by Boltzmann in 1868; studies devoted to this hypothesis appeared only in the 20th century, beginning with a critical survey of statistical mechanics by P. and T. Ehrenfest (1911).

The work of Boltzmann in physics and the development of the natural sciences in general led to a radical break with the past. It turned out that some fundamental laws of nature are stochastic rather than rigidly deterministic. It should be added, however, that in the 19th century physics provided no impulse for the development of the theory of probability.[88]

Describing the use of the statistical method in the 19th century, A.Ya. Khinchin maintained that "Fairly vague and somewhat timid probabilistic arguments do not pretend here to be the fundamental basis, and play approximately the same role as purely mechanical considerations. ... far reaching hypotheses are made concerning the structure and the laws of interaction between the particles ... the notions of the theory of probability do not appear in a precise form and are not free from a certain amount of confusion which often discredits the mathematical arguments by making them either devoid of any content or even definitely incorrect. The limit theorems of the theory

---

85 Boltzmann, L., *Zu Hrn. Zermelos Abhandlung "Über die mechanische Erklärung irreversibeler Vorgänge"*, 1897. Wiss. Abhandl., Bd. 3. Leipzig 1909, pp. 579–586. English translation: "On Zermelo's paper 'On the mechanical explanation of irreversible processes'". In: Brush, S.G., (see note 82), pp. 238–245.

86 We have mentioned (p. 216) that this model, whose appearance is considered as the beginning of the history of random processes, had been already introduced (in essence) by Daniel Bernoulli.

87 Russian translation: Moscow – Leningrad 1946.

88 However the theory of probability was undoubtedly used in concrete physical research. In this connection see, for example, Schneider, I., *Clausius' erste Anwendung der Wahrscheinlichkeitsrechnung im Rahmen der atmosphärischen Lichtstreuung*. Arch. hist. ex. sci., vol. 14, No. 2, 1974. Clausius' work was published in 1849. His role in creating the kinetic theory of gases, at least from the point of view of probability theory, has not yet been sufficiently studied.

of probability do not find any application as yet. The mathematical level of all these investigations is quite low, and the most important mathematical problems which are encountered in this new domain of application do not yet appear in a precise form."[89]

Of course, all this was written from the standpoint of statistical mechanics as it existed in the middle of our century, and perhaps it should be added that Boltzmann nevertheless attempted to introduce a logically rigorous notion of statistical probability. In this connection he considered in several instances, in pt. 2 of his "Lectures on the theory of gases" (*Vorlesungen über Gastheorie*, Tl. 1–2. Leipzig 1895–1899),[90] an infinity of "repetitions" of the same gas, thus effectively bringing into use an infinite parent population. Gibbs adopted the same point of view, which, incidentally, was already indirectly used by Laplace.[91]

In the section devoted to social and anthropometric statistics we described the development of population statistics, a traditional branch of applications of the theory of probability on the European continent. In England, statistics developed differently, viz., through applications to biology. The first who should be mentioned in this connection is F. Galton (1822–1911), a first cousin of C. Darwin. Being primarily a psychologist and an anthropologist, Galton attempted to base his scientific work in these and in other branches of science on measurements and on their mathematical treatment. After the appearance of Darwin's *Origin of species* (1859) he approached the problem of heredity in exactly the same manner.

Galton's book *Natural inheritance* (London – New York 1889), based on the elements of his theory of correlation,[92] attracted the attention of the mathematician K. Pearson and the zoologist W.F.R. Weldon (1860–1906) who aimed at justifying Darwin's theory of natural selection by a statistical study of animal and plant populations.

---

89 Khinchin, A.Ya., *Mathematical foundations of statistical mechanics*. New York 1949, p.2. Originally published in Russian (Moscow – Leningrad 1943).

90 Russian translation: Moscow 1956.

91 Later on, at the beginning of the 20th century, R. von Mises (1883–1953) constructed the entire theory of probability on a definition of probability based, in turn, on an infinite number of trials and, consequently, on the existence of a certain infinite parent population. As initially formulated, Mises' constructive ideas proved unacceptable. However, both these ideas and his sharp criticism of the elements of the classical theory of probability exerted a fruitful influence on the development of this scientific discipline.

92 Of course, the existence of correlative relations between certain events was known long ago. Thus, Aristotle, in his book *Problemata*, noted the existence of such relations in biology, and Kepler, in his *Harmonices Mundi*, Book 4, Chapter 7 (German translation: *Welt-Harmonik*. München – Berlin, 1939; reprint: 1967) mistakenly believed that in the absence of 'aspects' (i.e., of 'remarkable' arrangements of the heavenly bodies) the weather was usually calm. However, it seems that quantitative measurements of the closeness of correlation were not made before Galton; it was he who, in 1888, introduced the coefficient of correlation.

At the very end of the 19th century, by founding the still existing periodical "Biometrika",[93] Galton, Pearson, and Weldon established the Biometric school, that is, a scientific school aiming at the creation of methods of treating biological observations and of studying statistical regularities in biology.

The first issue of "Biometrika" (1902) contained two editorial papers. The first stated:

A very few years ago all those problems which depend for their solution on a study of differences between individual members of a race or species, were neglected by most biologists. ...

The starting point of Darwin's theory of evolution is precisely the existence of those differences ... the first step in an enquiry into the possible effect of a selective process upon any character of a race must be an estimate of the frequency with which individuals, exhibiting any given degree of abnormality with respect to that character, occur ... These, and many other problems, involve the collection of statistical data on a large scale ...[94]

And, further, in the second article:

... the problem of evolution is a problem in statistics ... we must turn to the mathematics of large numbers, to the theory of mass phenomena, to interpret safely our observations. ... may we not ask how it came about that the founder of our modern theory of descent made so little appeal to statistics? ... The characteristic bent of C. Darwin's mind led him to establish the theory of descent without mathematical conceptions; even so Faraday's mind worked in the case of electromagnetism. But as every idea of Faraday allows of mathematical definition, and demands mathematical analysis ... so every idea of Darwin — variation, natural selection ... — seems at once to fit itself to mathematical definition and to demand statistical analysis.[95]

The authors of the article then quote Darwin's statements — statements they managed to discover in his works — on the importance of the statistical method, and repeat his remark

I have no faith in anything short of actual measurement and the Rule of three.

---

93 We note in passing that "Biometrika" is one of the very few mathematical periodicals that regularly publishes papers on the history of its subject (more precisely, of the history of probability and mathematical statistics).

94 *The scope of Biometrika.* Biometrika, vol. 1, pt 1, 1901–1902, pp. 1–2.

95 *The spirit of Biometrika.* Ibid., p. 4.

K. Pearson

These words, the authors continue, appealing for an alliance of biologists, mathematicians, and statisticians, "may well serve as a motto for 'Biometrika' and for all biometricians".[96]

From the very beginning, Karl Pearson (1857–1936) was the head of the Biometric school. In particular, until the end of his life, he was the chief (and for many years the only) editor of "Biometrika". Pearson received his mathematical education at Cambridge where his teachers were G. Stokes, Maxwell, Cayley, and W. Burnside, and he also studied physics at Heidelberg. In 1884 he became professor of applied mathematics and mechanics at London University and, in 1911, professor of eugenics. In 1896 he was elected to the Royal Society.

Pearson was a very versatile scholar. In addition to work on mathematical statistics he wrote on applied mathematics and on philosophy.[97] In his youth he studied law and history, and his views tended towards moderate socialism

---

96 Ibid.

97 In principal philosophical problems Pearson was close to E. Mach. Lenin, in his *Materialism and empiriocriticism* (1909), sharply criticized Pearson's views expressed in his book *The grammar of science* (1892).

274

of the English variety. Pearson's first note devoted to mathematical statistics appeared about 1893 and, although he published more than twenty works before 1901, his main results belong to the 20th century. Pearson essentially advanced the theory of correlation and issued a large number of important statistical tables. The principal topic of his work, however, was the study of a number of distributions, partly introduced by himself, and the estimation of their parameters by observations. The development of this, the central trend of mathematical statistics at the turn of the 19th century and at the beginning of this century, led to the rejection of the obsolete method of determining the "true values" of observed quantities in the classical theory of errors. It was the mathematically precise demand to estimate the parameters of distributions under one or another additional condition that enabled statisticians to raise concrete problems more clearly and, in fact, to create the relevant branch of mathematical statistics. Pearson's works, at least those published up to the 1920s, were characterized by a rather low theoretical level — a circumstance that hindered their recognition and further development outside England.[98]

However, England's scientific community was prepared to accept the ideas and methods of the Biometric school by the previous work of F.Y. Edgeworth (1845–1926), also a versatile scientist, who occupied himself not only with the theory of probability, statistics and its applications, and the theory of errors, but also with political economy. Edgeworth was also a pioneer in the application of mathematics to economics. However, because of the pecularities of his style, his mathematical works remained little known.

In concluding this section one may say that the birth of mathematical statistics as a distinct mathematical discipline, which took place in the 1920s and 1930s, owed much both to the Biometric school and to the "Continental" school of population statistics. It is likely that the unification of these two schools and, at the same time, the penetration of the statistical method into a number of "new" branches of science and its applications in industrial production created mathematical statistics.

---

98 See, for example, Tschuprow, A.A., *Theorien för statistika räckors stabilitet*, 1926 (Russian translation in Chuprov, A.A., *Voprosy statistiki*. Moscow 1960). Note that Chuprov himself promoted the recognition of Pearson's works in Russia; witness also his *Ocherki po teorii statistiki* ("Essays on the theory of statistics"), 1909, and his correspondence with Markov (1910–1917). During the initial period of this correspondence Markov spoke of Pearson very negatively. Later on, however, he essentially changed his point of view under Chuprov's influence; see *On the theory of probability and mathematical statistics*; correspondence of A.A. Markov and A.A. Chuprov, ed. by Kh.O. Ondar. Moscow 1977. English translation: New York a.o. 1981. Generally speaking, the ideas of the Biometric school became known in Russia at the beginning of the 20th century through the works of L.K. Lakhtin (1904), R.M. Orzhentskiĭ (1910) and E.E. Slutskiĭ (1912).

Works of the Second Half of the 19th Century in Western Europe

At the end of the 19th century treatises written by J. Bertrand and H. Poincaré were devoted to probability theory proper.

Joseph Louis François Bertrand (1822–1900) was mentioned in Chapter 3, in connection with Chebyshev's work on analytic number theory. In his childhood he displayed unusual faculties and raised great hopes which, however, were not fully realized. An alumnus of the *Ecole Polytechnique*, he published his first work, on the mathematical theory of electricity, in 1839, and two years later he began to teach, first in a lyceum and then in other academic institutions. In 1856 Bertrand was appointed professor at the *Ecole Polytechnique* and, in 1862, he became professor at the *Collège de France* as well. A member of the Paris Academy of Sciences, he became its permanent secretary in 1874. His works embraced many branches of mathematics but they were not of supreme importance. Named after him are the Bertrand postulate in number theory, the Bertrand problem in group theory (p. 185), and curves of double curvature in differential geometry. As permanent secretary of the Academy of Sciences, he published many brief, elegant biographies of members who died during his term of office. These biographies were previously delivered orally as *Eloges* (Eulogies). Bertrand also penned many popular articles and books, for example on the history of mathematics and astronomy, as well as manuals. His "Calculus of probability" (*Calcul des probabilités.* Paris 1888), of which there exist several reprints, is exceptionally interesting even now, since it contains an abundance of important problems whose examination at times leads to unexpected results, and also apt and witty remarks in regard to many opinions and utterances of his predecessors.[99] Witness, for example, this problem, which became classical: A chord of a circle is chosen at random (*au hasard*). It is required to determine the probability that its length exceeds that of the side of an inscribed equilateral triangle. It turned out that the expression "at random" was insufficiently definite and that therefore the problem admitted many different solutions, of which the author gave three. This fact compelled mathematicians not only to formulate the conditions of problems more precisely but, in addition, to ponder the very foundations of the theory of probability.

Nevertheless, the general contents of Bertrand's treatise could hardly be called original in the mathematical sense. The author's own contribution to the theory of probability is small.

In the first edition of his "Calculus of probability" (*Calcul des probabilités.* Paris 1896) Poincaré studied the cause of equal probability of the various outcomes in the game of roulette and of the uniform distribution of the longitudes of the minor planets. Consider the first of these problems. A ball moves along the circumference of a circle, gradually slows down and stops.

---

99 We have quoted one such statement on Quetelet's average man on p. 244.

J.L.F. Bertrand

The law of distribution of the initial velocity of the ball is assumed to be continuous. If the circle is divided into congruent sectors, the probability that the ball stops in any given sector proves to be constant irrespective of the type of the law of distribution.

Poincaré returned to problems of this kind in his popular writings,[100] in an article dated 1906 and, finally, in the second edition of his treatise (1912; reprinted in 1923). But he also considered problems in pure physics. Thus the second edition of the *Calcul des probabilités* concludes with a remark on the diffusion of liquids. The formation of a single homogeneous "liquid", Poincaré notes, should be theoretically justified in one way or another. This problem is connected with the so-called ergodic hypothesis; in 1912, the series of works devoted to this hypothesis had barely begun to appear.

---

100 *La science et l'hypothèse*. Paris 1902; *Science et méthode*. Paris 1908. English translations of both books are in Poincaré, H., *Foundations of science*. Washington 1982. Note that in both these books the celebrated mathematician also showed himself a philosopher basically close to Machism. In his *Materialism and empiriocriticism* (1909), Lenin gave a minute critical analysis of Poincaré's interpretation of the problems in the theory of knowledge.

In the same edition of the *Calcul des probabilités* we find a passage on random events borrowed from *Science et méthode*. According to Poincaré, a distinctive feature of a random event is that insignificant causes lead to considerable changes in outcomes.[101]

In roulette, for example, a small change in the initial velocity of the ball leads to an essential change in the result — to the ball's stopping in another sector of the circle. However, Poincaré also pointed out a second form of randomness, the form that, given the numerous causes and the complexity of their combinations, operates within the bounds of the setup "small causes — small effects". The first kind of randomness is characterized by a uniform distribution, the second by the normal distribution that holds because of the "usually" fulfilled conditions of the Central Limit Theorem.

For modern mathematics, formalization of a concept of randomness suitable for application in the natural sciences remains a difficult and important problem.

Nonetheless, taken as a whole, the contents of Poincaré's treatise conform to those of older writings. For example, Poincaré, like Bertrand before him, did not mention Chebyshev or Lyapunov; for him, as for Laplace, the theory of probability still remained a branch of applied mathematics. This may have been connected with peculiarities of Poincaré's scientific work, and with his activities as head of the chair of mathematical physics[102] and the theory of probability at the Sorbonne: the first edition of his *Calcul des probabilités* corresponds to the course of lectures delivered by its author in the academic year 1893–1894.

We must still mention the works of the Austrian mathematician E. Czuber (1851–1925) and of the English logicians and mathematicians A. De Morgan, G. Boole, W.S. Jevons, and J. Venn who devoted attention to the problem

---

101 Such an interpretation of a random event does not contradict the reasoning expressed even by Aristotle. Thus, in his "Metaphysics", Aristotle maintains that discovering a buried treasure while digging a hole is 'random'. A small change in choosing the spot for the hole would have led to an essential change of the outcome: the buried treasure would have been left in the ground. Another illustration from Aristotle's "Physics" pertains to a 'random' meeting of two persons. A small change in the chain of preceding events would have resulted in preventing the meeting. A number of philosophers of the modern era, beginning with T. Hobbes and G.W. Leibniz, held a similar opinion about random events. In the 18th century, J.H. Lambert engaged in interpreting randomness. While attempting to introduce the notion of infinite random sequences he came close, on an intuitive level, to the idea of normal numbers that, in the 20th century, roused the interest of some prominent researchers (E. Borel and others). See Lambert J.H. *Anlage zur Architectonic*, Bd. 1. Riga, 1771, §324. Roughly speaking, a number in a given number system is called normal if all digits of its expansion into an infinite fraction occur equally often. Moreover, when the number of digits in its expansion increases infinitely, the limits of the relative frequencies of the occurrence of each combination of digits one, two, etc. at a time should be respectively equal.

102 Mathematical physics was understood to consist of those branches of physics in which mathematical computations were used.

of substantiating the theory of probability (see Chapt. 1 for their work in logic).

Czuber was the author of several works on the theory of probability and on the theory of errors. Without leaving any considerable scientific heritage, he was an honest compiler and his works enjoyed well-deserved fame for a long time. In this connection we mention his "Theory of errors of observations" (*Theorie der Beobachtungsfehler.* Leipzig 1891) and his "Calculus of probability" (*Wahrscheinlichkeitsrechnung),* the latter being a chapter in the "Encyclopedia of mathematical sciences" (*Encyklopädie der mathematischen Wissenschaften,* Bd. 1. Leipzig 1901). In addition, Czuber's works contain very valuable information on the history of the theory of probability.

De Morgan attempted to substantiate the initial probability-theoretic propositions by logical calculi. His book, *Formal logic, or the calculus of inference, necessary and probable* (London 1847), contains chapters devoted to the theory of probability, stochastic reasoning and induction. De Morgan mainly recognizes only subjective probability, treating it as the quantitative measure of knowledge. Having no direct practical applications, De Morgan's works remained little known. However, they were continued by Boole.

Aiming at translating Aristotle's logic into algebraic language, Boole also turned his attention to stochastic logic. In 1854, the year when he published his main work on mathematical logic, *An investigation of the laws of thought, on which are founded the mathematical theories of logic and probability,* Boole also published an article, *On the conditions by which the solution of questions in the theory of probabilities are limited,* in which he anticipated the necessity of substantiating the theory of probability axiomatically. The claim of the theory, he wrote,

> to rank among the pure sciences must rest upon the degree in which it satisfies the following conditions: 1st. That the principles upon which its methods are founded should be of an axiomatic nature. 2nd. That they should lead to results capable of exact verification, wherever verification is possible. 3rd. That they should be capable of a systematic development consistent in all its parts and processes, and neither acknowledging nor imposing any limitations but those which exist in the nature of things.[103]

Boole was ahead of his time also in formulating the aims of the theory of probability. In one of his articles of 1851 he repeated, in essence, Chebyshev's statement made in 1846 (see p. 256).

Like Boole, Jevons wrote his main work, *The principles of science. A treatise on logic and scientific method,* under De Morgan's considerable influence. In Chapter 10 of this book Jevons maintained that the theory of probability

---

103 Boole, G., *Studies in logic and probability.* London 1952, p. 288.

begins to matter where our ignorance begins. We have already pointed out (see p. 249)[104] that such statements are wrong. At the same time, Jevons attempted to subordinate the theory of probability to mathematical logic, and it is this direction of his works that is interesting. Venn had the same inclination. In the introduction to the first edition of his *Logic of chance* (London 1866) he even declared that "Of the province of Logic ... under its widest aspect Probability may ... be considered to be a portion".

The problem of the relations between mathematical logic and the theory of probability is still topical; however, it should probably be considered in the context of the more general problem of the relations between mathematical logic and mathematics.

In completing our account, we feel it appropriate to say a few words about the historian of mathematics and mechanics, and Fellow of the Royal Society I. Todhunter (1820–1884). In addition to being the author of a number of textbooks on elementary and higher mathematics, he published monographs on the history of the calculus of variations and on the mathematical theory of attraction and on the shape of the Earth (1873; reprinted 1962). Todhunter's monograph, *A history of the mathematical theory of probability* (Cambridge 1865; reprints 1949 and 1965), deals with scientists up to and including Laplace, and in part describes Poisson's writings. Todhunter was not a specialist in the theory of probability so that his comments can not be regarded as sufficiently expert, all the more so because more than a century has passed since his book appeared in print. But, owing to Todhunter's exceptional conscientiousness, this work is practically a complete source of information on the period he discusses.

Conclusion

The following features characterize the period from Laplace to the end of the 19th century:

1. The theory of probability was constructed as a discipline pertaining to natural science (Laplace). It used the tools of mathematical analysis including characteristic functions (Laplace), proofs of various forms of the law of large numbers (Laplace, Poisson, Chebyshev), and of the Central Limit Theorem (Laplace, Cauchy, Chebyshev, Markov).

---

104 J.S. Mill's opinion pronounced in his *System of logic* (London 1843) also proved methodically untenable. "... even when the probabilities are derived from observation and experiment", he wrote (p. 353 of the London edition of 1886), "a very slight improvement in the data by better observations, or by taking into fuller consideration the special circumstances of the case, is of more use than the most elaborate application of the calculus of probabilities founded on the data in their previous state of inferiority. The neglect of this obvious reflection has given rise to misapplications of the calculus of probabilities which have made it the real opprobrium of mathematics. It is sufficient to refer to the applications made of it to the credibility of witnesses, and to the correctness of the verdicts of juries".

2. The classical theory of errors was constructed (Laplace, Gauss).

3. The importance of population statistics increased sharply. Statistical services and statistical societies were created in the principal countries of the world; accordingly, public interest in the theory of probability (due, in addition, to other practical applications of the theory) intensified.

4. The theory of probability was shaped as a general mathematical discipline (Chebyshev).

5. Stochastic concepts began to be used in physics (Maxwell, Boltzmann); a most important fact, viz., that at least some fundamental laws of nature are stochastic in nature, was established.

6. The Continental school of population statistics (Lexis) and the Biometric school (Galton, K. Pearson, Weldon) were founded.

7. Publications substantiating the theory of probability from the standpoint of mathematical logic (De Morgan, Boole, Jevons, Venn) appeared.

Let us dwell on Item 4 and, moreover, on the relations between the theory of probability and mathematics in general. To be sure, during the 19th century the theory began to change from a collection of diverse and interesting particular problems into a mathematical theory having a rather strictly delimited field of concerns. In the process, however, neither Laplace, nor, later on, Bertrand or Poincaré reduced the theory of probability to a logically perfect mathematical discipline. In the 19th century it belonged primarily to applied mathematics. No wonder that D. Hilbert, in his celebrated report at the International Congress of Mathematicians in Paris (1900), assigned the theory of probability to the physical sciences. When stating one of his problems (problem No. 6) he said:

> To treat in the same manner [as the foundations of geometry], by means of axioms, those physical sciences in which mathematics plays an important part; in the first rank are the theory of probability and mechanics.[105]

It took much more time until, in the 1930s, there grew up a clear understanding that the theory of probability is a true mathematical discipline possessing in addition intimate and direct ties with the broad spectrum of the natural sciences and with the technological, sociological, and economic disciplines.

---

105 It is likely that another reason for Hilbert's assigning the theory of probability to the physical sciences was the remarkable progress of mathematical analysis in the second half of the 19th century, a fact that in essence transformed the whole of mathematics. The quotation above is from Hilbert's *Mathematische Probleme* (1901). Ges. Abh., Bd. 3. Berlin 1935, pp. 290–329 (p. 306) or, rather, from its English translation (Bull. Amer. Math. Soc., vol. 8, No. 10, 1902, pp. 403–479 (p. 454)).

The most important step in rooting mathematical structures in the theory of probability was due to Chebyshev, who actually included the teaching of random events in a wider study of random variables. The introduction of the concept of random variable as an object studied by the theory of probability was natural and necessary but, during the period under our consideration, it remained unnoticed. Only later, at the beginning of the 20th century, did this concept come to the fore, followed a long and complicated road of development, and was made mathematically precise. In particular, the introduction of this very notion resulted in distributions and characteristic functions, taken by themselves, becoming special objects of study.

The transition of the theory of probability to the study of random variables did not in the least weaken its tendency to acquire knowledge of random events, and did not prevent the appearance in primitive form of new problems that demanded the introduction of the new concept of a random process.

Addendum

1. French and German Quotations

Following are the original versions of the French and German quotations translated in the main text.

213 Laplace: *Cette inégalité [lunaire] quoique indiquée par les observations, était négligée par le plus grand nombre des astronomes, parce qu'elle ne paraissait pas résulter de la théorie de la pesanteur universelle. Mais, ayant soumis son existence au Calcul de Probabilités, elle me parut indiquée avec une probabilité si forte, que je crus devoir rechercher la cause.*

216/217 Laplace: *... l'irrégularité primitive de ces rapports disparait à la longue pour faire place à l'ordre le plus simple. ... On peut étendre ces résultats à toutes les combinaisons de la nature, dans lesquelles les forces constantes dont leurs éléments sont animés établissent des modes réguliers d'action, propres à faire éclore du sein même du chaos des systèmes régis par des lois admirables.*

217 Laplace: *principes éternels de raison, de justice et d'humanité.*

219 Laplace: *une nouvelle branche de la théorie des probabilités.*

221 Laplace: *un genre particulier de combinaisons du hasard.*

227/228 Gauss: *Ich müsse es nemlich in alle Wege für weniger wichtig halten, denjenigen Werth einer unbekannten Grösse auszumitteln, dessen Wahrscheinlichkeit die grösste ist, die ja doch immer nur unendlich klein bleibt, als vielmehr denjenigen, an welchen sich haltend man das am wenigsten nachtheilige Spiel hat; oder wenn $f(a)$ die Wahrscheinlichkeit des Werths a für die Unbekannte x bezeichnet, so ist weniger daran gelegen, dass $f(a)$ ein Maximum werde, als daran, dass $\int f(x)F(x-a)dx$ ausgedehnt durch alle möglichen Werthe des x ein Minimum werde, indem für F eine Function gewählt wird, die immer positiv und für grössere Argumente auf eine schickliche Art immer grösser wird.*

231 Klein: *Wie fruchtbar und vielseitig Poisson tätig war, möge man aus den vielen Einzelheiten ersehen, die sich immer noch an seinen Namen knüpfen: Poissons Klammerausdrücke in der Mechanik, Poissons Konstante in der Elastizitätslehre, Poissons Integral in der Potentialtheorie und schließlich die allgemein bekannte, viel verwendete Poissonsche Gleichung $\Delta V = -4\pi\rho$, die er im Innern eines anziehenden Körpers an die Seite der Laplaceschen $\Delta V = 0$ im äußeren Raum setzte.*

244 Bertrand: *Dans le corps de l'homme moyen, l'auteur belge place une âme moyenne. Il faut, pour résumer les qualités morales, fondre vingt mille caractères en un seul. L'homme type sera donc sans passions et sans vices, ni fou ni sage, ni ignorant ni savant, souvent assoupi: c'est*

la moyenne entre la veille et le sommeil; ne répondant ni oui ni non; médiocre en tout. Après avoir mangé pendant trente-huit ans la ration moyenne d'un soldat bien portant, il mourrait, non de vieillesse, mais d'une maladie moyenne que la Statistique révélerait pour lui.

244 Quetelet: *Il est un budget qu'on paie avec une régularité effrayante, c'est celui des prisons, des bagnes et des échafauds; c'est celui-là surtout qu'il faudrait s'attacher à réduire; et, chaque année, les nombres sont venus confirmer mes prévisions, à tel point, que j'aurais pu dire, peut-être avec plus d'exactitude: Il est un tribut que l'homme acquitte avec plus de régularité que celui qu'il doit à la nature ou au trésor de l'Etat, c'est celui qu'il paie au crime! ... Nous pouvons énumérer d'avance combien d'individus souilleront leurs mains du sang de leurs semblables, combien seront faussaires, combien empoisonneurs, à peu près comme on peut énumérer d'avance les naissances et les décès qui doivent avoir lieu.*

La société renferme en elle les germes de tous les crimes qui vont se commettre, en même temps que les facilités nécessaires à leur développement. C'est elle, en quelque sorte, qui prépare ces crimes, et le coupable n'est que l'instrument qui les exécute. Tout état social suppose donc un certain nombre et un certain ordre de délits qui résultent comme conséquence nécessaire de son organisation.

245 Marx: *Er hat großes Verdienst in der Vergangenheit, indem er nachwies, wie selbst die scheinbaren Zufälle des sozialen Lebens durch ihre periodische Rekurrenz und ihre periodischen Durchschnittszahlen eine innere Notwendigkeit besitzen. Aber die Interpretation dieser Notwendigkeit ist ihm nie gelungen. Er hat auch keine Fortschritte gemacht, nur das Material seiner Beobachtung und Berechnung ausgedehnt. Er ist heut nicht weiter, als er vor 1830 war.*

246 Rümelin: *Wenn mir die Statistik sagt, daß ich im Laufe des nächsten Jahres mit einer Wahrscheinlichkeit von 1 zu 49 sterben, mit einer noch größeren Wahrscheinlichkeit schmerzliche Lücken in dem Kreis mir theurer Personen zu beklagen haben werde, so muß ich mich unter den Ernst dieser Wahrheit in Demuth beugen; wenn sie aber, auf ähnliche Durchschnittszahlen gestützt, mir sagen wollte, daß mit einer Wahrscheinlichkeit von 1 zu so und so viel eine Handlung von mir der Gegenstand eines strafgerichtlichen Erkenntnisses sein werde, so dürfte ich ihr unbedenklich antworten: ne sutor ultra crepidam!*

256 Chebyshev: *Tout ingénieuse que soit la méthode employée par le célèbre Géometre, il reste à être impossible de montrer la limite de l'erreur que peut admettre son analyse approximative, et par cette incertitude de la valeur de l'erreur, sa démonstration n'est pas rigoureuse.*

261/262 Chebyshev: *Dans un Mémoire très intéressant, sous plus d'un rapport, que M. Bienaymé a lu à l'Académie des Sciences en 1833, et que* *l'on trouve imprimé dans les Comptes rendus, et reproduit dans le Journal de Mathématiques pures et appliquées de M. Liouville (2e série, t. 12, 1867), sous le titre: Considérations à l'appui de la découverte de Laplace sur la loi de[s] probabilité[s] dans la méthode des moindres carrés, l'illustre savant donne une méthode qui mérite une attention toute particulière. Cette méthode consiste dans la détermination de la valeur limite de l'intégrale ... d'après les valeurs des intégrals ... où $A > a$ et $f(x)$ une fonction inconnue, assujettie seulement à la condition de garder le signe $+$ entre les limites d'intégration. La démonstration simple et rigoureuse de la loi de Bernoulli, que l'on trouve dans ma Note, sous le titre Des valeurs moyennes,* n'est qu'un des résultats que l'on tire aisément de la méthode de M. Bienaymé, et d'après laquelle il est parvenu lui-même à démontrer une proposition sur les probabilités, d'où la loi de Bernoulli découle directement.*

note 82   Boltzmann: *Die Probleme der mechanischen Wärmetheorie sind daher Probleme der Wahrscheinlichkeitsrechnung.*

281   Hilbert: *Durch die Untersuchungen über die Grundlagen der Geometrie wird uns die Aufgabe nahe gelegt, nach diesem Vorbilde diejenigen physikalischen Disziplinen axiomatisch zu behandeln, in denen schon heute die Mathematik eine hervorragende Rolle spielt: dies sind in erster Linie die Wahrscheinlichkeitsrechnung und die Mechanik.*

2. Notes

More than ten years have passed since this contribution appeared in Russian and, clearly, some corrections are needed and some new facts must be brought to the readers' attention. These facts are found in the articles mentioned in the additional bibliography below. I have already supplemented the references in the main text by indicating English translations of Russian, French, and German sources. What follows are a few corrections. (I have already corrected a few misprints and/or mistakes.)

1) The method of maximum likelihood (p. 228). The first to introduce the principle, if not the method itself, was Lambert.

2) The term "normal distribution" (p. 241). It was coined not by Pearson but by Ch.S. Peirce [16, p. 99].

3) The theory of probability and the law. The conclusion on p. 242 is too harsh, cf. pp. 222–223 and 235–236.

4) The Average man has no passions or vices (p. 244): On the contrary, Quetelet's mean inclination to crime was related to the Average man.

5) The constancy of crime (p. 244). The number of crimes was not constant at all [33, pp. 299–300]. On the other hand, what Quetelet meant was constancy of crime under constant social conditions.

---

*   J. math. pures et appl., 2e sér., t. 12.

6) Bunyakovskiĭ on the theory of probability and ignorance (p. 249). Elsewhere, he effectively refuted his own statement [36, pp. 6–7].

7) The first quantitative measurement of correlation (note 92) is due to Seidel [32b, §§7.4.2–7.4.3] rather than to Galton.

3. Additional Bibliography

1. *American contributions to mathematical statistics in the 19th century*, vols 1–2, ed. by S.M. Stigler. New York 1980.

2. Bach, A., *Boltzmann's probability distribution of 1877*. Arch. hist. ex. sci., vol. 41, No. 1, 1990, pp. 1–40.

3. Bru, B., *Estimations Laplaciennes*. J. Soc. Stat. Paris, 129ᵉ année, No. 1–2, 1988, pp. 6–45.

4. Crepel, P., *De Condorcet à Arago: l'enseignement des Probabilités en France de 1786 à 1830*. Bull. Soc. Amis Bibl. Ec. Polyt. (Bull. SABIX), No. 4, 1989, pp. 29–81.

5. *Dictionary of scientific biography*, vols 1–16, ed. by C. Gillispie. New York 1970–1980.

6. Dutka, J., *The incomplete Beta-function — a historical profile*. Arch. hist. ex. sci., vol. 24, No. 1, 1981, pp. 11–29.

7. *Encyclopedia of statistical sciences*, vols 1–9, ed. by S. Kotz & N.L. Johnson. New York a.o. 1982–1988.

8. Ermolaeva, N.S., *On Chebyshev's unpublished course of lectures in probability*. Voprosy istorii estestvoznania i tekhniki, No. 4, 1987, pp. 106–112 (in Russian).

9. Farebrother, R.W., *The statistical estimation of the standard linear model, 1756–1853*. Proc. First Tampere Sem. Linear Models (1983). Tampere 1985, pp. 77–99.

10. Gillispie, C.C., *Mémoires inédits ou anonymes de Laplace sur la théorie des erreurs, les polynômes de Legendre, et la philosophie des probabilités*. Rev. hist. sci., t. 32, No. 3, 1979, pp. 223–279.

11. Gnedenko, B.V., *The works of A.A. Markov in the theory of probability* [13, pp. 223–237].

12. Good, I.J., *Some statistical applications of Poisson's work*. Stat. sci., vol. 1, No. 2, 1986, pp. 157–180.

13. Grodzenskiĭ, S.Ya., *A.A. Markov*. Moscow 1987 (in Russian).

14. Harter, H.L., *A chronological annotated bibliography on order statistics*, vol. 1. No place, 1978.

15. *International encyclopedia of statistics*, vols 1–2, ed. by W.H. Kruskal & J. Tanur. New York – London 1978.

16. Kruskal, W., *Formulas, numbers, words: statistics in prose* (1978). In: *New directions for methodology of social and behavioral science etc.*, No. 9, ed. by D. Fiske. San Francisco 1981, pp. 93–102.

17. Mackenzie, D.A., *Statistics in Britain 1865–1930*. Edinburgh 1981.

18. Markov, A.A., *On the solidity of glass*. Written in 1903, previously unpublished; pertains to the theory of errors. Comments by O.B. Sheĭnin. Istoriko-matematicheskie issledovania, vol. 32–33, 1990, pp. 451–467 (in Russian).

19. Nekrasov, P.A., *Theory of probability*. Petersburg 1912. (Three previous editions.)

20. Pearson, K., *Early statistical papers*. Cambridge 1948. Reprint: Cambridge 1956.

21. Pearson, K., *History of statistics in the 17th and 18th centuries*. Lectures delivered in 1921–1933, ed. by E.S. Pearson. London 1978.

22. Plackett, R.L., *The influence of Laplace and Gauss in Britain*. Bull. Intern. Stat. Inst., vol. 53, No. 1, 1989, pp. 163–176.

23. Poincaré, H., *On science*. Moscow 1983 (in Russian).

24. *The mathematical heritage of Henri Poincaré*. Proc. Symp. Indiana Univ. 1980, ed. by F.E. Browder. Proc. Symposia pure math. vol. 39, pt. 2. Providence RI, 1983.

25. *Poisson et la science de son temps*, ed. by M. Métivier et al. Paris 1981.

26. *The probabilistic revolution*, vols 1–2. Cambridge MA – London 1987.

27. *Probability and conceptual change in scientific thought*, ed. by M. Heidelberger & L. Krüger. Bielefeld 1982.

28. Seneta, E., *The central limit theorem and linear least squares in prerevolutionary Russia: the background*. Math. scientist, vol. 9, 1984, pp. 37–77.

29. Seneta, E., *A sketch of the history of survey sampling in Russia*. J. Roy. Stat. Soc., vol. A148, pt 2, 1985, pp. 118–125.

30. Schneider, I., (Hrsg.), *Die Entwicklung der Wahrscheinlichkeitstheorie von den Anfängen bis 1933*. Darmstadt 1988. (A source-book)

31. Sheĭnin, O.B., *C.F. Gauss and the theory of errors*. Arch. hist. ex. sci., vol. 20, No. 1, 1979, pp. 21–72.

32. Sheĭnin, O.B., [On the history of the statistical method in natural science, five papers]. Arch. hist. ex. sci., vol. 22 (1980) – vol. 33 (1985).

33. Sheĭnin, O.B., *Quetelet as a statistician*. Arch. hist. ex. sci., vol. 36, No. 4, 1986, pp. 281–325.

34. Sheĭnin, O.B., *C.F. Gauss and the χ^2-distribution*. Schriftenreihe Gesch. Naturwiss., Technik, Med., Bd. 25, No. 1, 1988, pp. 21–22.

35. Sheĭnin, O.B., *The concept of randomness from Aristotle to Poincaré.* Inst. hist. nat. sci. & technology, preprint No. 9. Moscow 1988 (in Russian).

36. Sheĭnin, O.B., *On V.Ya. Bunyakovskiĭ's works on the theory of probability.* Inst. hist. nat. sci. & technology, preprint No. 17. Moscow 1988 (in Russian). English version to appear in Arch. hist. ex. sci.

37. Sheĭnin, O.B., *A.A. Markov's work on probability.* Arch. hist. ex. sci., vol. 39, No. 4, 1989, pp. 337–377.

38. Sprott, D.A., *Gauss's contributions to statistics.* Hist. math., vol. 5, No. 2, 1978, pp. 183–203.

39. Stigler, S.M., *F.Y. Edgeworth, statistician.* J. Roy. Stat. Soc., vol. A141, 1978, pp. 287–322.

40. Stigler, S.M., *Mathematical statistics in the early States.* Ann. stat., vol. 6, No. 2, 1978, pp. 239–265.

41. Stigler, S.M., *The history of statistics.* Cambridge MA – London 1986.

42. *"Student". A statistical biography of W.S. Gosset based on writings by E.S. Pearson*, ed. by R.L. Plackett & G.A. Barnard. Oxford 1990.

43. *Studies in the history of statistics and probability*, vols 1–2, London 1970–1977. Ed. by E.S. Pearson & M.G. Kendall (vol. 1), M.G. Kendall & R.L. Plackett (vol. 2).

44. Tsykalo, A.L., *A.M. Liapunov.* Moscow 1988 (in Russian).

45. van der Waerden, B.L., *Über die Methode der kleinsten Quadrate.* Nachr. Akad. Wiss. Göttingen, Math. Phys. Kl., No. 8, 1977, pp. 75–87.

46. Walker, H.M., *Studies in the history of statistical methods* (1929). Reprints: Baltimore 1931; New York 1975.

Bibliography

Collected works and classical editions

Abel, N.H., *Œuvres complètes*, T. 1–2. Christiania 1881.

Betti, E., *Opere matematiche*, T. 1–2. Milano 1903–1914.

Boltzmann, L., *Wissenschaftliche Abhandlungen*, Bd. 1–2. Leipzig 1909.

Bolzano, B., *Gesammelte Schriften*, Bd. 1–12. Wien 1882.

Boole, G., *Collected logical works*, Vol. 1–2. Chicago-London 1940.

Boole, G., *Studies in logic and probability*. London 1952.

Cauchy, A.L., *Cours d'analyse de l'Ecole Royale Polytechnique*. Première partie: Analyse algébrique. Paris 1821.

Cauchy, A.L., *Œuvres complètes*, T. 1–27 (2 séries). Paris 1882–1974.

Cayley, A., *Collected mathematical papers*, Vol. 1–14. Cambridge 1889–1898.

Chebyshev, P.L., *Polnoe sobranie sochineniĭ* (Complete works), Vol. 1–5. Moscow-Leningrad 1944–1951.

Clifford, W.K., *Lectures and essays*, Vol. 1–2. London 1901.

Clifford, W.K., *Mathematical papers*. London 1882; New York 1968.

Clifford, W.K., *The common sense of the exact sciences*. New York 1885.

Cournot, A.A., *Exposition de la théorie des chances et des probabilités*. Paris 1848.

Dedekind, R., *Gesammelte mathematische Werke*, Bd. 1–3. Braunschweig 1930–1932.

Dirichlet, P.G. Lejeune, *Vorlesungen über Zahlentheorie*. Braunschweig 1863.

Eisenstein, F.G.M., *Mathematische Abhandlungen*. Berlin 1874.

Fedorov, E.S., *Nachala ucheniya o figurakh*. Moscow 1953.

Fedorov, E.S., *Simmetriya i struktura kristallov*. Moscow 1949.

Fourier, J.B.J., *Œuvres*, T. 1–2. Paris 1888–1890.

Frege, G., *Kleine Schriften*. Darmstadt 1967.

Frege, G., *Die Grundlagen der Arithmetik*, New York 1950.

Frege, G., *Funktion, Begriff, Bedeutung*. Fünf logische Studien. Göttingen 1962.

Frobenius, G., *Gesammelte Abhandlungen*, Bd. 1–3. Berlin 1968.

Frobenius, G., *De functionum analyticarum unius variabilis per series infinitas repraesentatione*. Berolini 1876.

Galois, E., *Œuvres mathématiques*. Paris 1897.

Gauss, C.F., *Werke*, Bd. 1–12. Göttingen 1863–1929. Reprint Hildesheim-New York 1973.

Gibbs, J.W., *The collected works*, Vol. 1–2. New York-London-Toronto 1928.

Grassmann, H., *Gesammelte mathematische und physikalische Werke*, Bd. 1–3. Leipzig 1894–1911.

Hadamard, J., *Œuvres*, T. 1–4. Paris 1968.

Hamilton, W.R., *The mathematical papers*, Vol. 1–3. Cambridge 1931–1967.

Hankel, H., *Vorlesungen über die complexen Zahlen und ihre Funktionen*, 2 Teile. Leipzig 1867.

Hermite, Ch., *Œuvres*, T. 1–4. Paris 1905–1917.

Hilbert, D., *Grundlagen der Geometrie*. Leipzig 1903.

Hilbert, D., *Gesammelte Abhandlungen*, Bd. 1–3. Berlin 1932–1935.

Hilbert, D., *Grundzüge der geometrischen Logik*. Berlin 1928.

Jacobi, C.G.J., *Gesammelte Werke*, Bd. 1–7. Berlin 1881–1891.

Jordan, C., *Œuvres*, T. 1–4. Paris 1961–1964.

Klein, F., *Gesammelte mathematische Abhandlungen*, Bd. 1–3. Berlin 1921–1923.

Korkin, A.N., *Sochineniya*. (Works), Vol. 1. SPb. 1911.

Kronecker, L., *Werke*, Bd. 1–5. Leipzig 1895–1930.

Lagrange, J.L., *Œuvres*, T. 1–14. Paris 1867–1892.

Laplace, P.S., *Essai philosophique sur les probabilités*. Paris 1816.

Laplace, P.S., *Exposition du système du monde*. Paris 1808.

Laplace, P.S., *Œuvres complètes*, T. 1–14. Paris 1878–1912.

Lejeune-Dirichlet, P.G., *Werke*, Bd. 1–2. Berlin 1889–1897.

Lie, S., *Gesammelte Abhandlungen*, Bd. 1–10. Leipzig-Oslo 1934–1960.

Lobachevskiĭ, N.I., *Polnoe sobranie sochineniĭ* (Complete works), Vol. 1–5. Moscow-Leningrad 1946–1951.

Lyapunov, A.M., *Sobranie sochineniĭ*. (Collected works), Vol. 1–4. Moscow 1954–1965.

Markov, A.A., *Izbrannye trudy* (Selected works). Moscow 1951.

Markov, A.A., *Izbrannye trudy po teorii nepreryvnykh drobeĭ i teorii funktsiĭ, naimenee uklonyayushchikhsya ot nulya*. Moscow-Leningrad 1948.

Minkowski, H., *Gesammelte Abhandlungen*, Bd. 1–2. Leipzig-Berlin 1911.

Ostrogradskiĭ, M.V., *Polnoe sobranie trudov* (Complete collected works), Vol. 1–3. Kiev 1959–1961.

Peano, G., *Formulario matematico*, Cremonese (Ed.). Roma 1960.

Peano, G., *Opere scelte*, Vol. 1–3, Cremonese (Ed.). Roma 1957–1958.

Pearson, K., *Early statistical papers*. Cambridge 1948.

Peirce, B.O., *Mathematical and physical papers*, 1903–1913. Cambridge MA 1926.

Peirce, Ch.S., *Collected papers*, Vol. 1–8. Cambridge MA 1931–1958.

Poincaré, H., *Œuvres de Henri Poincaré*. Paris 1916.

Poincaré, H., *Œuvres*, T. 1–11. Paris 1928–1956.

Riemann, G.F.B., *Gesammelte mathematische Werke und wissenschaftlicher Nachlass*. Leipzig 1876.

Ruffini, P., *Opere matematiche*, Vol. 1–3, Cremonese (Ed.). Roma 1953–1954.

Smith, H.J.S., *Collected mathematical papers*, Vol. 1–2. Oxford 1894.

Stieltjes, T.J., *Recherches sur les fractions continues*. Mémoires présentés par divers savants à l'Académie des Sciences, T. 32 (2e série). Paris 1909.

Stieltjes, T.J., *Œuvres complètes*, T. 1–2. Groningen 1914–1918.

Sturm, Ch., *Abhandlungen über die Auflösung der numerischen Gleichungen*. Ostwald's Klassiker, Bd. 143, 1904.

Sylvester, J.J., *Collected mathematical papers*, Vol. 1–4. Cambridge 1904–1911.

Voronoĭ, G.F., *Sobranie sochineniĭ* (Collected works), Vol.1–3. Kiev 1952–1953.

Weierstrass, K., *Mathematische Werke*, Bd. 1–7. Berlin 1894–1897.

Wroński, J. Hoëne, *Œuvres mathématiques*, T. 1–4. Paris 1925.

Zolotarev, E.I., *Polnoe sobranie sochineniĭ* (Complete works), Vyp. 1–2. Leningrad 1931.

Chapter 1
Sources

Al-Farabi, *Logicheskie traktaty*, Alma-Ata 1975.

Aristoteles, *Analytica priora*. Venetiis 1557.

Boole, G., *An investigation of the laws of thought*. London-Cambridge 1854.

Boole, G., *The mathematical analysis of logic, being an essay towards a calculus of deductive reasoning*. Cambridge-London 1847.

Church, A., *Introduction to mathematical logic*, Vol. 1. Princeton NJ 1956.

Couturat, L., *La logique de Leibniz*. Paris 1901.

Couturat, L., *L'algèbre de la logique*. Paris 1914.

Euler, L., *Lettres à une princesse d'Allemagne sur divers sujets de physique et de philosophie*, T. 2. St.-Pétersbourg 1768.

Hamilton, W., *Lectures on metaphysics and logic*, Vol. 1–4. Edinburgh-London 1860.

Ibn Sina, *Danish-mame* (Book of knowledge). Stalinabad 1957.

Jevons, W.S., *On the mechanical performance of logical inference*. Philos. Trans., 1870, **160**, 497–517.

Leibniz, G.W., *Die philosophischen Schriften*, Bd. 7. Berlin 1890.

Leibniz, G.W., *Opera philosophica quae exstant latina, gallica, germanica*, J.E. Erdmann (Ed.). Berlin 1840.

Leibniz, G.W., *Opuscules et fragments inédits de Leibniz*, L. Couturat (Ed.). Paris 1903.

Leibniz, G.W., *Philosophische Werke*, Bd. 3. Neue Abhandlungen über den menschlichen Verstand. Leipzig 1904.

Minto, W., *Inductive and deductive logic*. New York 1893.

Morgan, A. de, *Formal logic: or the calculus of inference, necessary and probable*. London 1847.

Morgan, A. de, *Trigonometry and double algebra*. London 1849.

Poretskiĭ, P.S., *Izlozhenie osnovnykh nachal logiki v vozmozhno bolee naglyadnoĭ i obshchedostupnoĭ forme*. Kazan' 1881.

Poretskiĭ, P.S., *Iz oblasti matematicheskoĭ logiki*. Moscow 1902.

Poretskiĭ, P.S., *Po povodu broshyury g. Volkova "Logicheskoe ischislenie"*. Kazan' 1884.

Poretskiĭ, P.S., *O sposobakh resheniya logicheskikh ravenstv i ob obratnom sposobe matematicheskoĭ logiki*. Kazan' 1884.

Schröder, E., *Der Operationskreis des Logikkalküls*. Leipzig 1877.

Schröder, E., *Vorlesungen über die Algebra der Logik*, Bd. 1–3. Leipzig 1890–1905.

Venn, J., *Symbolic logic*. London 1881; 2nd ed. 1894.

Venn, J., *On the diagrammatic and mechanical representations of propositions and reasoning*. The London, Edinburgh and Dublin Philos. Mag. and J. Sci., ser. 5, 1880, **10**.

Secondary literature

Berg, J., *Bolzano's logik*. Stockholm 1962.

Berka, K., Kreisel, L., *Logik-Texte; Kommentierte Auswahl zur Geschichte der Logik*. Berlin 1971.

Biryukov, B.V., *Krushenie metafizicheskoĭ kontseptsii universal'nosti predmetnoĭ oblasti v logike*. Moscow 1963.

Blanché, R., *La logique et son histoire. D'Aristote à Russell*. Paris 1971.

Bobynin, V.V., *Opyty matematicheskogo izlozheniya logiki. Sochineniya Ernesta Shredera*. Fis.-matem. nauki v ikh nastoyashchem i proshedshem, 1886–1894, **2**, 65–72, 173–192, 438–458.

Bocheński, J.M., *Formale Logik*. Freiburg-München 1962.

Boltaev, M.N., *Voprosy gnoseologii i logiki v proizvedeniyakh Ibn Siny i ego shkoly.* Dyushanbe 1965.

Carruccio, E., *Matematica e logica nella storia e nel pensiero contemporaneo.* Torino 1958.

Heijenoort, J. van, *From Frege to Gödel. A source book in mathematical logic, 1879–1931.* Cambridge MA 1967.

Jørgensen, J., *A treatise of formal logic: its evolution and main branches with relation to mathematics and philosophy,* Vol. 1–3. New York 1962.

Kneale, W., Kneale, M., *The development of logic.* Oxford 1962.

Kotarbiński, T., *Wykłady z dziejów logiki.* Łódź 1957.

Kuzichev, A.S., *Diagrammy Venna (Istoriya i primeneniya).* Moscow 1968.

Lewis, C.J., *A survey of symbolic logic.* New York 1960.

Łukasiewicz, J., *Aristotle's syllogistic from the standpoint of modern formal logic.* Oxford 1951.

Narskiĭ, I.S., *Gotfrid Leĭbnits.* Moscow 1972.

Scholz, H., *Geschichte der Logik.* Berlin 1931.

Sternfeld, R., *Frege's logical theory.* Illinois Univ. Press 1966.

Styazhkin, N.I., Silakov, V.D., *Kratkiĭ ocherk obshcheĭ i matematicheskoĭ logiki v Rossii.* Moscow 1962.

Styazhkin, N.I., *Formirovanie matematicheskoĭ logiki.* Moscow 1967.

Wang Hao, *A survey of mathematical logic.* Peking 1962.

Yanovskaya, S.A., *Osnovaniya matematiki i matematicheskaya logika.* In: Matematika v SSSR za tridtsat' let. Moscow-Leningrad 1948.

Chapter II and III
Sources and monographs

Ayoub, R., *An introduction to the analytic theory of numbers.* Providence 1963.

Bachmann, P., *Zahlentheorie,* Bd. 1–5. Leipzig-Berlin 1921–1927.

Bukhshtab, A.A., *Teoriya chisel.* Moscow 1966.

Cassels, J.W.S., *An introduction to the geometry of numbers.* Berlin 1959.

Chebotarev, N.G., *Teoriya Galua.* Moscow-Leningrad 1936.

Dedekind, R., Weber, H., *Theorie der algebraischen Funktionen einer Veränderlichen.* In: *R. Dedekind's Gesammelte mathematische Werke,* Bd. 1. Braunschweig 1930, pp. 238–348.

Euler, L., *Commentationes arithmeticae collectae,* T. 1–2. Petropoli 1849.

Gauss, C.F., *Disquisitiones arithmeticae.* Gottingae 1801; in: *Werke,* Bd. 1. Göttingen 1863.

Hamilton, W.R., *Lectures on quaternions.* Dublin 1853.

Hanceck, H., *Development of the Minkowski geometry of numbers.* New York 1939.

Jordan, C., *Traité des substitutions et des équations algébriques*, 2e éd. Paris 1957.

Khovanskiĭ, A.N., *Prilozhenie tsepnykh drobeĭ i ikh obobshcheniĭ k voprosam priblizhennogo analiza.* Moscow 1956.

Kogan, L.A., *O predstavlenii tselykh chisel kvadratichnymi formami polozhitel'nogo opredelitelya.* Tashkent 1971.

Klein, F., *Ausgewählte Kapitel der Zahlentheorie*, Bd. 1–2. Göttingen 1896–1897.

Landau, E., *Handbuch der Lehre von der Verteilung der Primzahlen*, Bd. 1–2. Leipzig 1909.

Landau, E., *Vorlesungen über Zahlentheorie*, Bd. 1–3. Leipzig 1927.

Legendre, A.M., *Théorie des nombres*, T. 1–2. Paris 1830.

Markov, V.A., *O polozhitel'nykh troĭnichnykh kvadratichnykh formakh.* SPb. 1897.

Perron, O., *Die Lehre von den Kettenbrüchen.* Leipzig-Berlin 1913.

Prachar, K., *Primzahlverteilung.* Berlin 1957.

Seeber, L.A., *Untersuchungen über die Eigenschaften der positiven ternären quadratischen Formen.* Freiburg 1831.

Shafarevich, I.R., *Basic algebraic geometry.* Berlin 1977.

Sokhotskiĭ, Yu.V., *Nachalo obshchego naibol'shego delitelya v primenenii k teorii delimosti algebraicheskikh chisel.* SPb. 1893.

Venkov, B.A., *Elementarnaya teoriya chisel.* Moscow-Leningrad 1937.

Secondary literature

Archimedes, Huygens, Lambert, Legendre; *vier Abhandlungen über die Kreismessung*, hrsg. v. F. Rudio. Leipzig 1892.

Bashmakova, I.G., *Diofant i diofantovy uravneniya.* Moscow 1974.

Bashmakova, I.G., *Obosnovanie teorii delimosti v trudakh E.I. Zolotarëva.* IMI, 1949, **2**, 233–351.

Bashmakova, I.G., *O dokazatel'stve osnovnoĭ teoremy algebry.* IMI, 1957, **10**, 257–304.

Bashmakova, I.G., *Sur l'histoire de l'algèbre commutative.* XIIe Congrès intern. d'histoire des sciences. Colloques. Textes des rapports. Paris 1968, p. 185–202.

Berman, G.N., *Chislo i nauka o nem.* Moscow-Leningrad 1949.

Bespamyatnykh, N.D., *Arifmetichsekie issledovaniya v Rossii v XIX veke.* Uchen. zap. Grodnen. ped. in-ta, 1957, **2**, 3–42.

Bortolotti, E., *Influenza dell'opera matematica di Paolo Ruffini sullo svolgimento delle teorie algebriche.* Modena 1903.

Bourbaki, N., *Note historique* (chap. I à III). In: *Groupes et algèbres de Lie,* Chapitres 2 et 3. Paris 1972.

Brill, A., Noether, M., *Die Entwicklung der Theorie der algebraischen Funktionen in älterer und neuerer Zeit.* Jahresber. Dtsch. Math.-Verein. 1894, 3, 107–566.

Bunt, L.N.H., *The development of the ideas of number and quantity according to Piaget.* Groningen 1951.

Bunyakovskiĭ, V.Ya., *Leksikon chistoĭ i prikladnoĭ matematiki.* SPb. 1839.

Chebotarev, N.G., *Novoe obosnovanie teorii idealov (po Zolotarevu).* Izv. fiz.-matem. obshch-va Kazani, 1925, 2, No. 25.

Chebotarev, N.G., *Obosnovanie teorii delimosti po Zolotarevu.* UMN, 1947, 2, No. 6 (22), 52–67.

Crowe, M.J., *A history of vector analysis. The evolution of the idea of a vector system.* Univ. Notre Dame Press 1967.

Delone, B.N., *German Minkowskiĭ.* UMN, 1936, **2**, 32–38.

Delone, B.N., *Peterburgskaya shkola teorii chisel.* Moscow-Leningrad 1947.

Delone, B.N., *Puti razvitiya algebry.* UMN, 1952, **7**, Vyp. 3 (49), 155–178.

Delone, B.N., *Raboty Gaussa po teorii chisel.* In: *Karl Fridrikh Gauss.* Moscow 1956, 11–112.

Delone, B.N., *Razvitie teorii chisel v Rossii.* Uchen. zap. MGU, 1947, Vyp. 91, 77–96.

Depman, I.Ya., *Istoriya arifmetiki.* Moscow 1965.

Dickson, L.E., *History of the theory of numbers,* Vol. 1–3. Washington 1919–1923.

Dieudonné, J., *Minkowski Hermann.* Dictionary of scientific biography, Ch.C. Gillispie (Ed. in chief.), Vol. 9, p. 411–414.

Dieudonné, J. *Cours de géométrie algébrique. Aperçu historique sur le développement de la géométrie algébrique.* Paris 1974.

Dubreil, P., *La naissance de deux Jumelles. La logique mathématique et l'algèbre ordonnée.* XIIe Congrès intern. d'histoire des sciences. Colloques. Textes des rapports. Paris 1968, p. 203–208.

Freudenthal, H., *L'algèbre topologique, en particulier les groupes topologiques et de Lie.* Ibid., p. 223–243.

Gauss, C.F., *Disquisitiones arithmeticae.* Gottingae 1801; in: *Werke,* Bd. 1. Göttingen 1863.

Gericke, H., *Geschichte des Zahlbegriffs*. Mannheim 1970.

Hensel, K., *E.E. Kummer und der grosse Fermatsche Satz*. Marburg 1910.

Kanunov, N.F., *Pervyĭ ocherk teorii algebry F.E. Molina*. IMI, 1975, **20**, 150.

Kiselev, A.A., Ozhigova, E.P., *K istorii èlementarnogo metoda v teorii chisel*. Actes du XI congrès intern. d'histoire des sciences (1965), Vol. 3. Warszawa 1967, 244.

Konen, H., *Geschichte der Gleichung $t^2 - Du^2 = 1$*. Leipzig 1901.

Kuz'min, R.O., *Zhizn' i nauchnaya deyatel'nost' E.I. Zolotareva*. UMN, 1947, **2**, Vyp. 6 (22), 21–51.

Liebmann, H., Engel, F., *Die Berührungstransformationen. Geschichte und Invariantentheorie*. Leipzig 1914.

Matvievskaya, G.P., *Postulat Bertrana v zapisnykh knizhkakh Eĭlera*. IMI, 1961, **14**, 285–288.

Mel'nikov, I.G., *V.Ya. Bunyakovskiĭ i ego raboty po teorii chisel*. Trudy In-ta istorii estestvoznaniya i tekhniki, 1957, **17**, 270–286.

Minin, A.P., *O trudakh N.V. Bugaeva po teorii chisel*. Matem. sb., 1904, 25, No. 2, 293–321.

Mitzscherling, A., *Das Problem der Kreisteilung. Ein Beitrag zur Geschichte seiner Entwicklung*. Leipzig-Berlin 1913.

Morozova, N.N., *V.Ya. Bunyakovskiĭ i ego raboty po teorii chisel*. Uchen. zap. mosk. obl. ped. in-ta, 1970, **282**, No. 8.

Muir, T., *The theory of determinants in the historical order of development*, Vol. 1–5, London 1906–1930.

Natucci, A., *Il concetto di numero e le sue estensioni. Studi storico-critichi intorno ai fondamenti dell'Aritmetica generale col oltre 700 indicazioni bibliografiche*. Torino 1923.

Nový, L., *Origin of modern algebra*. Prague 1973.

Nový, L., *L'Ecole algébrique anglaise* XIIe Congrès intern. d'histoire des sciences. Colloques. Textes des rapports. Paris 1968, p. 211–222.

Ore, O., *Number theory and its history*. New York-Toronto 1948.

Ozhigova, E.P., *Razvitie teorii chisel v Rossii*. Leningrad 1972.

Posse, K.A., *A.N. Korkin*. Matem. sb., 1909, **27**, No. 1, 1–27.

Posse, K.A., *Zametka o reshenii dvuchlennykh sravneniĭ s prostym modulem po sposobu Korkina*. Soobshch. khar'k. matem. ob-va, ser. 2, 1910, **11**, 249–268.

Ryago, G., *Iz zhizni i deyatel'nosti chetyrekh zamechatel'nykh matematikov Tartusskogo universiteta*. Uchen. zap. Tartuss. un-ta, 1955, **37**, 74–103.

Smith, H.J.C., *On the history of the researches of mathematicians on the series of prime numbers*. In: *Collected mathematical papers*, Vol. 1. Oxford 1894, p. 35–37.

Smith, H.J.C., *Reports on the theory of numbers*. Ibid., p. 38–364.

Smith, H.J.C., *On the present state and prospects of some branches of pure mathematics*. In: *Collected mathematical papers*, Vol. 2. Oxford 1894, p. 166–190.

Sorokina, L.A., *Raboty Abelya ob algebraicheskoĭ razreshimosti uravneniĭ*. IMI, 1959, **12**, 457–480.

Studnicka, F.J., *Cauchy als formaler Begründer der Determinantentheorie*. Prag 1876.

Sushkevich, A.K., *Materialy k istorii algebry v Rossii v XIX v. i v nachale XX v*. IMI, 1951, **4**, 237–451.

Torelli, G., *Sulla totalità dei numeri primi fino a un limite assegnato*. Atti Acad. sci. fis. e mat. Napoli, sez. 2, 1901, **1**, 1–222.

Uspen'skiĭ, Ya.V., *Ocherk nauchnoĭ deyatel'nosti A.A. Markova*. Izv. Ros. Akad. nauk, 1923, **17**, 19–34.

Vasil'ev, A.V., *Tseloe chislo*. Petrograd 1922.

Verriest, G., *Evariste Galois et la théorie des équations algébriques*. Louvain 1934.

Wieleitner, H., *Der Begriff der Zahl in seiner logischen und historischen Entwicklung*. Berlin 1918.

Wussing, H., *Die Genesis des abstrakten Gruppenbegriffes*. Berlin 1969.

Yushkevich, A.P., Bashmakova, I.G., *Algebra ili vychislenie konechnykh. N.I. Lobachevskogo*. IMI, 1949, **2**, 72–128.

Chapter IV
Sources

Adrain, R., *Research concerning the probabilities of the errors which happen in making observations*. Analyst or math. Museum, 1808, **1**, N 4.

Bertrand, J., *Calcul des probabilités*. Paris 1888.

Bienaymé, I.J., *Considerations à l'appui de la découverte de Laplace sur la loi des probabilités dans la méthode des moindres carrés*. C. r. Acad. sci. Paris 1853, **37**.

Bienaymé, I.J., *Mémoire sur la probabilité des erreurs d'après la méthode des moindres carrés*. J. math. pures et appl., 1852, **17**.

Boltzmann, L., *Vorlesungen über Gastheorie*, 2 vols. Leipzig 1896–1898.

Boole, G., *Studies in logic and probability*. London 1952.

Bunyakovskiĭ, V.Ya., *O prilozhenii analiza veroyatnosteĭ k opredeleniyu priblizhennykh velichin transtsendentnykh chisel*. Memuary Peterburg. akad. nauk, 1837, **1**, (3), No. 5.

Bunyakovskiĭ, V.Ya., *O soedineniyakh osobogo roda, vstrechayushchikhsya v voprose o defektakh.* Prilozh. No. 2 k T. 20. Zapisok Peterburg. akad. nauk za 1871.

Bunyakovskiĭ, V. Ya., *Osnovaniya matematicheskoĭ teorii veroyatnosteĭ.* SPb. 1846.

Chebyshev, P.L., *Teoriya veroyatnosteĭ (Lektsii 1879–1880).* Izdano A.N. Krylovym po zapisyam A.M. Lyapunova. Moscow-Leningrad 1936.

Chuprov, A.A., *Voprosy statistiki.* Moscow 1960.

Chuprov, A.A., *Ocherki po teorii statistiki.* Moscow 1959.

Czuber, E., *Theorie der Beobachtungsfehler.* Leipzig 1891.

Davidov, A.Yu., *Prilozhenie teorii veroyatnosteĭ k statistike.* In: *Ucheno-literaturnye stat'i professorov i prepodavateleĭ moskovskogo universiteta.* Moscow 1855.

Davidov, A.Yu., *Teoriya veroyatnosteĭ.* Moscow Litografirovannyĭ kurs lektsiĭ 1879–1880.

Ehrenfest, P., *Sbornik stateĭ.* Moscow 1972.

Galton, F., *Natural inheritance.* London-New York 1889.

Gibbs, J.W., *Elementary principles in statistical mechanics.* New York-London 1902.

Khinchin, A.Ya., *Matematicheskie osnovaniya statisticheskoĭ mekhaniki.* Moscow-Leningrad 1943.

Kuz'min, R.O., *Ob odnoĭ zadache Gaussa.* DAN SSSR, ser. A, 1928, No. 18–19, 375–380.

Lacroix, S.F., *Traité élémentaire du calcul des probabilités.* Paris 1816.

Legendre, A.M., *Nouvelles méthodes pour la détermination des orbites des comètes.* Paris 1805 et 1806.

Linnik, Yu.V., *Zamechaniya po povodu klassicheskogo vyvoda zakona Maksvella.* DAN SSSR, 1952, **85**, No. 6, 1251–1254.

Markov, A.A., *Ischislenie veroyatnosteĭ.* SPb. 1900, 4-e izd., Moscow 1924.

Mikhel'son, V.A., *Sobranie sochineniĭ*, T. 1. Moscow 1930.

Morgan, A. de, *Theory of probability.* In: Encyclopaedia metropolitana, Vol. 2. London 1845.

O teorii dispersii. Sb. st. V. Leksisa, V.I. Bortkevicha, A.A. Chuprova, R.K. Bauèra. Moscow 1968.

Pearson, K., *On a method of ascertaining limits to the actual number of marked members in a population of given size from a sample.* Biometrika, 1928, **20 A**, pt. 1–2.

Pirogov, N.N., *Osnovaniya kineticheskoĭ teorii mnogoatomnykh gazov.* Zhurn. rus. fis-khim. ob-va, 1886–1887, **18–19**.

Poincaré, H., *Calcul des probabilités*. Paris 1896.

Poisson, S.D., *Mémoire sur la proportion des naissances des filles et des garçons*. Mém. Acad. sci. Paris 1830, **9**.

Poisson, S.D., *Recherches sur la probabilité des jugements en matière criminelle et en matière civile*. Paris 1837.

Poisson, S.D., *Sur la probabilité des résultats moyens des observations*. Conn. des tems., 1827 et 1832 (publ. 1824 et 1829).

Poisson, S.D., *Sur l'avantage du banquier au jeu de trente-et-quarante*. Ann. math. pures et appl., 1825–1826, **16**.

Quetelet, A., *Sur l'homme et le développement de ses facultés ou essay de physique sociale*, 2 vols. Paris 1835.

Recherches statistiques sur la ville de Paris et de département de la Seine. Sous la direction de J.B.J. Fourier, T. 1–4. Paris 1821–1829.

Sleshinskiĭ, I.V., *K teorii sposoba naimen'shikh kvadratov*. Zap. matem. otdeleniya Novoros. ob-va estestvoispytateleĭ. Odessa 1892, **14**.

Tikhomandritskiĭ, M.A., *Kurs teorii veroyatnosteĭ*. Khar'kov 1898.

Venn, J., *Logic of chance*. London 1866.

Secondary literature

Adams, W.J., *The life and times of the central limit theorem*. New York 1974.

Bowley, A.L., *F.Y. Edgeworth's contributions to mathematical statistics*. London 1928.

Brashman, N.D., *O vliyanii matematicheskikh nauk na razvitie umstvennykh sposobnosteĭ*. Moscow 1841.

Czuber, E., *Wahrscheinlichkeitsrechnung*. In: *Encyclopädie der mathematischen Wissenschaft*, Bd. 1. Leizig 1904.

Druzhinin, N.K., *Razvitie statisticheskoĭ praktiki i statisticheskoĭ nauki v èvropeĭskikh stranakh*. In: *V.I. Lenin i soremennaya statistika*, Vol. 1, Moscow, "Statistika", 1970, 33–36.

Druzhinin, N.K., *K voprosu o prirode statisticheskikh zakonomernosteĭ i o predmete statistiki kak nauki*. Uchen. zap. po statistike, 1961, **6**, 65–77.

Freudenthal, H., Steiner, H.-G., *Aus der Geschichte der Wahrscheinlichkeitstheorie und mathematischen Statistik*. In: *Grundzüge der Mathematik*, Bd. 4. Göttingen 1966, S. 149–195.

Gnedenko, B.V., Gikhman, I.I., *Razvitie teorii veroyatnosteĭ na Ukraine*. IMI, 1956, **9**, 477–536.

Gnedenko, B.V., *Kratkiĭ ocherk istorii teorii veroyatnosteĭ*. In: *Kurs teorii veroyatnosteĭ*. Moscow 1954, 360–388.

Gnedenko, B.V., *Razvitie teorii veroyatnosteĭ v Rossii*. Trudy In-ta istorii estestvoznaniya, 1948, **2**, 390–425.

Gnedenko, B.V., *O rabotakh A.M. Lyapunova po teorii veroyatnosteǐ.* IMI, 1959, **12**, 135–160.

Gnedenko, B.V., *O rabotakh K.F. Gaussa po teorii veroyatnosteǐ.* In: *K.F. Gauss.* Moscow 1956, 217–240.

Gnedenko, B.V., *O rabotakh M.V. Ostrogradskogo po teorii veroyatnosteǐ.* IMI, 1951, **4**, 99–123.

Grattan-Guiness, I., *Fourier's anticipation of linear programming.* Operat. Res. Quarterly, 1970, **21**, 361–364.

Heyde, C.C., Seneta, E. *I.J. Bienaymé.* New York-Heidelberg-Berlin 1977.

Kolmogorov, A.N., *Rol' russkoǐ nauki v razvitii teorii veroyatnosteǐ.* Uchen. zap. MGU, Ser. matem. nauk, 1947, 91, 53–64.

Koren, J., *The history of statistics. Their development and progress in many countries.* New York 1970.

Maǐstrov, L.E., *Teoriya veroyatnosteǐ. Istoricheskiǐ ocherk.* Moscow 1967.

Ondar, Kh.O., *O rabotakh A.Yu. Davidova po teorii veroyatnosteǐ i ego metodologicheskikh vzglyadakh.* Istoriya i metodologiya estestvennykh nauk, 1971, **11**, 98–109.

Pavlovskiǐ, A.F., *O veroyatnosti. Rechi, proiznesennye v torzhestvennom sobranii Khar'kovskogo universiteta.* Khar'kov 1821.

Plackett, R.L., *The discovery of the method of least squares.* Biometrika, 1972, 59 N 2, 239–251.

Ptukha, M.V., *Ocherki po istorii statistiki v SSSR,* T. 2. Moscow 1959.

Schneider, I., *Beitrag zur Einführung wahrscheinlichkeits-theoretischer Methoden in die Physik der Gase nach 1856.* Arch. hist. exact. sci., 1974, **14**, N 3, 237–261.

Schneider, I., *Clausius' erste Anwendung der Wahrscheinlichkeitsrechnung im Rahmen der atmosphärischen Lichtstreuung.* Arch. hist. exact. sci., 1974, **14**, N 2, 143.

Sheǐnin, O.B., *D. Bernoulli's work on probability.* Rete. Strukturgesch. Naturwiss., 1972, **1**, N 3–4, 273–300.

Sheǐnin, O.B., *Laplace's theory of errors.* Arch. hist. exact. sci., 1977, **17**, N 1, 1–61.

Sheǐnin, O.B., *Laplace's work in probability.* Arch. hist. exact. sci., 1976, **16** N 2, 137–187.

Sheǐnin, O.B., *S.D. Poisson's work in probability.* Arch. hist. exact. sci., 1978, **18**, N 3.

Sheǐnin, O.B., *Teoriya veroyatnosteǐ P.S. Laplaca.* IMI, 1977, **22**, 212–224.

Sheǐnin, O.B., *O poyavlenii del'ta-funktsii Diraka v trudakh P.S. Laplaca.* IMI, 1975, **20**, 303–308.

Sheǐnin, O.B., *O rabotakh R. Edreǐna v teorii oshibok.* IMI, 1965, **16**, 325–336.

Todhunter, I., *A history of mathematical theory of probability from the time of Pascal to that of Laplace*. New York 1965.

Truesdell, C., *Early kinetic theory of gases*. Arch. hist. exact. sci., 1975, **15**, N 1, 1–66.

Walker, H.M., *Studies in the history of statistical methods*. Baltimore 1929.

Westergaard, H., *Contributions to the history of statistics*. New York 1969.

Abbreviations

Abhandl. Preuss. Akad. Wiss.	Abhandlungen der Preussischen Akademie der Wissenschaften. Mathematisch-Naturwissenschaftliche Klasse
Amer. J. Math.	American Journal of Mathematics
Ann. Ecole Norm.	Annales scientifiques de l'Ecole Normale Supérieure
Ann. Math.	Annales de mathématiques de M. Gergonne
Ann. math. pues et appl.	Annales des mathématiques pures et appliquées
Ann. Phys. und Chem.	Annalen der Physik und der Chemie
Ann. Soc. sci. Bruxelles	Annales de la Société sientifique de Bruxelles
Arch. Hist. exact. sci.	Archive for History of Exact Sciences
Atti Accad. sci. fis. e mat. Napoli	Atti della Accademia delle scienze fisiche e matematiche di Napoli
Bericht. Verhandl. Akad. Wiss.	Berichte über die Verhandlungen der Sächsischen Akademie der Wissenschaften zu Leipzig. Mathematisch-Physikalische Klasse
Bericht. Königl. Akad. Wiss. zu Berlin	Bericht der Königlichen Akademie der Wissenschaften. Mathematisch-Naturswissenschaftliche Klasse(Berlin)
Bull. Acad. Sci. St.-Pétersbourg	Bulletin de l'Académie des Sciences de St.-Pétersbourg
Bull. sci. math. et astron.	Bulletin des sciences mathématiques et astronimiques
Bull. Sci. math.	Bulletin des sciences mathématiques de M. Férussac
Bull. Soc. math. France	Bulletin de la Société mathématique de France
C. r. Acad. sci. Paris	Comptes rendus hebdomadaires des séances de l'Académie des Sciences (de Paris)
Gött. Nachr.	Nachrichten von der Gesellschaft der Wissenschaften zu Göttingen. Mathematisch-Physikalische Klasse
IMI	Istoriko-matematicheskie issledovaniya
J. Ec. Polyt.	Journal de l'Ecole Polytechnique
J. für Math.	Journal für die reine und angewandte Mathematik (Crelle's)
J. math. pures et appl.	Journal de mathématiques pures et appliquées
Math. Ann.	Mathematische Annalen
Mém. Acad. Sci. St.-Pétersbourg	Mémoires de l'Académie des sciences de St.-Pétersbourg
Messenger of Math.	Messenger of Mathematics
Nouv. Ann. Math.	Nouvelles des Mathématiques
Operat. Res. Quarterly	Operational Research Quarterly

Philos. Mag.	Philosophical Magazin and Journal of Science
Prace mat.-fiz.	Prace Matematyczno-Fizyczne
Proc. Nat. Acad. USA	Proceedings of the Nationale Academy of Science (Washington)
Proc. Roy. Soc.	Proceedings of the Royal Society. Series E. Mathematical and Physical Sciences (London)
Sitzungsber. Akad. Wiss. Wien	Sitzungsberichte der Kaiserlichen Akademie der Wissenschaften. Mathematisch-Naturwissenschaftliche Klasse (Wien)
Trans. of the Cambridge Philos. Soc.	Transacions of the Cambridge Philosophical Society
Trans. Roy. Irish Acad.	Transactions of the Royal Irish Academy
Trans. Roy. Soc. Edinburgh	Transactions of the Royal Society of Edinburgh

Index of Names

Abel, N.H. 36, 37, 40, 55, 56, 110, 126, 139, 190, 253
Abelard, P. 2
Adrain, R. 228, 268
Alberti, L.B. 243
Alexander of Aphrodisiäs 2
Amizura, A.L. 200
Appell, P. 139
Arago 286
Aristotle 1, 2, 2, 272, 278, 288
Aronhold, Ś.H. 80, 85, 86
Axer, A. 166
Ayoub, R.G.D 175, 190
Babbage, C. 242
Bach, A. 286
Barnard, G.A. 288
Bashmakova, I.G. 120
Baskakov, S.I. 200
Bayes, T. 219, 263
Bell, E.T. 200
Bennett, J. 3
Bentham, G. 10
Bernoulli, D. 213, 216, 223, 242, 251, 271
Bernoulli, J. 105, 166, 213, 214, 232, 233, 241, 256, 259, 262, 267, 285
Bernshtein, S.N. 216, 253, 267
Bertrand, J.L.F. 161, 185, 186, 188, 244, 276, 278, 281, 283
Bervi, N.V. 200
Bessel, F.W. 227
Bienaymé, I.J. 238, 247, 258, 261, 262, 285, 174
Biermann, K.R. 174
al-Bīrūnī 222
Blichfeldt, H.F. 151
Boethius 2
Boltzmann, L. 268, 269–272, 281, 285, 286
Bolyai, J. XI
Bolzano, B. XI
Bonnet, P.O. 161
Boole, G. 9, 13–19, 24, 25, 27, 33, 82, 278, 279, 281
Borchardt, K.B. 117
Borel, E. 278
Borevich, Z.I. 94
Borisov, E.V. 153
Bortkevich V.I. (L. von Bortkiewicz) 247

Boscovich, R. 217
Bouquet, J.C. 161
Bourbaki, N. 108, 117, 128
Bourget, H. 140
Brashman, N.D. 197, 248, 251, 252
Brouwer, L.E.J. 33
Browder, F.E. 287
Bru, B. 286
Brun, V. 196
Brush, S.G. 269, 271
Buffon, G.L. 243
Bugaev, N.V. 181, 195–200
Bukhshtab, A.A. 196
Bunitskii, E.L. 33
Bunyakovskii, V.Ya. 143, 169, 181, 182, 195, 225, 248–251, 253, 286, 288
Burnside, W. 274
Cahen, E. 193
Cantor, G. 101, 131, 208
Capelli, A. 71
Cardano, G. 213
Cartan, E. 134
Cassels, J.W.S. 163, 170
Cauchy, A. XI, 38, 49, 51, 65, 68, 100, 106, 134, 141, 174, 186, 189, 225, 229–231, 236, 238–241, 280
Cayley, A. 38–40, 63, 66, 70, 77, 79, 80, 82–85, 195, 274
Cesàro, E. 181, 200
Chebyshev, P.L. 108, 110, 111, 138, 143, 144, 152, 160, 165, 171, 178, 181, 183, 184, 186–188, 196, 200, 212, 213, 216, 217, 241, 247, 251–263, 266, 267, 276, 278–282, 284, 286
Chetverikov, N.S. 247
Chevalier, A.C. 61, 62
Chrisippus of Soli 1
Christoffel, E.B. 117
Chuprov, A.A. 246, 247, 275
Cipolla, M.U.L. 200
Clausius, R. 268, 271
Clebsch, R.F.A. 39, 40, 80, 86, 206
Clifford, W.K. 78, 79
Condorcet, J.A. 236, 286
Corput, J.G. van der 195
Cournot, A.A. 222, 248
Couturat, L. 5, 7, 33
Cramer, G. 260

Crelle, A.L. 55, 102, 105, 132
Crepel, P. 286
Czuber, E. 278, 279
D'Alembert, J. 41
Darboux, G. 139
Darwin, Ch. 212, 272, 273
Davenport, H. 153, 170
Davidov, A.Yu. 246, 251, 252
Dedekind, R.J.W. 40, 70, 74, 91, 95, 105–
    108, 116, 118–131, 133, 134, 181
Delone, B.N. 153, 154, 164, 169, 170
Demidov, P.G. 256
Demidov, P.N. 256
De Moivre, A. 213–215, 217, 233, 241,
    256
De Morgan, A. 2, 9–14, 31, 72, 73, 248,
    278, 279, 281
Descartes, R. X, 3, 41
Dirac, P. 235
Dirichlet, P.G. Lejeune 37, 40, 91–93, 95,
    98–101, 106, 116–118, 120, 122,
    138, 139, 152, 155–159, 161, 163,
    164, 167–171, 173, 174, 177–182,
    189, 190, 195, 198, 199, 218, 264
Dobrolubov, N.A. 267
Dodgson, Ch.L. 71
Du Bois-Reymond, P. 101
Duhamel, J.M.C. 197
Dutka, J. 286
Edgeworth, F.Y. 275, 288
Edwards, H.M. 100, 192
Egorov, D.F. 198
Ehrenfest, P. 216, 271
Ehrenfest, T. (Afanasieva-
    Ehrenfest T.A.) 216, 271
Eisenstein, F.G.M. 37, 39, 84, 85, 98, 106,
    137, 143, 153, 161
Erdös, P. 188, 194
Ermolaeva, N.S. 286
Eubulides of Miletus 1
Euclid of Alexandria 3, 76, 95, 96, 100,
    129
Euclid of Megara 1
Euler, L. X, 5, 27, 41, 43, 44, 49, 50, 62,
    63, 87, 88, 91, 138, 143, 168, 171,
    174–176, 178–183, 185, 187, 190,
    196, 201, 208
Faddeev, D.K. 169
al-Fārābī 2
Faraday, M. 273
Farebrother, R.W. 286
Fedorov, E.S. 169

Fermat, P. X, 37, 63, 88, 91, 94, 97, 99–
    101, 105,
Fischer, R.A. 228
Fiske, D. 287
Fourier, J.B. 173, 191, 213, 230, 238, 241,
    243
Fraenkel, A. 131
Frege, G.F.L. 13, 34
Frobenius, G. 117, 134, 152
Frullani, G. 262
Fuss, P.N. 182
Galileo 223
Galois, E. XI, 36, 37, 39, 42, 49, 50, 55,
    57–67, 202
Galton, F. 272, 273, 281, 286
Gauss, K.F. X, XI, 36–38, 40, 43, 44–54,
    57, 62, 63, 65, 68, 76, 80, 81, 86–
    92, 94, 96–99, 101, 104, 105, 116,
    137, 138, 141–145, 154–156, 159,
    161, 163, 172–174, 178, 180, 182,
    189, 195, 224, 226–231, 241, 262,
    281, 283, 287
Gegenbauer, L.B. 199, 200
Gel'fand, I.M. 128
Gel'fond, A.O. 203–205, 208
Gerhardt, C.I. 5
Germain, S. 99
German, K.F. 246
Gibbs, J.W. 271, 272
Gillispie, C.C. 286
Gnedenko, B.V. 286
Goldbach, C. 196, 209
Good, I.J. 286
Gordan, P. 39, 40, 80, 85, 101
Gorkii, A.M. 267
Gram, J.P. 200
Grant, E. 249
Grassmann, H. XI, 38, 70, 78, 79
Grattan-Guinness, I. 238
Grave, D.A. 254
Graves, J.T. 77
Gregory, D.F. 72, 73
Grodzenskii, S.Ya. 286
Hadamard, J. 193
Hamilton, W. 10
Hamilton, W.R. XI, 38, 70, 74–76
Hardy, G.H. 195
Harter, H.L. 286
Haüy, R.J. 154
Heidelberger, M. 287
Helmholtz, H. 125, 161

Hensel, K. 38, 103, 108, 112, 116, 134
Hermite, Ch.H. 80, 82, 98, 108, 110, 137, 139–145, 147, 148, 150, 153, 159, 161, 164, 165, 168, 170, 186, 192, 205–208, 216, 260
Herschel, J. 268
Hesse, L.O. 39, 80, 82, 84, 86
Hilbert, D. X, 80, 86, 105, 106, 134, 162, 208, 281, 285
Hobbes, T. 278
Hofreiter, N. 150, 170
Hölder, O.L. 67
Humboldt, A. von 173, 174
Hurwitz, A. 117, 161, 205
Huygens, C. 213
Ibn Rushd (Averroes) 2
Ibn Sīnā (Avicenna) 2
Ivanov, I.I. 188
Jacobi, C.G.J. 37–40, 62, 69, 71, 80, 82, 86, 98, 101, 138, 139, 141, 142, 152, 161, 164, 167, 168, 171, 174, 181, 182, 195, 196
Jevons, W.S. 13, 20–27, 33, 278, 279, 281
Johnson, N.L. 286
Jordan, C. 39, 63, 66, 67, 72, 150, 161
Kant, I. 211
Keller, E.H.O. 170
Kendall, M.G. 288
Kepler, J. 272
Khinchin, A.Ya. 271, 272
Killing, W. 134
Kirchhoff, G.R. 108, 125, 161
Klein, F. 39, 51, 66, 70, 77, 91, 117, 125, 131, 134, 139, 159, 189, 231, 283
Kogan, L.A. 154
Koksma, J.F. 170
Kolmogorov, A.N. 212, 254, 255
Korkin, A.N. 108, 144–152, 159, 164, 166, 169, 170, 254, 266
Kotz, S. 286
Kronecker, L. 40, 47–50, 71, 99–101, 106–108, 117, 127, 131–134, 153, 161, 168, 181, 196, 197
Kruger, L. 287
Kruskal, W. 287
Krylov, A.N. 145, 254, 258, 268
Kummer, E.E. 37, 38, 40, 95, 99–106, 108, 111, 116, 117, 120, 131, 132, 161, 181, 197
Kuz'min, R.O. 230
Lacroix, S.F. 248

Lagrange, J.L. 14, 41, 42, 44, 49, 50, 57, 62, 63, 68, 87, 89–91, 103, 105, 138, 143, 144, 154, 167, 182, 203, 212, 213, 217, 230
Lakhtin, L.K. 198, 275
Lambert, J.H. 8, 201, 223, 278, 285
Lamé, G. 99–101, 197,
Landau, E. 188, 193, 195
Laplace, P.S. 42–45, 49, 211–226, 230, 231, 233, 236, 239, 241–243, 247–249, 251, 256, 262, 264, 272, 278, 280, 281, 283, 285– 287
Legendre, A.M. 57, 87, 88, 99, 142, 143, 152, 160, 171, 174, 178, 180–184, 195, 196, 201, 228, 229, 286
Leibniz, G.W. X, 1–8, 278
Lenin, V.I. 274, 277
Levy, P. 212
Lexis, W. 247, 281
Lie, S. 134
Lindemann, F. 161, 205–208
Linnik, Yu.V. 196, 268
Liouville, J. 58, 100, 102, 110, 139, 154, 161, 181, 196–198, 201–204, 253, 285
Lipschitz, R. 79
Littlewood, J.E. 195
Lobachvskii, N.I. XI, 211, 212, 251
Loschmidt, J. 269
Lullius, R. 2
Lyapunov, A.M. 212, 241, 258, 260, 263–265, 278, 288
Mackenzie, D.A. 287
Maclaurin, C. 41, 215
MacMahon, P.A. 195
Mahler, K. 153, 170
Mangoldt, H.C.F. von 200
Markov, A.A 144, 151–153, 166, 208, 215, 216, 228, 241, 242, 254, 259, 261, 263, 265, 266, 267, 275, 280, 287–288
Markov, V.A 153
Marx, K. 245, 284
Maxwell, J.C. 268, 269, 271, 274, 281
Merlin, J. 196
Mertens, F.K.J. 187
Metivier, M. 287
Michelson, V.A. 270
Mill, J.S. 280
Minin, A.P. 200
Minkowski, H. 117, 151, 159, 161, 163–165, 170

Minto, W. 11
Mises, R. von 272
Möbius, A.F. 191, 196, 199
Molin, F.E. 134
Mordell, L.J. 153, 170
Nazimov, P.S. 195, 198, 200
Nekrasov, P.A. 287
Neumann, E. 86
Newton, I. X, 14, 106, 120
Noether, E. 131, 135
Nogués, R. 100
Ondar, Kh.O. 275
Oppenheim, A. 170
Oresme, N. 249
Orzhentskii, R.M. 275
Osipovskii, T.F. 247
Ostrogradskii, M.V. 250, 251, 253
Painlévé, P. 139
Pascal, B. 221, 222
Pasteur, L. 139
Pavlovskii, A.F. 247
Peacock, G. 72, 73, 76
Peano, G. 13
Pearson, E.S. 287, 288
Pearson, K. 221, 241, 272–275, 281, 285, 287
Peirce, B. 79
Peirce, Ch.S. 13, 285
Pellegrino, F. 200
Petrus Hispanus 2
Philo of Megara 1
Picard, E. 139, 140
Pirogov, N.H. 270
Pisarev, D.I. 267
Plakett, R.L. 228, 287, 288
Plank, M. 270
Ploucquet, G. 8
Poincaré, H. 134, 139, 150, 162, 188, 222, 276–278, 281, 287
Poisson, S.D. 217, 229, 230–237, 239, 241, 243, 246, 247, 248, 251, 252, 256–258, 280, 286, 287
Polygnac, A. 187
Popken, I. 200
Poretskii, P.S. 23, 24, 27, 31, 32, 196
Porphirius 2
Posse, K.A. 205, 254
Ptolemy 222
Puiseux, V. 134
Quetelet, L.A.J. 242–247, 251, 276, 284, 287
Raabe, J.L. 116

Ramanujan, S. 195
Remnant, P. 3
Riccati, J.F. 203
Riemann, B. 70, 74, 126–130, 171, 178, 181, 189–193
Rogers, C.A. 170
Rothe, P. 41
Ruffini, P. 63
Rümelin, G. 246, 284
Russell, B. 13
Salmon, G. 39, 80, 82
Scharve, L. 153, 162
Schneider, I. 271, 287
Schneider, T. 208
Schröder, E. 13, 27, 29, 30, 33, 131
Schur, I. 152
Schwarz, H.A. 101, 117, 225
Scotus, J.D. 2
Seeber, L.A. 137, 154–156
Segner, J.A. von 8
Seidel, 286
Selberg, A. 188, 194, 196, 200
Selling, E. 153
Seneta, E. 287
Serret, J.A. 63, 66, 186
Shafarevich, I.R. 94, 106
Shatunovskii, S.O. 33
Sheĭnin, O.B. 287, 288
Siegel, C.L. 153, 192, 208
Sierpiński, W. 171, 195
Simpson, T. 212, 217, 223,
Sleshinskii, I.V. 24, 33, 260
Smith, H.J.S. 98, 159, 160, 162, 172
Smoluchowski, M. 216
Socrates 1
Sokhotskii, Yu.V. 254
Sonin, N.Ya. 171, 198, 200, 254
Sprott, D.A. 288
Stanevich, V.I. 188
Steinitz, E. 131
Steklov, V.A. 216, 254
Stevin, S. 127
Stieltjes, T.J. 140, 192
Stigler, S.M. 286, 288
Stirling, J. 177, 185
Stokes, G. 274
Sturm, J.Ch.F. 141
Sylvester, J.J. 10, 39, 71, 80, 82, 83, 188, 195
Tait, P. 268
Tannery, P. 139

Tartakovskii, V.A. 196
Thomson, W. (Lord Kelvin) 268
Tichomandritskii, M.A. 260
Titchmarsh, E.C. 192
Todhunter, I. 10, 280
Torelli, G. 188
Tsinger, N.Ya. 268
Tsykalo, A.L. 288
al-Tūsī 2
Uspenskii, Ya.V. 143, 153, 170, 195, 195
Vaidyanathaswamy, R. 200
Val'fish, A.Z. 170, 195
Vallee Poussin, C.J. 193
Vandermonde, A.T. 50, 63, 68
Vantzel, P.L. 100
Vasil'ev, A.V. 228, 254
Venkov, B.A. 152, 170
Venn, J. 24–27, 278, 281
Viète, F. 3
Vinogradov, I.M. 171, 194, 195, 201
Voronoï, G.F. 143, 151, 159, 166, 168–
      170, 179, 195, 201, 254

Voynich, E.L. 14
Waerden, B.L. van der 131, 135, 288
Walker, H.M. 288
Wallis, J. 177, 201
Waring, E. 120
Weber, H. 40, 105, 106, 117, 125–128,
      132, 134, 161
Weierstrass, C.T.W. 40, 72, 80, 101, 108,
      117, 131, 161, 189, 197, 208
Weldon, C.F.R. 272, 273, 281
Weyl, H. 80, 133, 170
Wiener, N. 194
William of Occam 2
Yamamoto, C. 200
Yanovskaya, S.A. 129
Zeno of Citium 1
Zermelo, E. 270, 271
Zernov, N.E. 249
Zhegalkin, I.I. 17
Zolotarev, Ye.I. 38, 40, 95, 103, 107–116,
      133, 134, 143–151, 159, 164, 169,
      170, 266